**Communication Systems Engineering with GNU Radio**

# Communication Systems Engineering with GNU Radio

A Hands-on Approach

*Jean-Michel Friedt*
University of Besançon
France

*Hervé Boeglen*
University of Poitiers
France

Copyright © 2025 by John Wiley & Sons, Inc. All rights reserved, including rights for text and data mining and training of artificial technologies or similar technologies.

Published by John Wiley & Sons, Inc., Hoboken, New Jersey.
Published simultaneously in Canada.

No part of this publication may be reproduced, stored in a retrieval system, or transmitted in any form or by any means, electronic, mechanical, photocopying, recording, scanning, or otherwise, except as permitted under Section 107 or 108 of the 1976 United States Copyright Act, without either the prior written permission of the Publisher, or authorization through payment of the appropriate per-copy fee to the Copyright Clearance Center, Inc., 222 Rosewood Drive, Danvers, MA 01923, (978) 750-8400, fax (978) 750-4470, or on the web at www.copyright.com. Requests to the Publisher for permission should be addressed to the Permissions Department, John Wiley & Sons, Inc., 111 River Street, Hoboken, NJ 07030, (201) 748-6011, fax (201) 748-6008, or online at http://www.wiley.com/go/permission.

Trademarks: Wiley and the Wiley logo are trademarks or registered trademarks of John Wiley & Sons, Inc. and/or its affiliates in the United States and other countries and may not be used without written permission. All other trademarks are the property of their respective owners. John Wiley & Sons, Inc. is not associated with any product or vendor mentioned in this book.

Limit of Liability/Disclaimer of Warranty: While the publisher and author have used their best efforts in preparing this book, they make no representations or warranties with respect to the accuracy or completeness of the contents of this book and specifically disclaim any implied warranties of merchantability or fitness for a particular purpose. No warranty may be created or extended by sales representatives or written sales materials. The advice and strategies contained herein may not be suitable for your situation. You should consult with a professional where appropriate. Further, readers should be aware that websites listed in this work may have changed or disappeared between when this work was written and when it is read. Neither the publisher nor authors shall be liable for any loss of profit or any other commercial damages, including but not limited to special, incidental, consequential, or other damages.

For general information on our other products and services or for technical support, please contact our Customer Care Department within the United States at (800) 762-2974, outside the United States at (317) 572-3993 or fax (317) 572-4002.

Wiley also publishes its books in a variety of electronic formats. Some content that appears in print may not be available in electronic formats. For more information about Wiley products, visit our web site at www.wiley.com.

*Library of Congress Cataloging-in-Publication Data applied for:*
Hardback ISBN: 9781394218882

Cover Design: Wiley
Cover Images: Courtesy of Jean-Michel Friedt, © GNU Radio logo - Wikimedia Commons/public domain

Set in 9.5/12.5pt STIXTwoText by Straive, Chennai, India

SKY10100498_031825

*To Jean-Michel for his unfailling enthusiasm and for convincing me to join the world of freedom (Linux).*

# Contents

**About the Authors** *xi*
**Foreword** *xiii*
**Acknowledgments** *xvii*
**Acronyms** *xix*
**About the Companion Website** *xxi*
**Introduction** *xxiii*

**1**     **Getting Started with GNU Radio: Synthetic Signals** *1*
1.1     Evolution of Radio Frequency Electronics Toward SDR *1*
1.2     The Complex Envelope and the Justification of the IQ Structure *2*
1.3     Complex Number Manipulation *7*
1.4     GNU Radio and GNU Radio Companion *7*
1.5     Sample Rates, Decimation and Aliasing *13*
1.6     Low-pass Filtering or Working on Upper Nyquist Zones *19*
1.7     ADC and DAC Resolution *24*
1.8     Power Spectral Density Display with the Frequency Sink *27*
1.9     Conclusion *29*
       References *29*

**2**     **Using GNU Radio with Signals Collected from SDR Hardware** *31*
2.1     SDR Hardware Architecture *32*
2.2     Using Readily Available Processing Tools *33*
2.3     Amplitude Modulation and Demodulation *36*
2.4     Frequency Modulation and Demodulation *41*
2.4.1   Commercial FM Broadcast: Demodulation and Audio Output *46*
2.4.2   Stereo Sound and RDS *48*
2.4.3   FSK Modulation *50*
2.5     Phase Modulation and Demodulation *53*
2.5.1   Phase Modulation *53*
2.5.2   Phase Demodulation *55*
2.5.3   Radio Data System (RDS) *55*
2.5.4   The Global Positioning System (GPS) *57*
2.6     Spectral Occupation of the Various Modulation Schemes *63*
2.7     Local Oscillator Leakage Issue *63*
2.8     Conclusion *67*
       References *67*

**viii** | *Contents*

**3**      **Communicating with External Software (Python, Networking, ZeroMQ, MQTT)** *69*

3.1      Connecting to an External Sentence Decoding Tool Using Named Pipes   *70*

3.1.1      POCSAG Single Channel Decoding   *70*

3.1.2      POCSAG Multichannel Decoding   *72*

3.2      TCP/IP Server Running in a Separate Thread   *75*

3.3      XML-RPC   *81*

3.4      Zero MQ (0MQ) Streaming   *84*

3.5      MQTT   *93*

3.5.1      MQTT for Python, Bash and Octave   *94*

3.6      Conclusion   *96*

     References   *96*

**4**      **Correlating: Passive and Active Software-Defined Radio (SDR) – RADAR**   *99*

4.1      SDR–RADAR Requirements and Design   *99*

4.2      Correlation: GNU Radio Implementation   *103*

4.3      Passive RADAR Principle and Implementation   *113*

4.4      Active RADAR Principle and Implementation   *113*

4.5      Measurement Principle   *114*

4.6      From Theory to Experiment: Ranging by Frequency Stacking   *115*

4.7      Results   *120*

4.8      Conclusion on Range Measurement   *122*

4.9      Azimuth Resolution Through Spatial Diversity: Synthetic Aperture RADAR   *122*

4.9.1      OFDM RADAR (WiFi)   *126*

4.10      Acquisition for Azimuth Measurement   *129*

4.11      Suppressing Direct Coupling Interference   *132*

4.12      Signal Processing   *132*

4.13      Result Analysis   *135*

4.14      Interferometric Measurement   *136*

4.15      Reproducible Positioning of the Receiving Antenna: Motorized Rail   *138*

4.16      The Radio Frequency Corner Reflector   *139*

4.17      Fine Displacement Measurement   *141*

4.18      Impact of the Atmosphere   *142*

4.19      Time of Flight Measurement with Sub-sampling Period Resolution and the Use of a Surface Acoustic Wave Cooperative Target for Reproducible Range Simulation   *144*

4.20      Conclusion   *149*

     References   *150*

**5**      **Digital Communications in Action: Design and Realization of a QPSK Modem** *153*

5.1      Digital Communication Concepts   *153*

5.1.1      What Is Digital Information?   *153*

5.1.2      From Digital Data to Electrical Pulses   *155*

5.1.3      Occupied Bandwidth and Spectral Efficiency   *157*

5.2      Building a QPSK Modulator with GNU Radio   *160*

5.3      Building a QPSK Demodulator with GNU Radio   *165*

5.3.1      Synchronization   *167*

| | | |
|---|---|---|
| 5.3.1.1 | The Digital PLL *168* | |
| 5.3.1.2 | Maximum Likelihood Estimation and the Costas Loop *174* | |
| 5.3.1.3 | QPSK Timing Recovery *176* | |
| 5.3.2 | Automatic Gain Control (AGC) *181* | |
| 5.3.3 | Assembling All the Components: The Ultimate QPSK Receiver Flowgraph *182* | |
| 5.4 | Conclusion *187* | |
| | References *188* | |

**6 Messages, Tags, and Packet Communications** *189*

| | | |
|---|---|---|
| 6.1 | Introduction *189* | |
| 6.2 | Polymorphic Types *190* | |
| 6.3 | Messages *193* | |
| 6.4 | Tags *202* | |
| 6.5 | Case Studies *206* | |
| 6.5.1 | Improving the `gr-nordic` OOT Module *206* | |
| 6.5.2 | Converting the QPSK Modem to Packet Mode *212* | |
| 6.6 | Conclusion *216* | |
| | References *217* | |

**7 A Digital Communication Standard: The DAB+ Radio Broadcasting System** *219*

| | | |
|---|---|---|
| 7.1 | Introduction *219* | |
| 7.2 | The DAB+ Standard *220* | |
| 7.2.1 | Foundations of Digital Audio Coding *220* | |
| 7.2.1.1 | The Absolute Threshold of Hearing *220* | |
| 7.2.1.2 | Critical Bands *220* | |
| 7.2.1.3 | Masking *222* | |
| 7.2.2 | Audio Coding Standards and Their Usage in DAB and DAB+ *224* | |
| 7.2.2.1 | The MPEG/ISO/IEC International Audio Standards *224* | |
| 7.2.2.2 | The DAB and DAB+ Audio Coders *225* | |
| 7.2.3 | Digital Transmission over Time and Frequency-Selective Channels: The Need for COFDM *228* | |
| 7.2.3.1 | Time and Frequency-Selective Wireless Channels *228* | |
| 7.2.3.2 | COFDM: Digital Communication Techniques for the Wireless Channel *232* | |
| 7.2.3.3 | The ETSI EN 300 401 DAB Standard *236* | |
| 7.3 | Building a DAB+ Transmitter *241* | |
| 7.4 | Building a DAB+ Receiver with GNU Radio *244* | |
| 7.4.1 | Basic Usage of `gr-dab` *244* | |
| 7.4.2 | `gr-dab` in More Details *246* | |
| 7.5 | Conclusion *249* | |
| | References *249* | |

**8 QPSK and CCSDS Packets: Meteor-M 2N Satellite Signal Reception** *251*

| | | |
|---|---|---|
| 8.1 | Introduction *251* | |
| 8.2 | When Will the Satellite Fly Overhead? *255* | |
| 8.3 | Why Such a Complex Protocol? *258* | |
| 8.4 | How to Tackle the Challenge? *260* | |

x | *Contents*

| | | |
|---|---|---|
| 8.5 | From the Radio frequency Signal to Bits | *261* |
| 8.5.1 | Data Format | *262* |
| 8.5.2 | Decoding Data | *264* |
| 8.5.3 | Convolutional Encoding of the Synchronization Word | *265* |
| 8.5.4 | Convolutional Code Representation as State Machines | *267* |
| 8.5.5 | Decoding a Convolutional Code: Viterbi Algorithm | *269* |
| 8.5.6 | Constellation Rotation | *271* |
| 8.5.7 | From Bits to Sentences: Applying the Viterbi Algorithm Decoding | *273* |
| 8.6 | From Sentences to Paragraphs | *279* |
| 8.7 | So Much Text ... Pictures Now | *281* |
| 8.8 | JPEG Image Decoding | *286* |
| 8.9 | Conclusion | *291* |
| | References | *296* |

**9 Custom Source and Sink Blocks: Adding Your Own Hardware Interface** *297*

| | | |
|---|---|---|
| 9.1 | Python Block | *298* |
| 9.2 | Out-of-Tree Blocks | *300* |
| 9.3 | Cross-compiling for Running on Headless Embedded Systems | *310* |
| 9.4 | Conclusion | *312* |
| | References | *312* |

**10 Conclusion** *315*

References *318*

**Index** *319*

## About the Authors

**Jean-Michel Friedt** was trained as a physicist at École Normale Supérieure in Lyon (France). He completed his PhD on scanning probe microscopy in 2000 before joining IMEC (Leuven, Belgium) as a postdoctoral researcher working on surface acoustic wave (SAW)-based biosensors. He joined the company SENSeOR in 2006 as a systems engineer, developing short-range RADAR systems for probing SAW resonators acting as wireless passive cooperative targets with sensing capability. Before becoming associate professor at Franche-Comté University in Besançon (France) in 2014 with his research activities hosted by the Time & Frequency department of the FEMTO-ST Institute, he became intrigued by the field at the intersection of computer science, radio frequency, and digital signal processing with access to the physical properties of the electromagnetic waves, namely software-defined radio (SDR), and its opensource implementation GNU Radio. Visiting the radio-silent research station of Ny-Ålesund (Norway, see cover) had the most profound impact on his personal and research and development activities, from satellite communication to remote sensing during field trips since 2007. He has been a regular contributor to the French GNU/Linux Magazine/France and related journals since 2005 whose publications motivated most of this research, with an emphasis to present results toward the general public and curious readers willing to reproduce experiments with affordable and readily available hardware and opensource software. Current investigations focus on the use of SDR for time and frequency distribution including GNSS (with anti-jamming and spoofing strategies), RADAR and spectrum spreading for time dissemination, remote sensing, and spaceborne communications.

**Hervé Boeglen** graduated from the University of Haute Alsace in Mulhouse, France with an MSc degree in electrical engineering in 1994. He worked as a full-time lecturer in electronics in the telecommunications and networks department at the IUT of Colmar, University of Haute Alsace, France, from 1995 to 2013. Between 2006 and 2008, while working, he pursued a PhD in digital communications. In 2013, he joined the University of Poitiers, France, as an associate professor in electrical engineering. He currently teaches graduate-level courses in embedded systems and digital communications using GNU Radio. He is also a member of the XLIM lab at the Futuroscope site in France. His research focuses on wireless channel modeling and digital communication systems, both radio and optical, using software-defined radio. He is also a radio transmission enthusiast and holds an amateur radio license (callsign F4JET).

# Foreword

Communications Engineering solves one of the central problems of mankind: making sure that what is known in one place becomes known in another. This is a book instructing the reader how to implement that, wirelessly, with naught but a computer, free software, and a radio frontend.

Sufficiently obvious is the need for education in the practicalities of this communication technology, given how transformative they have been: The geopolitical situation of the twenty-first century is hard to imagine without the TV; neither the Arab Spring nor the public perception of Russia's invasion of Ukraine would have assumed their shape without ubiquitous mass communication. However, this is not a guide on how to make a call using a cell phone, or how to upload a video to social media.

Instead, its purpose lies in enabling the reader to work on a more fundamental, the *physical* layer, themselves, with nothing between them and the radio waves that need to be modified to communicate information. The power of *Software-Defined Radio* (SDR) lies in its ability to make the mathematical description of the physical phenomenon that is radio available for analysis and manipulation in a computer, and thus gives its users the ability to control what and how information is transported to the fullest extent.

Conversely, understanding how communication is actually done using SDR allows for a deeper insight into the nature of wireless communications. As the presence of a chapter on passive and active radar shows, the same techniques enabling us to exchange information with a communication partner allow us to retrieve information about things far away. This serves to illustrate one of the strengths of teaching concepts from communications engineering based on working, practical implementations: The theory taught on one page to establish the working of an aspect of communications serves as an explanation for the technology taught a few pages further down.

The authors elected to use GNU Radio to teach these concepts, why? GNU Radio is *Free and Open Source Software* (FOSS). This means three things for its usage as an educational tool:

1. Its openness allows for introspection: An interested user could always look inside and discover "how it's done."
2. Its wide availability across platforms, free of cost, makes it a desirable platform to work on, while not sacrificing on its professionality: The very same tools used by researchers, companies, and hobbyists worldwide are at the fingertips of the learner, giving them the opportunity to grow continuously from a beginner to an expert in designing communication systems.
3. The community built around GNU Radio is largely motivated by the idea that everyone should have access to both the knowledge and the tools needed to build communication systems. This leads to intense sharing of knowledge, both in the shape of software and its source code, as well as of methodology and theory.

**xiv** | *Foreword*

Surprisingly to me, as former leader of the software and its development process, openness is *not* the strongest argument (but still important to many) for using GNU Radio to teach communications engineering.

More valuable is the community arising from the FOSS nature. GNU Radio understands itself not only as a software project, but as the unifying part of a whole ecosystem of SDR applications, libraries, research groups, and users. This has made it the most popular open source environment for SDR development, with an annual conference in the United States and in Europe, a pervasive presence in research articles, the amateur radio community, demonstrators, and common use case for the vendors of SDR frontends. This comes with a network of educators, users, developers, supporters, and learners. But GNU Radio did not start out as a large project. However, its function as tool to understand wireless communications is part of the very foundations of it, which is a short look back as its history shows:

When Eric Blossom, a computer-science-friendly electrical engineer in the United States was designing cryptographically secure telephones around 1993, he found commercial and official understanding of secrecy to be lacking. The freshly evolving digital wireless, cellular communication standards, in the shape of IS-95, had serious cryptographical flaws.

Considering the technological and financial entry hurdles of working with the waveforms on the air, Blossom talked to John Gilmore, of fame for sponsoring the Electronic Frontier Foundation's (EFF) efforts to build a demonstrator to prove the Data Encryption Standard (DES) encryption standard to be weak, the two enter into a patronage, where Blossom gets to be paid on a Free and Open Source SDR framework – leading to the birth of the GNU Radio idea in 1998; the early code was based on MIT's Pspectra SDR framework.

The progression from a small project to the most popular SDR framework in existence was fostered by the Moving Picture Association of America throwing in their weight when the United States moved analog TV to the digital Advanced Television Systems Committee (ATSC) – and enforcing a "copy-protection bit," to be respected by video recorders. That not sitting well with EFF ideals of free access to technology, a reason emerged to write a complex receiver in what would become GNU Radio – something that can receive ATSC TV, get the video, and not care at all about any copyright bits.

The EFF proving the ATSC copy-protection is a publicity success, but a project of that size showed the need to overhaul the code. Clearing this milestone, Blossom convinces Gilmore to let him rewrite GNU Radio from scratch to remove its limitations resulting from Pspectra legacy.

At this point, about 2003, Matt Ettus gets involved and starts building his SDR frontend – what he coined the "Universal Software Radio Peripheral" (USRP) – a device that attaches to PCs through USB, allowing anyone to work with electromagnetic spectrum with off-the-shelf, relatively affordable hardware. Ettus becomes a contributor to GNU Radio, and hard- and software co-evolve. Early versions of the driver for the hardware are part of GNU Radio; only later on, a more generally useful driver for the Ettus hardware is written. Academic interest is massively picking up – numerous PhD students work with GNU Radio, contribute code, and most importantly: They (and the students they supervise) form a lively community. At the same time, the project becomes commercially relevant enough to support consultants.

One of these PhD students is Thomas Rondeau – who later becomes the lead to replace Blossom; the consultant, to become the maintainer of GNU Radio, is Jonathan Corgan, both who shape the project into a software project that is good at accepting contributions. At the same time, community events start to emerge. Students – the author of this foreword not being an exception – get highly invested. A very active mailing list forms the glue of an international community of users and developers.

Things go smoothly, and GNU Radio 3.7 becomes the stable release found in nearly all Linux distributions; there's binary installers for Windows, the annual GNU Radio conferences attract more than 300 people a year. Development moves from a self-hosted git server with trac as project management to github and mediawiki for documentation. However, stability is a double-edged sword; it means contributions begin to dry up in development branches that stand little chance to actually get deployed to users. As Corgan leaves the position of maintainer, GNU Radio has excellent support on all relevant desktop operating systems, but the "next" branch, which was destined to become GNU Radio 3.8, is not in a shape immediately ready for release.

This is when I'm asked to step in and switch from a very unofficial role, where I try to stay atop of what is happening on the mailing list, explain the code, the communications engineering base and the occasional bug to users, to a more official role; from 2018 to 2021, I become architect and maintainer of GNU Radio. What an honor! Getting the 3.8 release out of the door with the help of a lot of friends, we gave GNU Radio a new velocity (and, as you can imagine, we broke some poor people's applications in the process of making sure GNU Radio still works on machines with Python 3). Releases become more regular, and the number of contributions surges strongly.

Handing my responsibility off to two people – Josh Morman as the new architect, and Jeff Long as release maintainer concerned especially with the stable releases – was excellent for the project, too. It allows the evolution of the developer code base with less worries about things breaking on the machines of users, without sacrificing on the ability to spin reliable new releases.

However, it goes without saying that a software framework having an easy 20 years of development history comes with some baggage. Not all things are as intuitive, or as fast, or as safe, or as consistent as they could be, even in GNU Radio 3.10, the current release series as of writing. The GNU Radio project continues to evolve; what I believe will stay the same is the dedication to one central principle:

Through offering a very accessible FOSS framework for SDR application, with which everybody can get access to this great shared resource, the electromagnetic spectrum, GNU Radio will continue to foster, and live off, an active community of coders, researchers, hackers, users, and operators that drive the project forward.

I hope this book motivates generations of people willing to learn to tackle something as rich in facets as communications engineering – using tools that allow and encourage them to go beyond what is available as what companies are willing to sell them as wireless devices.

June 2024

*Marcus Müller*

# Acknowledgments

This book would not exist without ... the GNU Radio development team. Over the years, numerous contributors have improved the software, sometimes with challenging decision on the software architecture (e.g. SWIG to PyBIND transition), but always to achieve utmost quality and performance. Next to the development, the opensource spirit has driven not only software development but also sharing the complex field of digital signal processing. By tackling practical applications, GNU Radio provides a fun and attractive framework for becoming familiar with such obscure concepts as an imaginary voltage or a negative frequency.

In this vein, the GNU Radio related conferences have been the opportunity for developers not only to share results as is usual in "scientific" conferences, but most importantly the means to achieve and reproduce these results. The American GNU Radio Conference with its proceedings available at https://pubs.gnuradio.org and YouTube channel at https://www.youtube.com/@GNURadioProject, the European Free Open-Source DEveloper Meeting (FOSDEM) software defined radio devroom and the European GNU Radio Days – with its YouTube channel at https://www.youtube.com/@europeangnuradiodays1445/ that both authors have helped co-organize since 2018 – have been the source of inspiration and motivation with technical discussions with speakers and attendees.

*Jean-Michel Friedt and Hervé Boeglen*

## Acronyms

| | |
|---|---|
| ACARS | aircraft communication addressing and reporting system |
| ADC | analog-to-digital converter |
| AGC | automatic gain control |
| BPSK | binary phase shift keying $\varphi \in \{0, \pi\}$ |
| CCSDS | consultative committee for space data systems |
| CDMA | code division multiple access used in particular for identifying which GPS satellite is broadcasting |
| CGRAN | comprehensive GNU Radio archive network at https://cgran.org |
| COTS | commercial off the shelf |
| CRC | cyclic redundancy check |
| DAB | digital audio broadcasting |
| DAC | digital-to-analog converter |
| FDMA | frequency division multiple access |
| FFT | fast Fourier transform an $N \log_2(N)$ complexity implementation of the Fourier transform |
| FM | frequency modulation |
| FSK | frequency-shift keying (FM digital modulation) |
| FSPL | free space propagation loss, the logarithmic expression of Friis energy conservation |
| GNU | GNU is not Unix |
| GPS | global positioning system |
| GRAVES | Grand Réseau Adapté à la VEille Spatiale is the French space surveillance RADAR emitting a continuous wave at 143.05 MHz |
| IF | intermediate frequency |
| IQ | in-phase/quadrature |
| ISI | inter-symbol interference |
| ISS | International Space Station whose amateur service is broadcasting on 145.8 MHz |
| LEO | low Earth orbit |
| LO | local oscillator |
| LOS | line of sight |
| MEO | medium Earth orbit |
| NCO | numerically controlled oscillator |
| OFDM | orthogonal frequency division multiplexing |
| PDU | Protocol Data Unit |
| PMT | polymorphic types |

| | |
|---|---|
| POCSAG | **P**ost **O**ffice **C**ode **S**tandardisation **A**dvisory **G**roup pager protocol for emergency services |
| QPSK | quad phase shift keying $\varphi \in \{0, \pi/2, \pi, 3\pi/2\}$ |
| RADAR | radio detection and ranging |
| RDS | radio data system |
| RRC | root raised cosine |
| RF | radio frequency |
| RTL-SDR | a set of low-cost SDR receivers based on a RF front end and the Realtek RTL2832U analog-to-digital converter to USB |
| SDR | software-defined radio |
| SNR | signal-to-noise ratio |
| TED | timing error detector |
| WBFM | wideband frequency modulation the modulation scheme used by commercial FM broadcasters |

# About the Companion Website

This book is accompanied by a companion website:

**www.wiley.com/go/friedtcommunication**

The website https://gitlab.xlim.fr/gnuradio_book/flowcharts mirrored at https://gitlab.com/gnuradio_book/flowcharts includes:

- the flowcharts, included in the book as static figures, to be executed with GNU Radio 3.10 for assessing the output of the signal processing chains and tuning the parameters to observe the impact on the output signals. All figures are included in the folder with the name of the associated chapter for easy matching. Specific information needed to perform some of the experiments is also provided.

# Introduction

What is GNU Radio? GNU Radio is a toolkit providing the means to address discrete-time digital signal processing chains oriented toward radio frequency (RF) communication, but not limited to it. GNU Radio is *not* a readily functional decoding software for a given communication protocol: understanding the principles of signal representation, frequency transposition, synchronization, and digital information extraction will be needed before implementing functional communication systems. Thanks to its flexibility, GNU Radio is not restricted to digital communication over radio frequency channels but can be used for instrumentation, RADAR and radio frequency channel characterization, time and frequency transfer, beamforming and null steering for e.g. jamming and spoofing suppression, or any application benefiting from accessing the raw radio frequency wave characteristics.

To make the learning curve less steep, a graphical interface for assembling signal processing blocks is provided: GNU Radio Companion. It should be emphasized, though, that the graphical interface is for development ease only and is not needed for executing the resulting flowgraph; hence, GNU Radio is perfectly suited for running on embedded systems not fitted with graphical interfaces. Indeed, GNU Radio is included in the embedded Linux-built frameworks, Buildroot and OpenEmbedded, allowing to use the Python-generated processing scripts on headless embedded systems running the operating system and the associated C++ libraries.

The reader is encouraged to test all processing sequences and assembling processing chains step by step: even though some of the examples are a lengthy sequence of processing steps, assessing the impact of each block by displaying the frequency domain or time domain characteristics after each processing step is mandatory. In order to help the reader test various processing sequences, records of relevant signals are made available on https://iqengine.org in the GNU Radio repository.

A word of caution before starting to experiment with software-defined radio (SDR) and storing huge files: make sure to remember the experimental setup leading to these records, and most significantly the carrier frequency, the sampling rate, and data format. Many times have these authors run days of records to forget after a few days how data had been collected and hence how to read them for post-processing. To avoid such hassle, the SigMF (signal metadata format) standard has been proposed, specified at https://github.com/sigmf/SigMF. This format associates with each data file (`sigmf-data`), a format description (`sigmf-meta`) which provides the receiver characteristics including carrier frequency, bandwidth (sampling rate), or data format (floating point or integer, and integer size of each sample). All records at https://iqengine.org comply with the SigMF format, hence providing the necessary information for processing the collected data.

The GNU Radio Companion processing sequences described throughout the book are available at https://gitlab.xlim.fr/gnuradio_book and also mirrored at https://gitlab.com/gnuradio_book, and each chapter starts by referring to the relevant IQEngine record. For optimal layout of the proposed flowcharts, it will be assumed that GNU Radio Companion is configured (View menu) with the Show parameter value in block. All flowcharts have been tested with the 3.10 version of GNU Radio and GNU Radio Companion.

This discussion on using GNU Radio aims at a balance between processing synthetic signals and live signals collected from hardware. Indeed, the philosophy of SDR is to minimize the specificity of hardware and move most processing steps to software.

The book is organized as follows. The first chapter introduces GNU Radio and GNU Radio Companion as tools for becoming familiar, through simulations, with basic concepts needed when processing radio frequency signals, including the manipulation of complex numbers and baseband versus radio frequency bands. The second chapter extends the processing to real signals collected from hardware, with records available to readers for reproducing the processing steps if such signals are not available at their location or if the relevant hardware is not available. The third chapter tackles the communication between GNU Radio and external tools, either through network sockets or filesystems, introducing concepts needed in the following chapters. The fourth chapter benefits from all these knowledge to demonstrate how to assemble various SDR RADAR architectures, whether passive or active, and how accessing the raw radio frequency samples allows for target range and velocity detection as well as azimuth when combined with a moving antenna for spatial diversity of the sources in the synthetic aperture RADAR implementation. The fifth chapter returns to some basic concepts of GNU Radio for synchronizing processing tasks and propagating tags marking some features detected in the processed signal. The newly acquired knowledge is used in the sixth chapter to develop a complete digital communication system. The seventh chapter extends the custom digital communication system to decoding all layers of a satellite communication system using the same underlying modulation scheme: being a low-earth orbiting (LEO) satellite flying every day over every area in the world, the signal is accessible to all readers irrelevant of their geographical setting. While spaceborne communication benefits from ideal propagation conditions in free space, ground-based communication is plagued with multipath interferences and fading, an issue tackled in the eighth chapter using orthogonal frequency division multiplexing (OFDM) as implemented in the digital audio broadcast standard. Finally, the ninth chapter develops how the open-source GNU Radio framework can be complemented with custom processing blocks written in Python or C++, with an emphasis on custom source blocks for adding new hardware to the processing chain or new sinks for implementing processing or decoding protocols not yet supported by the standard GNU Radio processing blocks.

# 1

## Getting Started with GNU Radio: Synthetic Signals

This first chapter aims at achieving three outcomes; introducing the general concepts of software-defined radio (SDR) and how to reduce to a minimum the hardware dependence of the processing to move all digital signal processing steps after the analog-to-digital conversion; justifying the handling of complex numbers with a real and imaginary by GNU Radio; and becoming familiar with the GNU Radio Companion graphical user interface (GUI). These goals will be reached by using GNU Radio to process synthetic signals so that no hardware is needed to complete this first chapter. All GNU Radio Companion flowgraphs presented in this and the subsequent chapters are available from the GitHub repository at https://gitlab.xlim.fr/gnuradio_book, also mirrored at https://gitlab.com/gnuradio_book. When opening these flowcharts, it is assumed that `View → Show parameter expressions in block` and `Show parameter value in block` as well as `Show Block comments` are active for best layout experience.

## 1.1 Evolution of Radio Frequency Electronics Toward SDR

The term "software radio" was coined by J. Mitola in the early 1990s [Mitola, 1993]. The main idea of this technology is to reduce the analog electronics of a radio receiver to the part near the antenna, namely the radio frequency (RF) front end (RF amplifier + filters), and use a high-speed analog-to-digital converter (ADC) to get the baseband signal in the digital domain. All the usual operations (demodulation, filtering, etc.) are then performed digitally with the help of a digital signal processor (DSP). This would represent the ultimate SDR solution which is achievable today, though still quite expensive, for example, AMD-Xilinx Zynq Ultrascale+ RFSoC field programmable gate arrays (FPGAs) [AMD-Xilinx, 2024]. In this case, 5.9 GSps ADCs and 10 GSps digital to analog converters (DACs) are available, allowing to directly sample signals with a carrier frequency up to about 3 GHz. It is widely acknowledged that in telecommunications, a carrier frequency is employed to transmit information from one point to another. The carrier itself does not convey information; thus, it is removed at the receiving end. This is achieved by multiplying the signal received from the antenna by a local replica of the carrier. Consequently, this process results in a baseband spectrum containing the transmitted information, centered around 0 Hz. In the not-so-distant past, these operations were carried out using analog electronic components. In the case of the ultimate SDR system, the concept is to execute all demodulation and decoding functions in the digital domain. To gain a comprehensive understanding of the SDR hardware to be utilized in Chapter 2 (i.e. Adalm-Pluto and RTL-SDR), it is essential to grasp the fundamental concepts that have driven modern SDR architectures, particularly those constructed around the **zero intermediate frequency** structure (zero-IF or ZIF). Let us concentrate on the receiving side. It is widely recognized that the heterodyne

*Communication Systems Engineering with GNU Radio: A Hands-on Approach,*
First Edition. Jean-Michel Friedt and Hervé Boeglen.
© 2025 John Wiley & Sons, Inc. Published 2025 by John Wiley & Sons, Inc.
Companion website: www.wiley.com/go/friedtcommunication

structure is the most commonly used in analog receivers. Therefore, to achieve efficient reception of various channels, the demodulation stage operates at a fixed **intermediate frequency** (IF), typically 10.7 MHz for the frequency modulation (FM) broadcast band. All the electronic parts of this section are optimized for this fixed IF frequency (amplifiers, filters, etc.). The frequency translation of the signal coming from the antenna (e.g. 88–108 MHz for FM broadcast) is performed by a mixer and a tunable local oscillator structure. In the case of SDR, having an IF is not desirable as our aim is to process only the baseband signal. Therefore, the direct-conversion or ZIF architecture is what we seek. Although this structure has been known for quite some time, it was not until the 1990s, with advancements in integrated electronics, that it became a viable option. Let us now illustrate this evolution with two different SDR hardware platforms that we will be utilizing in Chapter 2. The first one, whose structure is illustrated in Figure 1.1, is the RTL-SDR receiver, initially designed for digital TV reception. It adopts a so-called zero second IF structure and uses an analog front end, which down-converts the RF signal to a user-defined IF frequency. This IF signal is then sampled by an 8-bit ADC at 28.8 MHz. Interestingly, the ADC is followed by a two-channel structure called an **IQ (in-phase and quadrature)** demodulator. The IQ structure is a fundamental constant found in any SDR system, and we will come back to it shortly.

The second hardware example is the Adalm-Pluto from Analog Devices Inc. (ADI) [Analog Devices, 2024] (Figure 1.2). It integrates a radio chip, which is a complete $2 \times 2$ transceiver (AD9363, see Figure 1.3), working on carrier frequencies between 325 MHz and 3.8 GHz and an instantaneous bandwidth of 56 MHz. This is a complete ZIF transceiver and one of the first efficient implementations of this apparently simple structure. For completeness, it is important to specify to the reader that the ZIF structure is not without its issues, which partly explains why its introduction is relatively recent. These issues include local oscillator (LO) leakage, DC offsets, and IQ mismatch, which may require compensation during a relatively complex calibration stage. Interested readers can refer to Razavi's book [Razavi, 2012], which delves into this subject in great detail.

## 1.2 The Complex Envelope and the Justification of the IQ Structure

The fundamentals of signal processing teach us that the Fourier transform of a real-valued signal is complex symmetric. This implies that in the frequency domain, negative frequency components emerge redundantly alongside the positive frequencies. However, negative frequencies do not exist in the physical domain; this phenomenon can be considered as an annoying mathematical artifact that often disrupts most students! It is, however, possible to build a time signal known as an **analytic signal**, whose Fourier transform exclusively reveals positive frequencies. This analytic signal, being complex valued, encompasses the original real-valued signal. The concept of the analytic signal originates from the contributions of Gabor [1946], Ville [1948, 1958], as stated by Viswanathan [2017]. Interested readers can explore classical signal processing literature for a more comprehensive understanding. The application of the analytic signal holds particular significance in numerous signal processing contexts, especially within the software-defined radio (SDR) domain. The theory concerning the analytic signal is intricately connected to the concept of the **complex envelope**, a well-established notion within the telecommunication community. For further insight into constructing the complex envelope of a bandpass signal $s(t)$ from its analytical representation, readers are encouraged to consult [Guimaraes, 2020]. Let us now consider a modulated **bandpass signal** $s(t)$ having a center frequency of $f_c$. It can be shown [Guimaraes, 2020] that

$$s(t) = Re\left[\tilde{s}(t)\exp(j2\pi f_c t)\right] \tag{1.1}$$

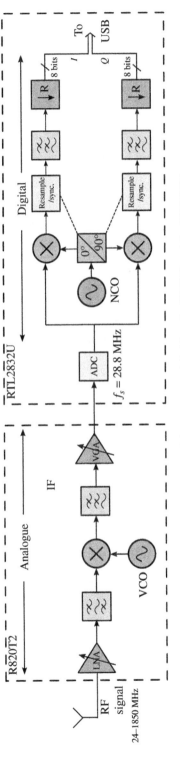

**Figure 1.1** Block diagram depicting the RTL-SDR receiver, comprising two essential chips: the R820T2 and the RTL2832U.

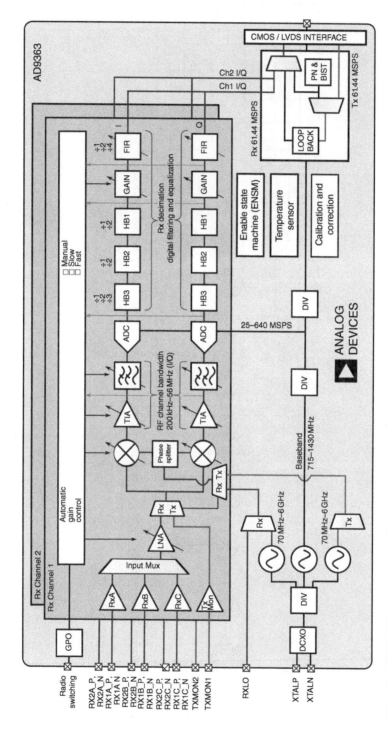

**Figure 1.2** ADI AD9363 IC receiver section. This radio chip is integrated into the Adalm-Pluto device.

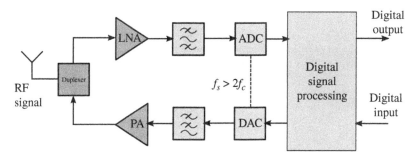

**Figure 1.3** The ultimate software-defined radio.

where $\tilde{s}(t)$ is the **baseband signal** defined as the **"complex envelope"** of $s(t)$. Considering that the baseband signal is of complex form (this is the case for digital modulation schemes), it can be written as

$$\tilde{s}(t) = s_I(t) + js_Q(t) \tag{1.2}$$

where $s_I(t)$ and $s_Q(t)$ are usually referred to as the **in-phase** (or direct) component and the **"quadrature"** component of $\tilde{s}(t)$, respectively. Using Euler's formula $\exp(jx) = \cos(x) + j\sin(x)$ and substituting 1.2 into 1.1 yields the important relation which allows to build the well-known **IQ modulator** structure:

$$s(t) = s_I(t)\cos\left(2\pi f_c t\right) - s_Q(t)\sin\left(2\pi f_c t\right) \tag{1.3}$$

The complex envelope of 1.1 can also be written in polar form as

$$\tilde{s}(t) = A(t)\exp(j\vartheta(t)) \tag{1.4}$$

So that $s(t)$ can be written in the general form:

$$s(t) = Re\left[A(t)\exp(j\vartheta(t))\exp(j2\pi f_c t)\right] = A(t)\cos\left[2\pi f_c t + \vartheta(t)\right] \tag{1.5}$$

with being $A(t)$ the envelope of the passband signal $s(t)$. The general form $s(t) = A(t)\cos\left[2\pi f_c t + \vartheta(t)\right]$ can be used to represent usual modulated signals. As an example, if $A(t)$ is constant and $\vartheta(t)$ defines $M$ discrete phases, then $s(t)$ yields an $M$-ary phase-shift keying (MPSK) signal.

Understanding the complex envelope representation, which leads to the IQ modulator/demodulator structures, is crucial for comprehending the principles underlying the design of SDRs. This importance stems from the fact that the complex envelope, denoted as $\tilde{s}(t)$, serves as an alternative baseband version of the passband signal $s(t)$, thereby necessitating lower sampling rates. Consequently, this reduction in sampling rate also decreases the computational workload. In the SDR context, the $s_I(t)$ and $s_Q(t)$ baseband signals enable efficient processing in the digital domain, utilizing DSPs, FPGAs, or computers. As an illustration, a specific modulated signal can be digitally generated and subsequently converted into the analog domain by utilizing its complex envelope represented as $s_I(t)$ and $s_Q(t)$. These signals are then fed into an **IQ modulator**, which is the electronic structure responsible for generating $s(t)$ from $s_I(t)$ and $s_Q(t)$, via equation 1.3. Figure 1.4a depicts the structure of the IQ modulator used in SDR. The analogy also extends to the passband received signal. The in-phase and quadrature baseband signals obtained from the received signal, which are then processed in the digital domain, are acquired by passing $s(t)$ through an **IQ demodulator** followed by appropriate low-pass filters (LPFs). Figure 1.4b illustrates this typical IQ demodulator structure. The reader can also refer to Figures 1.1 and 1.2 representing two common SDR hardware where these structures are readily identifiable.

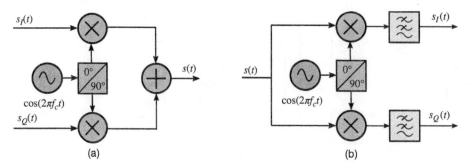

**Figure 1.4** Simplified block diagram of an IQ modulator (a) and IQ demodulator (b).

In addition to the complex envelope approach, the IQ structure can also be derived intuitively by considering a baseband signal that has been digitized by an ADC. Interestingly, this approach can be linked to the "self-corruption" phenomenon encountered during the down-conversion process to a zero-IF of an "asymmetrically modulated" signal, as explained in section 4.2.2 of [Razavi, 2012]. Unlike AM signals, which produce a symmetric spectrum around the carrier frequency, FM and most digital modulation schemes (e.g. Frequency-Shift Keying [FSK], MPSK) do not, and are therefore subject to self-corruption in the mixing process to baseband. The only way to circumvent this issue is to utilize the IQ structure, which is derived intuitively in what follows.

We remind that the **radio frequency carrier is never involved** in SDR digital processing since it has been removed during the preliminary analog step, justifying the spectra displayed later by GNU Radio Companion as centered on 0 Hz rather than on the radio frequency carrier.

With a single mixer used to transpose the time $t$ dependent radio frequency signal $s(t) = A\cos(\omega t + \varphi)$ at angular frequency $\omega = 2\pi f$ with $f$ as the radio frequency carrier frequency and $\varphi$ the phase and amplitude $A$, to baseband $b(t)$, then the output of the mixer tuned to a nearby frequency $f'$ would be expressed as $b(t) \propto \cos((\omega + \omega')t + \varphi) + \cos((\omega - \omega')t + \varphi)$, and a low-pass filter rejects the frequency sum so that only the frequency difference at baseband $\cos((\omega - \omega')t + \varphi)$ is sampled. If the radio frequency receiver is properly tuned, then $\omega = \omega'$, and we are only left with $\cos\varphi$. This phase component includes two wave propagation terms: the time-dependent part and the space-dependent part, classically expressed as $\varphi = \vec{k}\cdot\vec{d}$ with $\vec{k}$ as the wavevector with magnitude $2\pi/\lambda$, where $\lambda$ is the electromagnetic wavelength, and $d$ the movement of the receiver with respect to the emitter. Rather than expressing $\lambda = c/f$ with $c = 3\times 10^8$ m/s and $f$ a frequency in hertz, it is most convenient to normalize both numerator and denominator of the wavelength expression by $10^6$ to remember that

$$\lambda = 300/f$$

when $f$ is expressed in MHz and $\lambda$ in meters. Thus, a 100 MHz signal is associated with a 3 m wavelength, which induces a phase cancellation every 1.5 m: were a single mixer used, the broadcast commercial frequency-modulated signal commonly used in most of the world would cancel as the receiver is moving every 1.5 m, a very inconvenient listening experience.

Hence, a second signal must be generated that is maximized when $\cos\varphi$ vanishes, and most intuitively the trigonometric complement $\sin\varphi$ will solve the problem, generated by mixing the incoming signal $s(t)$ with a second copy of the local oscillator tuned to $\omega'$ but this time shifted by 90° since $\cos(\vartheta + 90) = \sin(\vartheta)$ and $s(t)\cdot\sin(\omega't) \propto \sin((\omega+\omega')t + \varphi) + \sin((\omega-\omega')t + \varphi)$ with again the frequency sum component filtered out by a low-pass filter. Hence we have created two outputs from the analog processing of the real signal $s(t)$ received by the antenna, $A\cos\varphi$ and $A\sin\varphi$

when $\omega = \omega'$. Both these signals will be sampled by two synchronous ADCs, and the resulting signals will be called $I$ and $Q$ for in-phase and quadrature. These two signals are handled with the complex number algebra by considering $I$ as the real part and $Q$ as the imaginary part so that $I + jQ$ provides the expected properties of extracting the amplitude as $A = |I + jQ| = \sqrt{I^2 + Q^2}$ and $\varphi = arg(I + jQ) = atan2(I, Q)$ with $atan2$ being the arc-tangent function considering the sign of $I$ and $Q$, where $j^2 = -1$.

The two conclusions from these developments are as follows:

1. SDR will naturally handle complex numbers characterized by a real part $I$ and an imaginary part $Q$, and the default configuration of most GNU Radio processing blocks is to handle such quantities.
2. There is no such thing as an imaginary voltage or a negative frequency: these concepts will be addressed experimentally in Section 1.3 to illustrate the teaching skills of the GUI of GNU Radio Companion. Moreover, the concept of complex envelope we presented in Section 1.2 should also have helped to clarify these elements.

## 1.3 Complex Number Manipulation

As we are convinced that complex numbers naturally represent at baseband the radio frequency signal $s(t)$ after transposition, let us remind some of the basic properties of complex number algebra.

A complex number $I + jQ$ is also expressed in its Euler notation $A \exp(j\varphi)$ with $A$ as the amplitude and $\varphi$ the phase, which then becomes $A \cos \varphi + jA \sin \varphi$. This expression is most convenient when considering mixers or, in their digital representation, multiplications since $A \exp(j\varphi) \times A' \exp(j\varphi') = AA' \exp(j(\varphi + \varphi'))$.

From these considerations, the real trigonometric functions are expressed as $\cos \vartheta = \frac{1}{2}(\exp(\vartheta) + \exp(-\vartheta))$ and $\sin \vartheta = \frac{1}{2j}(\exp(\vartheta) - \exp(-\vartheta))$, emphasizing how the magnitude of the spectrum of a real signal is even, $\cos(\omega t)$ being represented by two spectral components at $\omega$ and $-\omega$. On the other hand, a non-even spectrum is necessarily represented by complex quantities.

Finally, since the amplitude and the phase – the frequency $f$ being the time derivative of the phase $\varphi$ as $f = \frac{1}{2\pi}\frac{d\varphi}{dt}$ – are the core characteristics of a signal, a representation in the plane where the real part is along the horizontal axis, and the imaginary part along the vertical axis allows for an intuitive representation of the amplitude as the distance to the origin and the phase as the angle to the abscissa. Such a diagram is called a **phasor diagram** or, in SDR, a **"constellation diagram"** and will be used whenever digital modulation is considered.

## 1.4 GNU Radio and GNU Radio Companion

Having reminded the reader of the basic mathematical relations needed to get started, these rather theoretical concepts are used to become familiar with the GNU Radio Companion GUI. Indeed, the core of GNU Radio is a scheduler transferring data from one processing block to another, each block having access to the IQ stream. Most GNU Radio processing functions are written in C++ for efficiency, although GNU Radio can accept processing blocks written in Python. Connecting processing blocks with each other to define the dataflow is achieved by programming either in C++ or in Python. To get started, though, a GUI allows for selecting processing blocks from a list of available functions and connect blocks to route signals: **GNU Radio Companion** is a Python or C++ code generator so that all arguments in the fields for configuring the blocks should be syntactically correct Python commands. The generation of a Python script not necessarily relying on a GUI to be executed will be most welcome when addressing embedded systems, as will be shown in some of the coming Chapters 4 and 9. Once the reader has become familiar with GNU Radio block connection, programming in Python or C++ becomes straightforward, although GNU Radio Companion always remains an enjoyable prototyping tool. Remember, though, that the IQ samples are **only available to each processing block** and not the program connecting the blocks together: new signal processing features do require writing a new processing block. Although we should emphasize that most core processing functions are available, it is worth noting that if some functions do not exist, they can also be managed externally with dedicated tools (Python, GNU Octave,[1] etc.).

GNU Radio is a complex piece of software requiring multiple dependencies, and we strongly advise the new user to benefit from the packaging system of their favorite GNU/Linux distribution, taking care of meeting all dependencies. However, this solution has the drawback that distribution package versions lag significantly behind the current version. This can pose a real problem in cases where the package version contains bugs, a situation that has occurred frequently with the GNU Radio packages in the Ubuntu distribution in the past. On Ubuntu, this issue is now partially solved as the latest GNU Radio builds are maintained as PPAs (see [GNU Radio Linux Install, 2024] for more details). The advanced user will benefit from the latest releases by manually installing GNU Radio from the source (available on GitHub). Please note that we do not recommend the use of PyBOMBS. Indeed, as indicated on the GNU Radio Wiki, "we are no longer including PyBOMBS as a recommended method of installing GNU Radio, unless you want to play around with old versions (i.e. 3.7 or 3.8 versions)." Our experience with installing GNU Radio from source has been positive, as it typically takes less than five minutes to build on a recent GNU/Linux machine by cloning the source from Git and executing the standard CMake commands to build and install the software. While ensuring the necessary dependencies are met is essential, this process is well documented on the dedicated page of the GNU Radio Wiki [GNU Radio Linux Install, 2024].

Furthermore, an efficient alternative method for installing GNU Radio is by using a specific version of the Anaconda distribution called `radioconda` [GNU Radio Conda Install, 2024]. This approach has the advantage of being platform agnostic and has allowed many Microsoft Windows users to discover GNU Radio. Radioconda utilizes Python virtual environments, enabling the use of different GNU Radio versions on the same machine without affecting the native system. It is even possible to build and integrate out-of-tree (OOT) modules within these virtual environments. However, building OOTs on a Windows system can be challenging due to the need to install and configure a specific version of the bulky Microsoft Visual Studio-based compiling tools. The process is significantly simpler on GNU/Linux, leveraging the native GNU Compiler Collection (GCC) compiler inherent to all GNU/Linux-based systems, in the original tradition of "*the fact that non-Unix operating systems don't come bundled with development tools meant that very little source was passed over them. Thus, no tradition of collaborative hacking developed*" [Raymond, 2001].

Finally, we draw the reader's attention on the fact that *mixing distribution binaries and custom compilation will be the source of most problems when executing GNU Radio*. One has to make sure to *either* use distribution binaries *or* manually compiled libraries and executables.

---

1 Throughout this book, the free open-source implementation of the Matlab language will be referred to as "Octave" as found at octave.org, despite the full project name of GNU Octave as part of the GNU project and licensed as GNU General Public License (GPL). All processing scripts referring to this language were developed and tested with GNU Octave version 8.4.0.

After the installation of GNU Radio, the graphical assistant GNU Radio Companion is launched by executing the following command from a terminal:

```
gnuradio-companion
```

Please note that all examples in this book have been tested with GNU Radio 3.10 running on GNU/Linux. Upon start-up, a clean sheet is displayed with two windows: one named "Options" and a Variable named "samp_rate." The Variable block is a Python assignment where the name of the block is the name of the variable, and its argument is the value assigned to the name. The samp_rate variable is the sampling rate that will define time throughout the processing chain. Indeed, SDR is not aware of the concept of time since it handles discrete samples: only at the analog-to-digital conversion step is time a relevant quantity, with each sample collected at time intervals inverse of the sampling rate. Once the data have been collected, assuming no sample is lost during data transfer and processing, then all signal handling could be asynchronous, using first in first out (FIFO) memories, pipelined or by blocks, each sample index uniquely defined from the beginning of the stream, the time at which it was sampled. The second Option window defines whether the processing chain uses GUI features (QT GUI) or if it can run on a system with no such feature (No GUI), or if the processing chain is a hierarchical block grouping multiple processing steps (Hier Block).

After launching GNU Radio Companion, the list of all processing blocks is displayed on the right menu. Enumerating all possible processing blocks would be a time-consuming activity, so we can benefit from the Search function of the magnifying glass in the top icons: searching for the keyword "Source" allows for identifying the various virtual or hardware signal generators. Indeed, the GNU Radio processing chain considers Source as the new data to be processed through a set of signal processing blocks until the data reach a Sink such as a sound card, a graphical interface, or an output (file, network socket, etc.). Double-clicking on the Signal Source provides a virtual signal generator.

This first processing block output is drawn in blue. The color encodes the type of data handled by the block. In the Help menu, Types lists all available data types (Figure 1.5), including the blue complex float 32 (single-precision floating-point number complex) but also the orange Float 32 (single-precision floating-point number real) and the various integers (8 bits in purple, 16 bits in yellow, 32 bits in green). In this first approach to handling signals with GNU Radio, we wish to address a real voltage as would be seen at the output of a physical signal generator, so by double-clicking on the signal source block, the data type is replaced from complex to float, and upon validation, the output color becomes orange. Displaying the signal on an oscilloscope in the time domain requires the appropriate sink: inquiring in the magnifying glass icon menu about QT, the QT Instrumentation menu appears, including the QT GUI Time Sink, the name of the oscilloscope. Again selecting this block assumes a complex input with a blue square, and double-clicking on the QT GUI Time Sink block allows for changing the data type from complex to float.

We might wish to connect the Signal Source output to the QT GUI Time Sink input to display the time dependence of the signal, but doing so would be unwise. The GNU Radio scheduler is not aware of the concept of time and only knows sample indices. If a hardware signal source (e.g. ADC or sound card microphone) or a hardware signal sink (e.g. DAC or sound card speaker) is included in the flowgraph, such peripherals will clock the samples and slow down the scheduler until all data have been produced or consumed. But in the current flowgraph with only a virtual Signal Source and virtual QT GUI Time Sink, no hardware clock defines the flow of time.

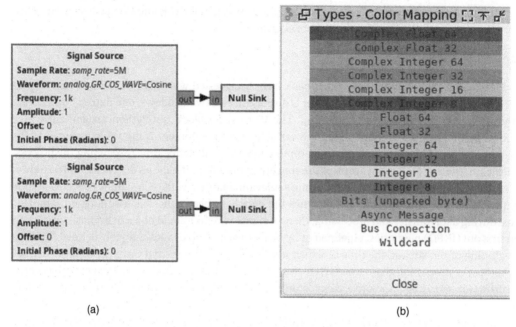

**Figure 1.5** (a) Each block output must be connected to an input of the same type. Here, the null sink just absorbs and destroys any input item. (b) The window created when requesting Help → Types with the color coding displayed by the GNU Radio Companion interface of the output and input type of each block, most usually blue for floating point complex (real and imaginary) numbers and orange for floating point real numbers. We will also use the 8-bit integers (purple) and 16-bit integers (yellow) when saving bandwidth or manipulating bytes and words. Source: GNU Radio project.

Under such circumstances, and **only if no hardware source or sink is included in the flowgraph,** a block defining a software time must be included: the Throttle block. Selecting this block exhibits again complex input and output, which are switched to float by double-clicking on the block to change configuration, and the first flowgraph should look like Figure 1.6. The comments under the blocks are added by double-clicking on a block and adding the comment under the Advanced tab in the Comment field. As with any software, including comments that help the reader understand the parameter selection of each block is always useful when returning later to a complex flowchart and trying to remember the design rationale.

A real signal (e.g. a voltage) can be displayed on an oscilloscope or in this virtual environment in a time sink. Only real quantities have a physical meaning, whereas the complex (real and imaginary) values of the IQ stream, naturally handled by SDR in general and GNU Radio here, will be an artifact to fully use the available bandwidth from $-f_s/2$ to $+f_s/2$ – including the negative frequency range – with $f_s$ the front-end analog-to-digital sampling rate. Indeed, the magnitude of the Fourier transform is even since $\cos(x) = \frac{1}{2}(\exp(jx) + \exp(-jx))$ so that the negative and positive parts of the spectrum include the same redundant information. This mathematical relation is again illustrated using GNU Radio, this time by adding a spectrum analyzed called the QT Frequency Sink (Figure 1.7)

Hence, a natural representation of baseband signal is a complex quantity made of an in phase part ($I$) and a quadrature part ($Q$), each resulting from mixing the radio frequency signal with a local oscillator LO expressed as $\cos(\omega_{LO}t)$, a copy of which in quadrature expressed as $\sin(\omega_{LO}t)$ generates $Q$ since $\cos \times \sin \propto \sin$ while $\cos \times \cos \propto \cos$. These two quantities meet the expected

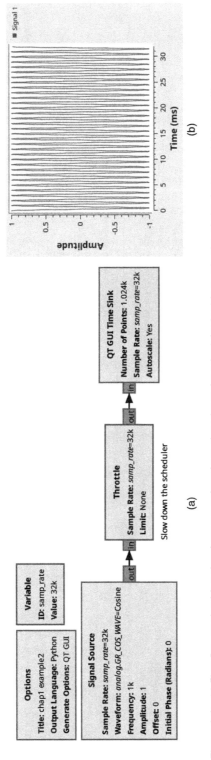

**Figure 1.6** (a) First flowgraph with a signal source, a throttle to slow down the scheduler in this purely virtual simulation, and (b) an oscilloscope output as the QT time sink.

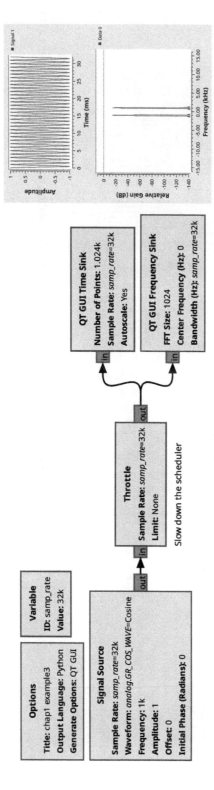

**Figure 1.7** Adding a spectrum analyzed called the QT frequency sink for displaying the real signal Fourier domain representation as a function of frequency.

complex number algebra that the magnitude of $|I + jQ|$ is the amplitude of the signal and $arg(I + jQ)$ is the phase of the signal. However, handling these complex quantities brings us into an abstract world best addressed with GNU Radio. The time domain now exhibits two curves: the real and imaginary part, with the former being late or early with respect to the second depending on whether the frequency is negative or positive (Figure 1.8).

However, the even symmetry of the spectrum is lost, and now the signal can fully spread from $-f_s/2$ to $+fs_2$ with no redundancy since the expression of $\exp(j\omega t)$ separates the contribution of positive angular frequency $\omega$ and negative angular frequency (Figure 1.9).

## 1.5   Sample Rates, Decimation and Aliasing

When a new flowgraph is created with GNU Radio Companion, a variable named "`samp_rate`" is created. This fundamental quantity representing the sampling rate of the ADC clocks the flowchart. GNU Radio is not aware of time but only of the sample index so that knowing any symbol index in the stream defines the time at which it was sampled. Despite sharing information from one processing block to the next as buffers asynchronously transferred with undetermined latency, knowing the sample index is enough to recover its collection time. Hence, pipelined stream processing is not necessary, but accumulating buffers for efficient memory transfer will be used for transferring data from one processing step to the next.

As a signal is being demodulated and processed, information can only be lost and never created. From Shannon's capacity theorem [Shannon, 1949], information is determined by bandwidth and signal-to-noise ratio: as information is removed during the reception steps, less and less bandwidth is needed. Reducing $f_s$ involves only keeping one every $D > 1$ sample, a processing step called **decimation**. Many GNU Radio processing blocks include such a decimation step **after** the signal processing is completed and excess samples can be dropped. Decimating by a factor $D$ means reducing the sampling rate from $f_s$ to $f_s/D$, lowering the computational load of more complex processing steps as we are moving forward in the reception chain. GNU Radio will not track the decimation factors throughout the processing chain, and the developer must make sure to inform all processing blocks following a decimation by $D$ that the new sampling rate is `samp_rate/D`. As in all programming language, processing consistency is maintained when a quantity is used multiple times in a program by creating a variable and using this variable whenever necessary. GNU Radio Companion is a Python code generator, and all fields are Python-compatible expressions, with the `Variable` block creating a new variable whose name is its ID and its value the associated field, possibly a Python expression.

Decimation is another opportunity to use GNU Radio to illustrate a basic concept of discrete-time signal processing, namely the periodicity of the spectrum and aliasing. Analyzing a data stream sampled at $f_s$ in the Fourier domain exhibits a signal spanning from $-f_s/2$ to $+f_s/2$ so that after a decimation by a factor $D$, the new frequency domain ranges from $-f_s/(2D)$ to $+f_s/(2D)$.

---

**Frequency range of the discrete Fourier transform**

While in continuous time signal processing, the frequency range of the Fourier transform spans from 0 to $f_s$, or from the spectrum periodicity, from $-f_s/2$ to $+f_s/2$, it is important to notice that in discrete time, a Fourier transform computed on $N$ samples will lead to $N$ complex

---

*(Continued)*

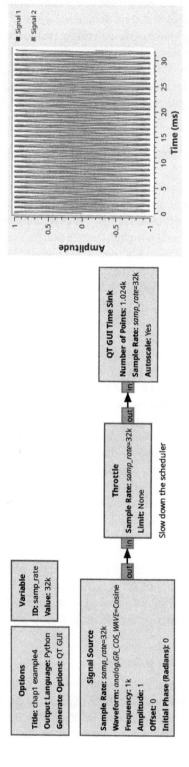

**Figure 1.8** SDR naturally handles complex numbers: replacing the real-valued quantities with complex quantities doubles the information content as emphasized on the time-domain (oscilloscope) plot now displaying two curves, a real part and imaginary part.

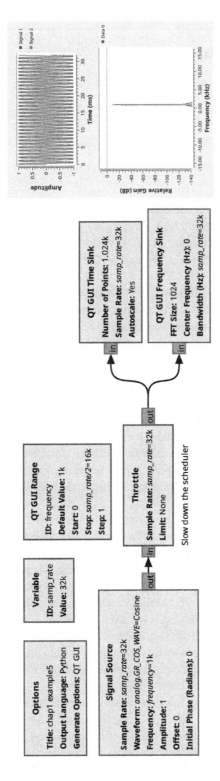

**Figure 1.9** The magnitude of the spectrum of the complex quantity is no longer an even function, and in the case of a periodic signal, the positive and negative frequencies are separated depending on whether the in-phase component is early or late with respect to the quadrature component.

> **(Continued)**
>
> coefficients in the spectral domain at frequencies ranging from $-f_s/2$ to $+f_s/2 - f_s/N$. Using the Python language, the abscissa is expressed as
>
> ```python
> import numpy as np
> fs = 32000;                          # sampling frequency
> t = np.linspace(0, 1, fs)            # sampling time
> s = np.sin(2*np.pi*2000*t)+0.1       # signal
> freq = np.linspace(-fs/2, fs/2-fs/len(s), len(s)) # frequency axis
> from matplotlib import pyplot as plt
> plt.plot(freq,np.fft.fftshift(np.fft.fft(s)))
> plt.show()
> ```
>
> which will exhibit the two spectral components of the real-valued sine wave at $-2000$ and $+2000$ Hz on a frequency plot ranging from $-f_s/2$ to $+f_s/2 - f_s/N$, with $N$ being the number of samples in the input signal. A constant DC offset was introduced in the signal to emphasize how this spectral component is correctly centered on 0 Hz with this frequency axis definition. Notice that Python (and Matlab or GNU/Octave) follow the convention that after computing the Fourier transform using the `fft()` function, the first sample is representative of the DC (0 Hz) component and the last sample is representative of the highest spectral component at $f_s - f_s/N$. Shifting to the baseband representation spanning from $-f_s/2$ to $+f_s/2$ requires swapping the upper and lower halves of the spectrum, as achieved with the `fftshift` function.

The spectrum periodicity of discrete-time signal processing is illustrated in Figure 1.10 by explicitly decimating the data stream by using the `Keep 1 in N` block. In order to make the demonstration interactive, rather than creating a static variable using the `Variable` block, we can create a QT widget for manipulating variable: a slider is called a `QT GUI Range`, which creates a variable with the name in the `ID` field and a default value between a minimum and maximum interval. This variable can be used in any processing block, here for example as the `Frequency` argument of the `Signal Source` block. Being a frequency definition, the bounds for the variable definition are `-samp_rate/2` to `+samp_rate/2`.

In Figure 1.10, the slider defining the sine wave frequency is moved to higher frequencies until reaching $f_s/(2D) = f_s/8$ since here D=4, leading the spectral component to leave the baseband first Nyquist zone through the right side of the chart to return by the left side of the chart. While aliasing is usually considered a hindrance as will be discussed in Section 1.6, it can be efficiently used on purpose. As an example, the continuous wave signal of the GRAVES space surveillance radio detection and ranging (RADAR) emitted from North-Eastern France at 143.05 MHz is sampled by the radio frequency front-end BasicRX of an Ettus Research SDR receiver at a rate of 200 Msamples/s so that the spectra range from $-100$ to $100$ MHz, and the 143.05 MHz appears at $-43.5$ or 56.5 MHz. Because we observe that no emission occurs on this frequency, only the RADAR signal is detected, low-pass filtered, and decimated for observing the Doppler shift introduced by moving planes illuminated by the RADAR beam and backscattered. Notice that the Doppler shift in a RADAR system is **twice** the classical carrier frequency times the target velocity projection to the observer divided by the speed or light so that in the case of GRAVES, the frequency offset induced by a target moving at velocity $v$ m/s happens to be about $v$ Hz, or frequency offsets in the $\pm 300$ Hz for civilian aircrafts. The flowchart in Figure 1.11 illustrates how the basic concepts developed so far can be used in a practical application.

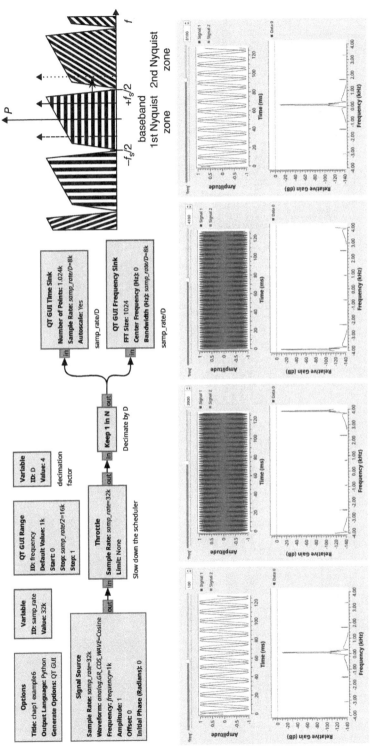

**Figure 1.10** Illustration of the discrete-time sampled periodic spectrum, with a decimation by $D=4$ of the sine wave leading to aliasing when the spectral component leaves the Nyquist zone by the right (frequency close to $f_s/2$) and returns from the left (frequency close to $-f_s/2$). Top left: the GNU Radio Companion flow chart, and bottom screenshots depict the frequency is moved out of the first Nyquist zone and into the second, with aliasing occurring when the spectral component returns in the negative side of the spectrum. Top right defines the periodicity of discrete-time sampled spectra; with horizontal stripes the baseband, also called the first Nyquist zone with upward tilted stripes the second and beginning of the third Nyquist zones respectively to the right of the baseband (dotted vertical arrow to dashed vertical arrow), and how a signal leaving baseband is aliased back to the first Nyquist zone. The asymmetry of the spectral shape emphasizes that complex values are considered.

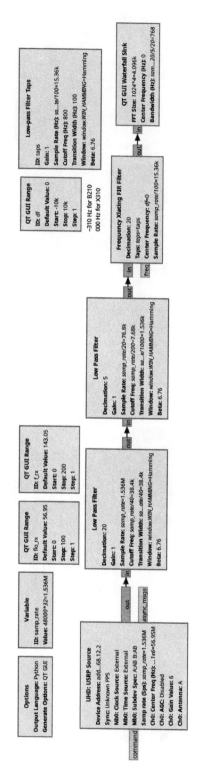

**Figure 1.11** GRAVES continuous wave RADAR flowchart using the second Nyquist zone of the Ettus Research X310 SDR receiver fitted with a BasicRX daughter board: the 143.05 MHz signal is aliased when sampled at 200 MHz to 56.95 MHz as shown in the `flo_rx` variable.

An important aspect of the flowchart shown in Figure 1.11 is the **absence** of a `Throttle` block. Indeed this block was included in simulation flowcharts when no hardware is included in the processing sequence to hint at timing constraints, attempting to slow down the scheduler to process `samp_rate` every second with no hardware clock other than the computer timers. In Figure 1.11, a hardware data source is included – the ADC and the FPGA of the Ettus Research X310 board – so that the dataflow is constrained by the production of samples by the hardware. Including a `Throttle` block with its coarse timing constraints would conflict with this hardware-accurate clock, and hence this timing block must be removed. The result of executing this flowchart is shown in Figure 1.12. Notice, though, that the straight line in the middle of the chart, generated by the direct current (DC) output of mixing the GRAVES continuous wave with the local oscillator, is only straight because this X310 was referenced to a stable frequency source, namely a hydrogen maser. Running this flowchart from the internal temperature compensated crystal oscillator (TCXO) clocking the X310 will exhibit frequency fluctuations of several Hz to tens of Hz ($x$-axis) over the several minutes shown along the $y$-axis of the waterfall, emphasizing the need for stable local oscillators when using SDR for Doppler RADAR applications (Figure 1.12).

Analog anti-aliasing filter on the analog-to-digital front end not only aims at rejecting unwanted out-of-band (or upper Nyquist zones) signals but also removes unwanted noise aliased to baseband by the fast track and hold located before the ADC. As can be shown in the simulation of Figure 1.13, decimating without aliasing by using a `Keep 1 in N` block will accumulate noise in the baseband from energy conservation, while low-pass filtering prior to decimation will remove this unwanted noise floor and help extract weak signals. The ratio between the unfiltered and low-pass filtered decimated signals is equal to $10 \log_{10}(D)$ when decimating by a factor $D$: in this case where $D = 8$, the ratio is 9.03 dB.

When using GNU Radio blocks with multiple inputs, such as the `Multiply` or `Add` blocks, the data rate at all inputs must be the same: make sure that the decimation factors in all branches leading to a block with multiple inputs are all the same.

## 1.6  Low-pass Filtering or Working on Upper Nyquist Zones

Aliasing is avoided by low-pass filtering prior to decimating. Indeed, since the baseband frequency range is decreased from `-samp_rate/2` to `+samp_rate/2` to `-samp_rate/(2D)` to `+samp_rate/(2D)` when decimating by D, a low-pass filter with a cutoff frequency set to `samp_rate/(2D)` removes unwanted spectral components prior to decimating. GNU Radio provides a `Low Pass Filter` block requiring three parameters, namely the `Gain` – we can keep this as unity gain – `Cutoff Frequency` and `Transition Width`. Understanding this last parameter is fundamental to efficiently use available processing power while keeping computational resources within reasonable limits.

A low-pass filter can be implemented either in the frequency domain or in the time domain. The two representations are related by the convolution theorem, which relates the Fourier transform (FT) and the convolution $*$ defined at

$$(x * y)(\tau) = \int x(t) \times y(\tau - t)dt$$

or in its discretized form

$$(x * y)(n) = \sum_k x_k \times y_{n-k}$$

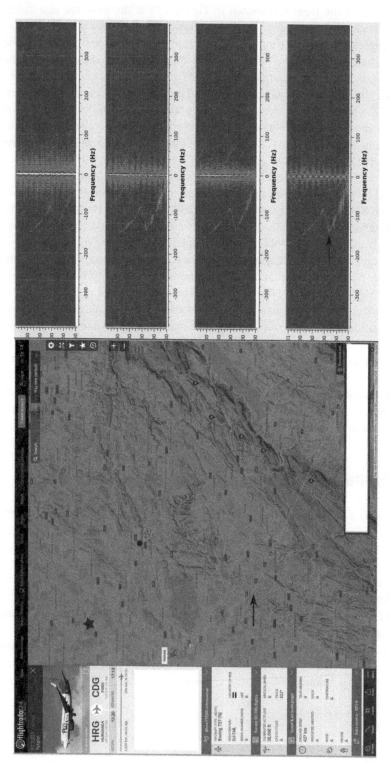

**Figure 1.12** Measurement result with four antennas connected to two Ettus Research X310 receivers, both fitted with two BasicRX boards and all synchronized by the same hydrogen maser frequency reference. The arrow on the right chart indicates a change in the Doppler shift from the expected arctangent shape due to the turning motion of the plane (left arrow) changing the velocity projection toward the receiver. Source: Flightradar24 AB.

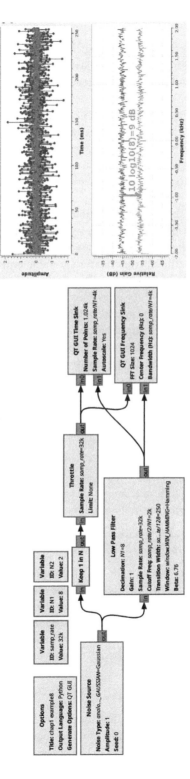

**Figure 1.13** Illustration of aliasing on the noise floor and the impact of the analog anti-aliasing filter on the analog-to-digital conversion stage to avoid the fast, wideband track and hold front end to alias broadband noise in the baseband.

**22** | *1 Getting Started with GNU Radio: Synthetic Signals*

as

$$FT(x * y) = FT(x) \times FT(y)$$

When defining the transition width of the filter $\delta f$, the spectral resolution of the Fourier transform must be better than the transition width, or the filter could not be defined sharply enough. A Fourier transform processed on $N$ samples will provide a frequency bin resolution of $f_s/N$ so that $f_s/N \leq \delta f \Leftrightarrow N \geq f_s/\delta f$. Hence, GNU Radio Companion filter design will aim at meeting the designer's request for a given transition width by selecting a number of filter coefficients, named taps, as `samp_rate/Transition Width`. Selecting too narrow a transition width will induce a filter with a unrealistically large number of taps requiring excessive computation load when calculating each output. Indeed, whether computed in the time domain or in the frequency domain, each output sample generated at a rate of the sampling rate divided by the decimation factor requires $N^2$ operations in the time domain or $N \cdot \log_2(N)$ in the frequency domain, thanks to the efficiency of the fast Fourier transform (FFT). If a sharp transition width is needed because an interfering noise is close to the wanted signal band and efficient approach is usually to first perform a coarse filtering followed by decimation, hence reducing the sampling rate and cascading low-pass filters with decimation until the sharp transition width only requires a reasonable number of taps. Reasonably it falls within the 32–128 coefficient range, so we define the transition width as `samp_rate/N` with N being the targeted number of taps.

The low-pass filter is instantiated in the flowgraph shown in Figure 1.14 and configured with an arbitrary cutoff frequency of a quarter of the sampling rate and a transition width of `samp_rate/128`, yielding an estimated 128-tap-long finite impulse response (FIR) filter. GNU Radio Companion is a Python code generator whose output can be read by editing the resulting Python program, as shown in the bottom window when running the `Generate the Flow Graph` icon (left of the arrow icon for executing). This program shows how the low pass filter is designed as

```
self.low_pass_filter_0 = filter.fir_filter_ccf(1, firdes.low_pass(1,
    samp_rate, (samp_rate/4), (samp_rate/256), window.WIN_HAMMING,
    6.76))
```

indicating that the filter name is `low_pass_filter_0`. Since the properties of the filter are dependent on a variable – the sampling rate `samp_rate` – a callback function updating the flowchart properties if this variable is dynamically modified during execution is defined as

```
self.low_pass_filter_0.set_taps(firdes.low_pass(1, self.samp_rate, (self
    .samp_rate/4), (self.samp_rate/256), window.WIN_HAMMING, 6.76))
```

hinting at a `set_taps()` method associated with the filter object. Displaying the coefficients is achieved by calling the `tap()` method of the filter class. However, modifying the Python code itself would be short-lived, since generating a new code from GNU Radio Companion would erase any update made in the code itself. GNU Radio provides, since version 3.8.1, a feature allowing for including Python code run at various steps of the flowchart initialization or execution: the `Python Snippet` [Müller, 2020] block allows for including a command executed after initialization. If we execute `After Init` the command

```
print(len(self.low_pass_filter_0.taps()))
```

we see in the terminal at the bottom of GNU Radio companion that selecting a transition width of `samp_rate/128` leads to a 309-long tap filter, while requesting a sharper transition of `samp_rate/256` doubles the number of taps to 617. The excessive computational burden of too sharp transition widths is hence demonstrated.

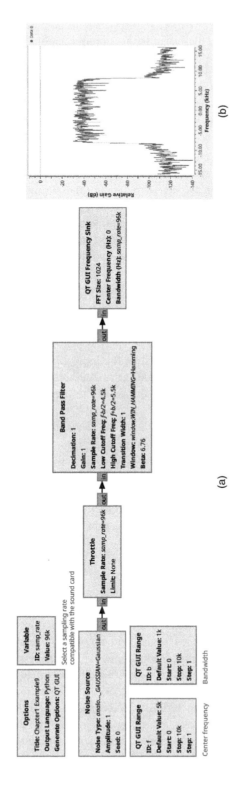

**Figure 1.14** (a) Band-pass filter block demonstration, fed by a noise source with a flat spectrum as the input, throttled in the absence of input or output hardware, restricting the dataflow to the GNU Radio scheduler, and displayed at the output of the spectrum representing the filter transfer function. (b) The sample rate of the QT GUI Frequency sink is used for defining the abscissa between minus half and plus half of the sampling rate.

**24** | *1 Getting Started with GNU Radio: Synthetic Signals*

Figure 1.14 also shows how after filtering, half of the allocated bandwidth is no longer necessary since no energy is seen from $\pm$samp_rate/4 to $\pm$samp_rate/2 so that a decimation factor of 2 will halve the data rate to be processed by the subsequent processing blocks with no loss of information.

We have introduced in Figure 1.11 another way of generating filter taps, namely the Low Pass Filter Taps block, whose parameters are the same as the Low Pass Filter block but whose output is not the ability to process a stream but to generate a vector of taps to be used in another processing block. In Figure 1.11, this processing block was a Frequency Xlating FIR Filter.

Indeed, frequency transposition followed by low-pass filtering to get rid of unwanted spectral components far from 0 Hz, followed by decimation to get rid of the unnecessary spectral components cancelled by the low-pass filter aimed at reducing the data rate, is so common that a single GNU Radio block performs all operations: the Xlating FIR Filter. This block multiplies the input data stream by $\exp(j2\pi f_{LO}t)$ with $f_{LO}$ being the frequency of a software numerically controlled oscillator with frequency $f_{LO}$ defined as the Center Frequency parameter of the block, followed by a filtering step by applying the taps and decimating. Doing so, the signal is frequency-shifted by $f_{LO}$ since an input signal expressed as $\exp(j2\pi ft)$ then becomes $\exp(j2\pi ft) \times \exp(j2\pi f_{LO}t) = \exp(j2\pi(f + f_{LO})t)$. In a discrete sampling sequence, the time is not continuous but expressed as a discrete sequence of $[0 : N - 1]/f_s$ with $N$ being the number of samples being processed and $f_s$ the sampling frequency. Once frequency-shifted, the resulting signal is filtered and then decimated. In summary, this block requires three sets of input parameters:

- the frequency of the local oscillator $f_{LO}$ with which the input signal is multiplied at its input sampling rate so that all input spectral components $f$ are shifted to $f - f_{LO}$ (Figure 1.15).
- the shape of the low pass filter defined by the taps of the convolution. In the example of Figure 1.15, the dedicated Low Pass Filter Taps block is used, with its ID defined with the same name as used to fill the Taps in the Xlating FIR block. Notice that alternatively, if no filtering was needed, then filling Taps of the Xlating FIR Filter with only the "1" coefficient would lead to a rotation of the spectrum without filtering. Indeed, in the expression of the convolution relating the output $y$ to the inputs $x$ weighted by the taps $b$ as $y_n = \sum b_k \cdot x_{n-k}$, then selecting $y_0 = 1$ and $y_k = 0$ for all $k > 0$ means that the output is expressed as $y_n = x_n$ or the identity.
- the decimation factor to be applied **after** the multiplication for spectrum rotation and low-pass filtering.

## 1.7 ADC and DAC Resolution

We consider here the statement about the finite resolution of the ADCs and the need for an analog frequency transposition from radio frequency to baseband. A $N$-bit ADC ideally exhibits a least significant bit step of $V_{ref} \times 2^{-N}$ when converting the incoming signal with respect to a fixed reference voltage $V_{ref}$. Once the pre-amplifier gain is tuned so that the full scale range is used, the ratio of the strongest to the weakest signal must be more than $10 \log_{10}(2^{-N})$ dB or $6.02 \times N$. For an 8-bit analog to digital convert as found in the low-cost RTL-SDR receivers, the dynamic range is hence only 48 dB. For a high-end SDR receiver, this dynamic range becomes 96 dB when $N = 16$ bits, which still limits access to weak signals if a strong interference source lies within the sampled bandwidth (Figure 1.16). This issue is even more acute in RADAR applications when the received signal drops as the fourth power of the distance to the target – rather than with the square of the distance for one-way communication links – as will be discussed in the chapter about SDR applications to RADAR.

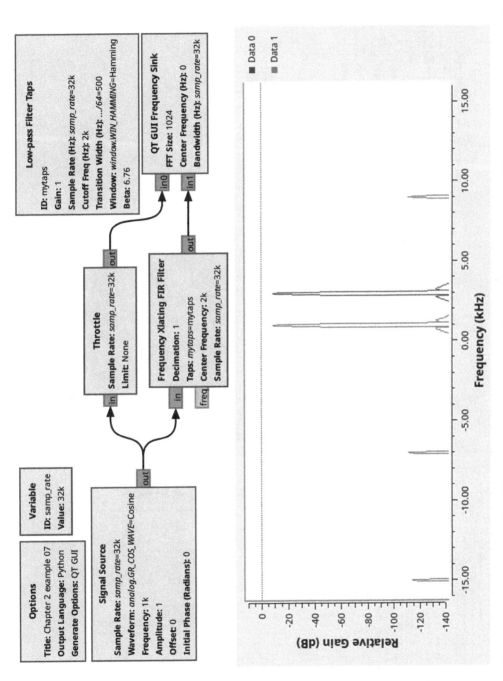

**Figure 1.15** Basic example of `xlating FIR Filter` usage demonstrating how the input signal frequency $f$ is shifted by the `xlating FIR Filter` local oscillator $f_{LO}$ so the output signal is characterized by a frequency $f - f_{LO}$. The low-pass filter taps are defined with the appropriate block whose ID is the same name filling the Taps field of the `xlating FIR Filter`.

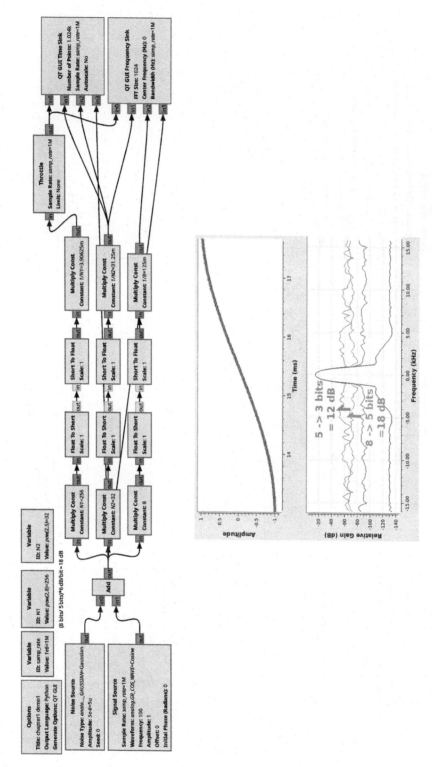

**Figure 1.16** Demonstration of the quantization limit and the impact on the signal to noise ratio when the signal is buried in noise.

> **Link budget and received power**
>
> The statement about limited dynamic range requires assessing the link budget between two transmitters at various distances $d$ from the receiver and how energy conservation impacts the received power. Energy conservation requires that an emitter generating a radio frequency power, $P_E$ spreads a spherical wave assuming an isotropic radiating element with area $4\pi d^2$ at distance $d$, hence a power density per unit area $P_E/(4\pi d^2)$. This wave reaches the receiver assumed to be fitted with an isotropic antenna with typical dimensions $\lambda^2$, and only a fraction of the $4\pi$ steradian sphere intersects the incoming power: Friis equation concludes that the received power
>
> $$P_R = P_E \left(\frac{\lambda}{4\pi d}\right)^2$$
>
> or when considering the logarithmic scale of decibels, a free space propagation loss
>
> $$FSPL = 20\log_{10}(f) + 20\log_{10}(d) - 147.55$$
>
> since the wavelength is $\lambda = c/f$ with $c$ as the speed of light and $f$ the carrier frequency, with the constant calculated as $20\log_{10}\left(\frac{c}{4\pi}\right)$. Hence, the 48 dB dynamic range of an 8-bit ADC is reached when two emitters operating on similar carrier frequency bands and emitting the same power are located at a distance ratio of $10^{48/20} = 251$ (e.g. one emitter 100 m away prevents reception of another emitter 25 km away if the gain of the ADC is tuned to avoid saturation, bringing the remote signal below the least significant bit) for a one-way communication, dropping to a range ratio of 80 if the nearest emitter emits 10 times more power than the weaker remote emitter (hence, e.g., a strong emitter 100 m away overwhelms a weak emitter only 8 km away). In case of a RADAR emitter illuminating a point-like target, acting itself as a new source of a spherical wave, hence replacing the $20\log_{10}$ of the sphere wave energy spreading with $40\log_{10}$, representing the received power as the fourth power of the emitter-to-target-to-receiver distance, this range ratio drops to $10^{48/40} = 16$.
>
> Practical applications are more complex by including the antenna gain on the emitter and receiver side, and interaction of the wave with potential reflecting sources in the so-called Fresnel zone when ground-based communication is considered.

## 1.8 Power Spectral Density Display with the Frequency Sink

The QT frequency sink is arguably the most useful display when analyzing signal characteristics in the frequency domain. The abscissa is defined from $-f_s/2$ to $fs/2 - f_s/N$ when analyzing a signal sampled at a rate $f_s$ with a Fourier transform computer over $N$ samples. The $y$-axis is given in decibel (dB), defined as a "power $P$" ratio between input and output $10\log_{10}\left(\frac{P_{output}}{P_{input}}\right)$ or between voltages (IQ samples) $20\log_{10}\left(\frac{IQ_{output}}{IQ_{input}}\right)$ since power is related to the square of the voltage. When comparing powers computed under the same conditions $f_s$ and $N$, this knowledge is sufficient. However, when comparing charts produced with different $f_s$, for example after decimations, or different $N$, it is worth mentioning that the displayed quantities are actually **power spectral densities** (PSD) defined as dB/Hz or power ratios **normalized to the bin width**. Unfortunately, in the current status of GNU Radio (3.10), the $y$-axis of the QT frequency sink is erroneously

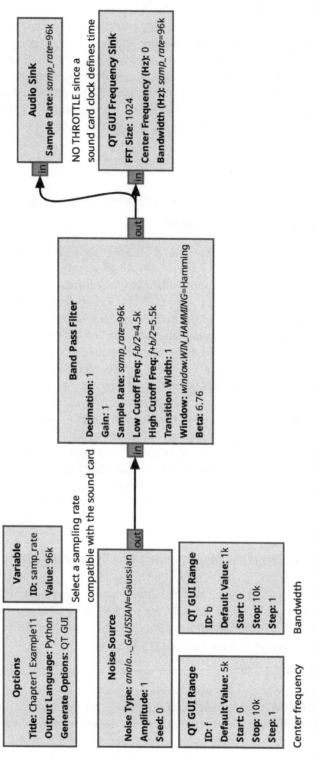

Figure 1.17 Characterizing a (real) bandpass filter and playing its output on the sound card.

normalized[2] with bin number and not bin frequency width. Hence, demonstrating noise power accumulation at baseband as decimation is performed without low-pass filtering, is not possible.

## 1.9 Conclusion

This first chapter was designed to introduce the basics of handling GNU Radio processing and generating code through its associated GUI GNU Radio Companion. In doing so, we introduced basic signal processing concepts needed throughout this book, including why SDR handles complex (real and imaginary parts) quantities, the abscissa of a discrete Fourier transform ranging from minus half the sampling rate to half the sampling rate, the concept of decimation and how it impacts the frequency range, and finally filtering and how to interact with custom Python code.

Although we have solely used synthetic signals throughout this chapter, except for the GRAVES data acquisition example, the reader might enjoy getting some physical sense of the signal processing steps. One widely available SDR interface is sound cards: despite their low sampling rate, such interfaces allow for collecting very low frequency (VLF) signals or narrowband signal transposed to baseband by external analog hardware on the microphone input and generate audible output on the headphone output. Accessing the sound card is achieved by using the `Audio Source` (microphone acting as data source) and `Audio Sink` (headphone acting as data sink), but most sound cards will only accept a few selected sampling rates, namely 48 kHz, 96 kHz, or for some of the higher grades sound cards 192 kHz. Remember though, that when using a hardware interface either as data input or output, the `Throttle` block must be **omitted** to avoid timing conflicts (Figure 1.17).

Using the flowchart of Figure 1.17, the impact of the bandwidth and central frequency of the filter is heard on the headphone output.

One final important note might be to remind the difference between the Python code generated by GNU Radio Companion, defining the links between processing block inputs and outputs, and the libraries (written in C++ or Python) processing the actual IQ streams as processing blocks. Indeed, the GNU Radio Companion-generated code **cannot access IQ samples** but only instantiates processing blocks which will process such samples. Hence, adding new processing capabilities for real-time processing in the context of GNU Radio requires writing new dedicated processing blocks – also called out of tree (OOT) blocks – as will be addressed in Chapter 10 of this book, although many processing functionalities will be readily found in existing libraries. These OOT blocks are then parameterized and instantiated by GNU Radio Companion to channel IQ samples to the newly defined processing capabilities.

## References

AMD-Xilinx. AMD Zynq™UltraScale+™RFSoCs, 2024. https://www.xilinx.com/products/silicon-devices/soc/rfsoc.html.

Analog Devices. ADALM-PLUTO Overview, 2024. https://wiki.analog.com/university/tools/pluto.

---

2 Marcus Müller points to https://github.com/gnuradio/gnuradio/issues/4827 for a description of the issue, best visible when reading the source code at `gr-qtgui/lib/freq_sink_c_impl.cc` and noticing that the sampling rate `samp_rate` is only involved in the definition of the frequency axis along $x$ but not in the calculation of the bin normalization factor of the PSD.

D. Gabor. Theory of communication. *Journal of the Institution of Electrical Engineers-Part III: Radio and Communication Engineering*, 93(26):429–457, 1946.

GNU Radio Conda Install. 2024. https://wiki.gnuradio.org/index.php?title=CondaInstall.

GNU Radio Linux Install. GNU Radio Linux install, 2024. https://wiki.gnuradio.org/index.php?title=LinuxInstall.

D. E. Guimaraes. Complex envelope based modems: A tutorial. *Journal of Communication and Information Systems*, 35(1):34–50, 2020.

J. Mitola. Software radios: Survey, critical evaluation and future directions. *IEEE Aerospace and Electronic Systems Magazine*, 8(4):25–36, 1993.

M. Müller. Re: GPIO lines on RPi4, 2020. https://lists.gnu.org/archive/html/discuss-gnuradio/2020-06/msg00125.html, accessed March 2024.

E. S. Raymond. *The Cathedral and the Bazaar: Musings on Linux and Open Source by an Accidental Revolutionary, Revised Edition*. O'Reilly, 2001.

B. Razavi. *RF microelectronics*, volume 2. Prentice Hall, New York, 2012.

C. E. Shannon. Communication in the presence of noise. *Proceedings of the IRE*, 37(1):10–21, 1949.

J. Ville. Théorie et application de la notion de signal analytique. *Câbles et transmissions*, 2(1):61–74, 1948.

J. Ville. Theory and application of the notion of complex signal. Technical report, RAND Corp, Santa Monica, CA, 1958.

M. Viswanathan. 2017. https://www.gaussianwaves.com/2017/04/analytic-signal-hilbert-transform-and-fft/, accessed March 2024.

# 2

## Using GNU Radio with Signals Collected from SDR Hardware

Having introduced the basics of GNU Radio operations and the design of processing flowcharts using GNU Radio Companion to analyze synthetic signals, we will address in this chapter the acquisition of live signals using hardware receivers. Doing so, we meet the challenge that the **emitter and receiver local oscillators are at different locations and at different frequencies** – if only because of the Doppler shift introduced by a moving emitter or receiver – and this frequency difference must be compensated for. Hence, the modulation scheme aims at imprinting the transmitted information on all possible carrier $s(t) = A(t)\cos(2\pi f(t)t + \varphi(t))$ variables, $A$ the amplitude, $f$ the frequency, or $\varphi$ the phase. As stated by Mitola [2000], increasing complexity is associated with each one of these modulation schemes due to increasing coherence, namely the need to recover on the receiver side increasingly accurate copies of the transmitted carrier. The increasing complexity however comes with the benefit of increased spectral efficiency and immunity to noise, and hence increased channel capacity. We first introduce some of the hardware used by the authors for collecting data from emitters before demonstrating the use of GNU Radio to analyze some of the widely available modulation schemes, with an emphasis on spaceborne signals accessible throughout the world.

A common issue with SDR signal recording and storage is remembering the data organization and acquisition condition. Indeed, one of the unique features of SDR over analog signal processing is the ability to store and share recordings for post-processing, applying multiple analysis steps on the same samples. We will advertise the use of the sample data found at iqengine.org, but reading these files in the GNU Radio Companion flowchart from a `File Source` block would not be possible without understanding the file organization, the format of each sample, and at least knowing the sampling rate, if not the acquisition conditions (hardware, antenna type, gain). Such information can be recorded in a meta file associated with each data file, as proposed by the SigMF format (https://github.com/sigmf/SigMF). Following this format, each `.sigmf-data` file holding the records is associated with a `.sigmf-meta` description file stating how each sample is represented – as a floating-point number or integer and on how many bytes each sample is encoded – and whether the samples are complex (interleaved real and imaginary part) or real (no imaginary part). This information will be needed to set the right `Output Type` of the `File Source` block. In the absence of hardware sink, the use of the `File Source` block will require a `Throttle` block to produce data at the appropriate sampling rate. Finally, readers willing to process samples using languages other than GNU Radio recorded from a `File Sink` or downloaded from `iqengine.org` will be able to load the data, thanks to these information: under Octave for example, the Matlab scripts `read_complex_binary.m` and `read_float_binary.m` found in `gr-utils/octave` of the GNU Radio source code will be used for loading complex or real data from a file. Since these scripts will always open the file, read a number of samples possibly

*Communication Systems Engineering with GNU Radio: A Hands-on Approach,*
First Edition. Jean-Michel Friedt and Hervé Boeglen.
© 2025 John Wiley & Sons, Inc. Published 2025 by John Wiley & Sons, Inc.
Companion website: www.wiley.com/go/friedtcommunication

## 2 Using GNU Radio with Signals Collected from SDR Hardware

provided as argument, and then close the file, large datasets are hardly accessible using this technique. For files too large to fit in memory, the file can be opened and then blocks of data are read to be processed before finally closing the file. Under Octave, such a program would be

```
fd =fopen(filename , 'rb ');
do
  d = fread(fd ,4*N,'int16 ');      % for example for 16-bit integers
  d = d(1:2:end)+j*d(2:2:end); % interleaved real/imaginary
  d1 = d(1:2:end);                   % channel1
  d2 = d(2:2:end);                   % channel1
  % here data processing on d1 and d2 (complex values)
until (length(d1)<N);
fclose(fd );
```

assuming the file was recorded from a dual-channel SDR receiver with complex samples saved as 16-bit integers interleaving the real and imaginary parts. Blocks of $N$ samples are read at each iteration of the loop, $N$ being selected to match available memory and allow for running the processing algorithm, e.g. the length of a Fourier transform or filtering. Since each block read from the file includes interleaved channels each represented by interleaved complex quantities, obtaining $N$ samples in each channel requires reading $4N$ samples from the file.

The signals associated with this chapter are available from https://iqengine.org in GNU Radio SigMF Repo → analog_FM_France, a 1.92 MHz wide acquisition in the analog frequency modulation (FM) band at a carrier frequency of 96.9 MHz removed by the analog hardware front end and saved as floating point numbers interleaving real and imaginary parts. This signal allows for demodulating analog FM signals and the associated radio data system (RDS) digital band broadcast with each FM station. Also, under GNU Radio SigMF Repo → space → 3genuine_69dB_11h15_1 a GPS L1 signal is recorded by connecting an active antenna through a bias-T to an Ettus Research B210 SDR at 1.023 MS/s.

## 2.1 SDR Hardware Architecture

The popularity of SDR in general, and GNU Radio in particular, is arguably associated with the availability of affordable hardware related, on the one hand, to low-cost radio frequency front end integrated circuits, and on the other hand, to the availability of huge computational power with fast communication buses.

On the hardware side, the now-defunct Elonics E4000, leading to the Rafael Micro R820T(2) radio frequency front end, coupled with the analog-to-digital converter (ADC) and USB transceiver Realtek RTL2832U, has provided an affordable radio frequency front end for SDR application under the generic naming RTL-SDR. At a unit price below US$20, the receiver is sensitive enough to receive signals from low earth orbiting (LEO) satellites and even from the global positioning system (GPS) at distances from 800 to 20 000 km from the receiver, despite being designed for receiving digital-video broadcast terrestrial (DVB-T) signals at a few tens of kilometers. Analyzing the various radio frequency front end performance versus cost demonstrates an obvious relation with bandwidth, while the carrier frequency range is only a matter of local oscillator tunability. With a carrier frequency range of 25 to 1750 MHz, the R820T2 front end allows accessing most very high frequency (VHF) and ultra high frequency (UHF) signals but excludes the popular super high frequency (SHF) 2.4 GHz unlicensed band used by protocols such as Bluetooth, Zigbee, or WiFi, but most significantly it limits the accessible bandwidth to 2.4 MHz. More expensive front ends, including the AD986x circuits by analog devices or LMS700x from Lime Microsystems, will achieve broader bandwidth and wider carrier frequency ranges at the expense of more expensive systems. We will later show that for discontinuous streams, as used for RADAR measurements, even radio frequency-grade oscilloscopes can be used as a signal source for GNU Radio processing. While historically super-heterodyne receivers were developed to compensate for the poor isolation of radio frequency mixers, with the double-frequency transposition allowing to remove the re-injected DC component leaking from the local oscillator to the receiver low noise amplifier, current integrated circuit technology allows for direct conversion (0-intermediate frequency) from radio frequency band to baseband by a single mixing stage feeding the ADC. Hence, removing the radio frequency carrier $f_c$ is taken care of at the analog level, and the ADC sampling frequency defines samp_rate. GNU Radio will display baseband spectra in its default configuration as centered on 0 Hz since removing $f_c$ was handled by the radio frequency front end, and SDR is not aware of frequency transposition to the radio frequency band. Nevertheless, for convenience, the QT frequency sink allows for defining a center frequency of $f_c$ rather than 0, despite the information being irrelevant to SDR processing (Figure 2.1).

## 2.2   Using Readily Available Processing Tools

Before getting into the manual processing of various modulation schemes, emphasizing how GNU Radio acts as a toolbox for tackling signal processing techniques not only on virtual signals (as discussed in the previous Chapter 1) but on real samples collected from hardware and processed in real time, let us mention that many authors are providing complete functional processing frameworks where GNU Radio is the underlying library but not necessarily exposed to the user. Among the most advanced, a non exhaustive list and limited to the interests of these authors is gnss-sdr [Fernández-Prades et al., 2011], the Global Navigation Satellite System (GNSS) free and open processing framework, but many other gr-* processing chains are listed on the Comprehensive GNU Radio Archive Network (CGRAN) repository at https://cgran.org. As an illustration, no detailed understanding of the Iridium satellite constellation [Bloom, 2016] modulation and communication scheme is needed when running gr-iridium [Sec and Schneider, 2015, 2022] – developed by Sec and Schneider from the Munich CCC group in Germany – from any location in the world and receiving e.g. plane communication messages (Figure 2.2). The hardware displayed on this picture Figure 2.2(a) varies in cost and bandwidth: the RTL-SDR (bottom left), communicating over USB2 can collect and stream a maximum of 2.4 MS/s but, in the case of Iridium, is limited to 2 MS/s; the Pluto+ can collect data up to 56 MS/s from its AD9361 front end, but despite the Ethernet connection (bottom right), the Zynq 7010 embedded processor limits the data-stream to about 5 MS/s, as is the case with the original Adalm-Pluto from Analog Devices, and (top) the Ettus Research B210 can stream up to 10 MS/s from the same radio frequency front end but at seven times the cost of the Pluto+.

Iridium modulation is differential encoding-quad phase shift keying (QPSK) at a baud rate of 25 kbauds with resource sharing in the time domain (time division multiple access – TDMA) and in the frequency domain (frequency division multiple access – FDMA) – following a scheme reminiscent of those described later in this book (Section 5.3.3) [ICAO, 2006].

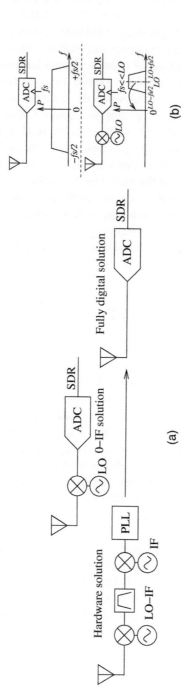

**Figure 2.1** Evolution of hardware architectures, from the fully hardware solution (a) with an intermediate frequency (IF) transposition and a band-pass filter to compensate for imperfections of the mixers, a hardware demodulation stage, for example, a phase locked loop (PLL) for FM reception. The dream of SDR is a fully digital solution where the antenna is directly connected to the radio-frequency grade analog-to-digital converter (b). Current solutions include an analog frequency transposition from radio frequency band to baseband, which is sampled by the analog-to-digital converter. In all cases, the sampled bandwidth is determined by the sampling rate, centered on 0 Hz in the first solution (right, top), and on the radio frequency local oscillator frequency LO (right, bottom) in the 0-IF solution.

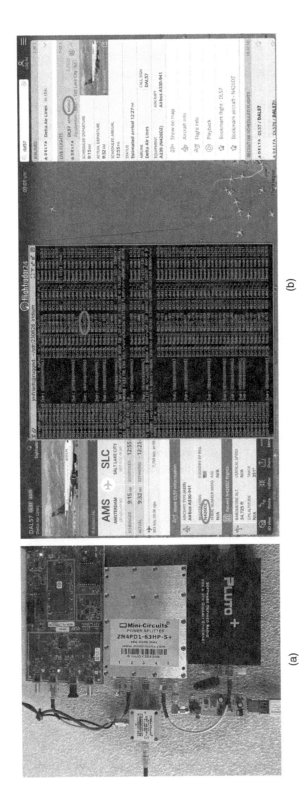

**Figure 2.2** (a) Family picture of the hardware connected to a modified GPS antenna whose band-pass filter was removed to access the 1622 MHz Iridium signals, powered by a bias-T, connected to a splitter with four outputs connected from top to bottom to the Ettus Research B210 (top), the Pluto+ (bottom-right), and the RTL-SDR dongle (bottom-left), all receiving the same radio frequency signal. (b) Example of sentences decoded using gr-iridium, matching the location of a plane well beyond the horizon of the receiver's location, here close to Amsterdam (Netherlands), 600 km north of Besançon (France), where the receiver is located. The antenna setup is shown in Figure 9.2. Source: Flightradar24 AB.

## 2.3 Amplitude Modulation and Demodulation

The first modulation scheme considered for transmitting information over a radio frequency carrier has historically been amplitude. Amplitude modulation (AM) and demodulation are considered noncoherent since the carrier does not need to be recovered on the receiver side. Only a rectifying circuit implemented as a diode bridge and low-pass filter in the analog domain, or absolute value and FIR filter in the digital domain, are needed to demodulate an AM modulated signal, with a filter bandwidth broader than the frequency offset between emitter and receiver local oscillators. Few users still use AM modulation other than aeronautical communications between planes and airports.

Formally, AM acts on the amplitude term outside the trigonometric argument, which only includes the fixed carrier frequency $f_c$ so that the signal $s(t)$ is expressed as

$$s(t) = A(t)\cos(2\pi f_c t)$$

The spectral occupation of the AM is analyzed by separating each spectral component of $A(t)$ as $A(t) = 1 + m\cos(2\pi f_m t)$, with $f_m$ being the modulation frequency and $m$ the modulation index ranging from 0 (no modulation) to 1. The spectral components are hence a carrier at $f_c$ and two sidebands at $f_c \pm f_m$, and most importantly, we observe that AM does not work at constant power since the emitter power is dependent on $A(t)$.

As a demonstration of AM demodulation and analysis, the digital communication mode ACARS (aircraft communications, addressing, and reporting system) is used as an example. Indeed, while AM has become obsolete in most applications, aeronautical communication remains the last field strongly attached to AM-modulated communication for reasons that will be explained at the end of Section 2.4. All communication in the airband ranging from 108 to 137 MHz worldwide is hence AM modulated: we shall not be interested in the audio communication between pilots and controllers on the ground but in the digital communication protocol ACARS, although the former is easily addressed by feeding the SDR receiver source to the AM Demod block of GNU Radio after low-pass filtering and decimating since AM signals are narrowband with respect to the bandwidth accessible to RF front ends.

The ACARS decoder, complementary to automatic dependent surveillance–broadcast (ADS-B), has been implemented in GNU Radio as found at https://sourceforge.net/projects/gr-acars, providing the software resource for the following description. Before getting into the details of the ACARS modulation and demodulation, let us consider why planes still communicate using AM: this modulation scheme is not subject to the FM-capture effect [Rondeau, 2016], namely the fact that the FM demodulator (a Phase Locked Loop (PLL)) locks on the strongest signal and rejects any weaker signal within the analyzed bandwidth. Hence, a weak plane transmitting an FM-modulated emergency message would not be heard over the stronger signal broadcast by a nearer plane, the airport tower only demodulating the strongest signal if it were frequency modulated. To demonstrate this concept using GNU Radio Companion, Figure 2.3 illustrates the synthesis of an amplitude $A(t)$ modulated (AM) signal as $s(t) = A(t) \times \exp(j\omega t)$. Since the rectifier and low-pass filter in the analog demodulator version, or the absolute value and low-pass filter in the digital implementation, will detect any spectral component of $A(t)$ during the demodulation process, both spectral components of the transmitted signals at arbitrarily selected frequencies of 480 and 1710 Hz of the carrier at 30 kHz are seen on the receiver (Figure 2.3 right chart, bottom).

Moving the sliders on the top of the output window allows for tuning the relative power of both spectral components, all of which are always visible and never vanish unless the corresponding slider driving the AM is brought to 0. We will see later (Figure 2.8) that the behavior is opposite in

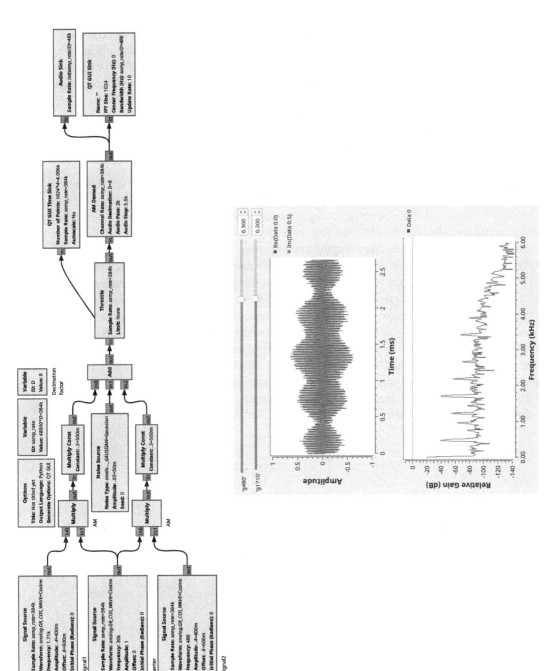

**Figure 2.3** Two tones at 480 and 1710 Hz are used to amplitude-modulate a 30 kHz carrier: the AM detector, implemented as a magnitude detector (absolute value and low-pass filter), detects both components with their respective relative power.

the case of FM modulation when only the strongest signal will be demodulated, hence justifying the use of AM for critical air safety communications.

Multiple protocols are used when communicating plane information to the ground, with ACARS being the main channel for sharing information between pilots and airport control towers. ACARS is a narrowband signal broadcast worldwide on 131.550 MHz, centered on 131.725 MHz as the primary channel in Europe and 131.525 MHz as the secondary channel, with all frequencies lying in the 129.125 to 131.850 MHz (2.725 MHz bandwidth) or 136.7 to 136.9 MHz with 50 kHz steps for the newer allocated channels (200 kHz bandwidth), easily accessible with most SDR receivers. ACARS is documented as a 2400 bits/s communication protocol encoding the two bit states as half a period of a 1200 Hz sine wave or a full period of a 2400 Hz sine wave. A header allows for tuning the receiver frequency and compensating for any offset with the broadcasting oscillator, as well as tuning an automatic gain control, before a synchronization word indicates when the first bit is aligned to form a byte, and following sentences are assembled.

Once a broadcast message has been detected, identifying the bit states is a matter of selecting the most probable value, an operation called *soft bit to hard bit* conversion (Figure 2.4). The soft bit is the output of the recorded signal running through two convolutional matched filters, one centered on 1200 Hz and the other one on 2400 Hz. The threshold decision for allocating one of the other bit states requires comparing the filter outputs and assigning the bit state of the strongest signal. However, this decision must be done only when a bit state has stabilized and not during a transition: searching for the strongest signal in quickly changing bit states allows for *symbol synchronization* where the sampling clock is aligned far from the transitions on states when the bits are stable.

When analyzing a new protocol, we found it most efficient to record a dataset including some communication and post-process the analysis on this sample using one of the prototyping languages such as Python or Octave. Having identified the processing sequence, the post-processing interpreted script is converted to C++ for computational efficiency and validated against the same dataset. Finally, the C++ code is included in the GNU Radio processing framework for real-time analysis of the continuous stream. As an example of Octave implementation of the soft bit generation and soft-to-hard bit analysis,

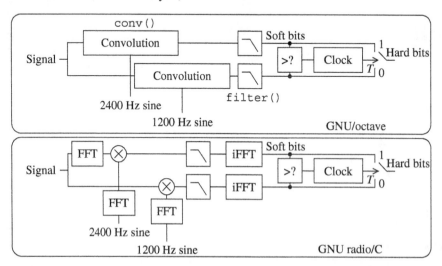

**Figure 2.4** ACARS decoding principle for converting the received signals to soft bits as the output of two convolutional filters – either implemented in the time domain (a) or the frequency domain (b) and converting soft bits to hard bits after the comparison of the two outputs. The symbol synchronization aims at sampling the comparison when the bit states are stable, far from transitions.

```
fs = 48000;
bps = 2400;
f2400 = exp(j*2*pi*2400*[0:1/fs:1/bps*2]');f2400 = f2400(1:end-1);
f1200 = exp(j*2*pi*1200*[0:1/fs:1/bps*2]');f1200 = f1200(1:end);
% s2400=abs(conv(y,f2400));s2400=s2400(200:end);  % convolution
% s1200=abs(conv(y,f1200));s1200=s1200(200:end);
% b=firls(256,[0 3400 4500 fs/2]*2/fs,[1 1 0 0]); % low pass filter taps
% s2400=filter(b,1,s2400);s2400=s2400(200:end);   % filter (remove 4800 Hz)
% s1200=filter(b,1,s1200);s1200=s1200(200:end);
sf = (fft(y));
f2400f = (fft(f2400, length(sf)));
f1200f = (fft(f1200, length(sf)));
sf2400f = sf.*f2400f;        % convolution
sf1200f = sf.*f1200f;
df = fs/length(sf);          % FFT bin width (in Hz)
fcut = floor(3500/df)        % cutoff bin position
sf2400f(fcut:end-fcut) = 0;  % low pass filter/Matlab FFT convention
     (0 left, fs/2 middle)
sf1200f(fcut:end-fcut) = 0;
s2400 = abs(ifft(sf2400f));  % back to time domain
s1200 = abs(ifft(sf1200f));
plot(s2400, '-');hold on; plot(s1200, '-'); legend('2400', '1200')
```

illustrates how the taps are generated in f2400 and f1200 as the two filters centered on 2400 and 1200 Hz respectively, and then how progressively the convolution is first implemented using the Octave function conv() before being converted to the Fourier domain, more readily accessible in C++ through the FFTW library [Frigo, 1999], performing the convolution as products in the Fourier domain, before returning to the time domain and comparing the relative magnitudes of both outputs. Whether considering GNU Radio or Octave implementation of discrete-time digital signal processing, the sampling rate remains the core quantity propagated throughout the processing sequence, as emphasized by the definition of fs at the beginning of the Octave script. Time is defined as a vector of $N$ elements with increments of $1/fs$ as [0:N-1]/fs. The filter taps are the impulse response of the filter or, in this case, the half wave at 1200 Hz or full wave at 2400 Hz stored in f1200 and f2400, respectively. One benefit of explicating the intermediate spectral information is that the low-pass filter is naturally implemented by setting to 0 the higher frequency spectral components (sf2400f(fcut:end-fcut)=0;) before returning to the time domain, so that both convolution and low-pass filtering are efficiently implemented in the Fourier domain.

Figure 2.5 illustrates the two possible bit states at the output of the filters centered on 1200 and 2400 Hz: the soft-to-hard bit decision is implemented as a comparison between the two outputs sampled at the nominal chip rate of 2400 bits/s or once every 20 samples assuming the signal was recorded at the audio-frequency rate of 48 kS/s. However, as can be seen around abscissa 800 to 850 of Figure 2.5, the strongest filter output does not always match the sampling time of the receiver. Indeed, since the transmitter and receiver do not share the same local oscillator, the sampling period on the receiver side is prone to drifting with respect to the emitter bit sequence, ending up sampling during a bit transition when its state is undetermined. This issue is corrected by searching for the strongest signal power around the theoretical sampling time and tuning the sampling clock to track the strongest filter output, as illustrated by shifting the sampling period

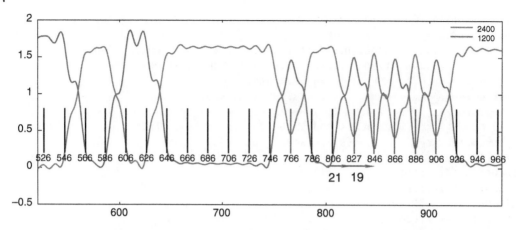

**Figure 2.5** Soft-bit output of the two filters centered on 1200 and 2400 Hz to identify the two possible bit states. Comparing these two outputs will allow selecting 1 or 0 hard-bit values, considering the sampling clock has been aligned to the stable bit position far from any transition (bold lines).

by 21 or 19 samples rather than the nominal 20 samples. Such a tracking is mandatory for decoding long ACARS sentences when the sampling clock would otherwise drift excessively and end up analyzing the bit state during a transition [Friedt and Laverenne, 2020].

Once bit states have been identified, bits are grouped into 8-bit-long bytes, which themselves are assembled as sentences according to https://www.scancat.com/Code-30_html_Source/acars.html, starting with a header 0x2B 0x2A 0x16 0x16 0x01, representing the 16-bit alternating bit sequence for decoder locking, pattern, and start of header (SOH or ASCII code 1), followed by the transmission type, aircraft registration, block identifier, start of text (STX or ASCII code 2), which itself starts with the flight number and text ending with the end of text (ETX or ASCII code 3), and completed with a checksum for message integrity validation. A typical ACARS message will read after decoding as:

```
Aircraft=.B-5938
STX
Seq. No=C30A
Flight=MU7080
#CFB.1/FLR/FR2004271711 24000006WHC1  1,,,,,,,POWER SUPPLY INTERRUPT,INTERMITTENT
ETX
```

although many airlines use proprietary undocumented message formats for sharing information.

Multiple ACARS communication channels are allocated to avoid message collision, spanning from 129.125 to 131.450 MHz so that a single RTL-SDR dongle is able to sample the whole band in which ACARS messages are broadcast. Each sub-band is brought to baseband using the Xlating FIR Filter, whose output each feeds an AM demodulator whose output is decoded by one of the multiple instances of the `gr-acars` decoder.

Notice that the classical FFTW3 library is *not* thread safe (https://www.fftw.org/fftw3_doc/Thread-safety.html), and using the GNU Radio wrapper of this library is mandatory to avoid memory collision during the FFT initialization, leading to memory corruption and software crash otherwise. As a demonstration of multiple instances of `gr-acars` running in parallel and decoding multiple channels, Figure 2.6 illustrates a flowchart with three adjacent channel decoding and the resulting sentences.

## 2.4 Frequency Modulation and Demodulation

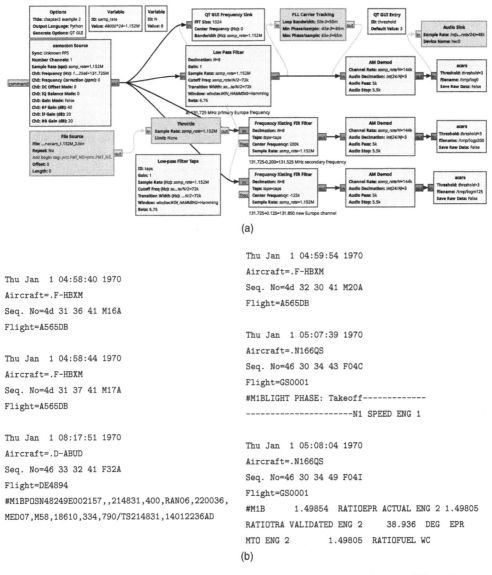

```
Thu Jan  1 04:58:40 1970
Aircraft=.F-HBXM
Seq. No=4d 31 36 41 M16A
Flight=A565DB

Thu Jan  1 04:58:44 1970
Aircraft=.F-HBXM
Seq. No=4d 31 37 41 M17A
Flight=A565DB

Thu Jan  1 08:17:51 1970
Aircraft=.D-ABUD
Seq. No=46 33 32 41 F32A
Flight=DE4894
#M1BPOSN48249E002157,,214831,400,RAN06,220036,
MED07,M58,18610,334,790/TS214831,14012236AD
```

```
Thu Jan  1 04:59:54 1970
Aircraft=.F-HBXM
Seq. No=4d 32 30 41 M20A
Flight=A565DB

Thu Jan  1 05:07:39 1970
Aircraft=.N166QS
Seq. No=46 30 34 43 F04C
Flight=GS0001
#M1BLIGHT PHASE: Takeoff-------------
----------------------N1 SPEED ENG 1

Thu Jan  1 05:08:04 1970
Aircraft=.N166QS
Seq. No=46 30 34 49 F04I
Flight=GS0001
#M1B      1.49854   RATIOEPR ACTUAL ENG 2 1.49805
RATIOTRA VALIDATED ENG 2       38.936  DEG   EPR
MTO ENG 2         1.49805    RATIOFUEL WC
```

(b)

**Figure 2.6** (a) Example of simultaneous decoding of multiple channels broadcast on different frequencies close enough to all lie within the bandwidth of the signal recorded by the hardware source, each brought to baseband using a frequency Xlating FIR filter and processed using its own decoding block, here for the ACARS protocol. (b) Output of the messages decoded at the European Primary frequency of 131.725 MHz (right) and 131.850 MHz (left).

## 2.4 Frequency Modulation and Demodulation

Frequency modulation has supplanted AM since the 1940s, thanks to its better immunity to broadband radio frequency interferences, at the expense of broader spectral occupation. FM is commonly known as the modulation scheme of analog commercial broadcast in the 88 to 108 MHz range in Europe or 76 to 95 MHz in Japan.

An FM-modulated signal $s(t)$ over a carrier at frequency $f_c$ is expressed as

$$s(t) = \sin\left(2\pi \cdot f_c \cdot t + 2\pi f\Delta \underbrace{\int_0^t x(\tau)d\tau}_{phase \to freq}\right)$$

where the integral on the transmitted signal $x(t)$ expresses the relation between the trigonometric argument of a phase and the frequency as derivate of the phase and $f_\Delta$ is the user-tunable frequency excursion of the modulation. The various spectral components are analyzed by using the decomposition of $x(t)$ as $x(\tau) = \cos(2\pi \cdot f_m t)$ with $f_m$ being the modulation signal frequency, leading to $s(t) = \sin\left(2\pi \cdot f_c \cdot t + \frac{f_\Delta}{f_m}\sin(2\pi f_m t)\right)$ emphasizing the two FM broadcast conditions, namely $\frac{f_\Delta}{f_m} \gg 1$ called Wideband FM (WBFM) broadcast as used in the commercial band, and $\frac{f_\Delta}{f_m} \ll 1$ called Narrowband FM (NBFM) as used by amateur radio communication including the downlink from the International Space Station (ISS) at 145.800 MHz broadcasting voice or slow scan television (SSTV) images from space, readily accessible using a RTL-SDR receiver fitted with a properly tuned dipole antenna – two 51-cm-long wires assembled as a 102-cm-long dipole – despite the hundreds of kilometers communication range.

GNU Radio Companion allows testing FM broadcast characteristics in simulated conditions to assess immunity to disturbances such as additive noise or frequency offset between transmitter and receiver, as well as the ability to receive signals, using flowcharts as shown in Figure 2.7.

As a follow-up to the previous discussion on FM capture and how only the strongest FM-modulated signal is demodulated, Figure 2.8 illustrates the impact of feeding a unique FM demodulator block with two incoming signals, one stronger than the other. An analog FM demodulator converts a frequency fluctuation to a voltage, as achieved with the classical PLL circuit in which the error signal between the incoming carrier and the local oscillator feeds the voltage control of the local oscillator to track any incoming frequency fluctuation. As such, the PLL is only able to track a single incoming signal, which is the strongest, practically filtering out the weaker signal despite being broadcast in the same frequency band. While the digital implementation of the FM quadrature demodulation is an open-loop processing and not a feedback control as achieved in the PLL [Bederov, 2015], the conclusion remains valid for the digital demodulator.

Returning to the analysis of signals recorded using hardware, the easiest yet most attractive demonstration of radio frequency reception is to analyze an analog commercial FM broadcast station signal and send the audio output to the sound card (Figure 2.9), despite this communication protocol slowly becoming obsolete when being replaced with digital audio broadcast (DAB), which will be investigated in depth in Chapter 7. This example is most simple since the received signal is powerful and broadcast continuously but nevertheless demonstrates the core concepts of SDR reception and will allow investigating FDMA when a unique FM channel broadcasts its information, namely the left channel audio, the right channel audio for stereo sound, and the radio data system (RDS) digital channel. Most importantly, this example illustrates that the consistency of data flow, as decimations are applied at various stages of the processing, must be handled manually since GNU Radio will not tune the data rate after decimation reduces the number of samples per unit time between input and output decimator blocks.

A **source** provides a complex I and Q data stream to feed the various processing blocks to finally reach a **sink** – in our case, the sound card. The data rate from the source defines the analysis

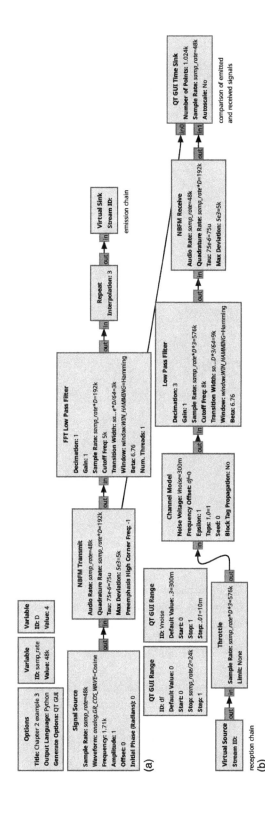

**Figure 2.7** Narrowband FM signal emission (a) and reception (b) simulation.

**44** | *2 Using GNU Radio with Signals Collected from SDR Hardware*

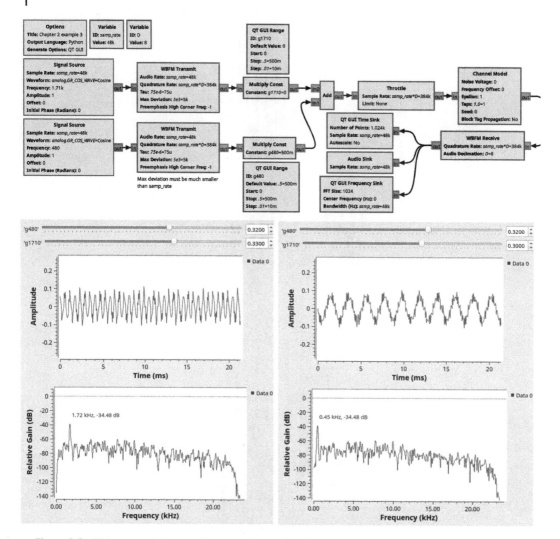

**Figure 2.8** FM capture demonstration on synthetic signals.

bandwidth and hence the amount of information we can collect, as stated by Shannon's capacity theorem. The bandwidth is limited by the sampling rate and the communication bandwidth between the acquisition peripheral and the personal computer or, in the case of the RTL-SDR receiver, the USB bus bandwidth.

In order to design the FM receiver:

- The RTL-SDR signal source is found by selecting the magnifying glass icon and typing osmo, which allows isolating the Osmocom Source block, assuming the gr-osmosdr package was installed on the GNU/Linux computer.
- The spectrum analyzer sink is found by selecting the magnifying glass icon and typing QT, which allows accessing the QT Frequency Sink.
- The sampling rate variable samp_rate is modified from its default value of 32 kHz to a value ranging from 800 kHz to 2.4 MHz as required by the RTL-SDR hardware characteristics.

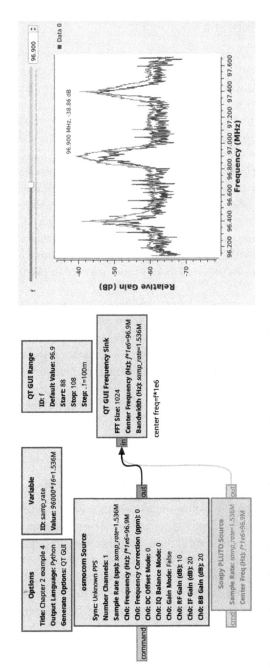

**Figure 2.9** Spectrum of part of the FM broadcast band as monitored by an RTL-SDR receiver connected through the osmocom source or the PluotSDR source, which has been commented out.

# 46 | 2 Using GNU Radio with Signals Collected from SDR Hardware

- The receiver local oscillator is tuned to the FM commercial broadcast band, and the gain is raised to 40 dB by double-clicking on the `Osmocom Source` block and tuning the `Frequency` and `RF Gain`, respectively. While the osmocom source can handle multiple hardware interfaces, including some providing Intermediate frequency gain and baseband gain, these parameters will not be used in the case of the RTL-SDR.

With these settings, a basic spectrum analyzer was assembled as shown in Figure 2.9, displaying the signal from multiple FM stations as allocated within the bandwidth around the carrier frequency. The maximum bandwidth is limited by the RTL-SDR hardware characteristics: hardware providing more bandwidth (and usually more expensive) allow to record more signals (Figure 2.10). In this example, rather than setting the central operating frequency when starting the acquisition, we might want to dynamically update such a parameter: the slider setting the frequency is selected by searching for the `QT Range` graphical user interface block. When selecting the block, its `ID` (top field when double-clicking on the selected block) defines the Python variable associated with the block, whose value is set by the slider position. The **same variable name** is used for setting the osmocom source frequency field, remembering by either selecting a slider in Hz and filling the frequency field with the variable name or more conveniently selecting a slider value in MHz and filling the frequency field with the variable name multiplied by $10^6$. Notice that GNU Radio Companion displays values prefixed with "m" for milli ($\times 10^{-3}$) or "k" for kilo ($\times 10^3$), but the developer must use Python-compatible syntax (e.g. `*1E6`) and is not allowed to use the prefix when filling the fields.

Although we have emphasized how SDR is only concerned with baseband signal processing after the radio frequency carrier was removed by the hardware front end mixer, some users might find it convenient to center the spectrum display on the carrier frequency rather than 0 Hz. This result is achieved by double-clicking on the QT Frequency Sink and setting the `Center Frequency` field to the same variable name as the one selected for filling the Frequency field of the osmocom source.

Whenever GNU Radio Companion blocks are selected and fields are filled with variables, commenting on the content of the block might be useful for future reference: comments are added by double-clicking on a block and filling the `Comment` field in the `Advanced` tab.

### 2.4.1 Commercial FM Broadcast: Demodulation and Audio Output

Once the frequency band including the FM broadcast signal has been identified, we must demodulate the signal (extract the information content from the carrier) and send the result to a meaningful peripheral, for example, the PC sound card, at the appropriate sampling rate.

The challenges lie in handling the data-flow, which must start with a radio frequency rate (several hundreds of ksamples/s) to an audio frequency rate (a few ksamples/s). GNU Radio does not automatically handle data flow rates and only warns the user of an inconsistent data rate with cryptic messages at first sight (but consistent once their meanings have been understood).

The data rate output from the RTL-SDR must lie in the 0.8 to 2.4 Msamples/s, the upper limit being given by the bandwidth of the USB communication link and the lower limit by hardware limitations. The sound card sampling rate must be selected from a few possible values up to 192 kHz, such as 48 000, 44 100, 22 050, or, for the older sound cards, 11 025 Hz. Handling the data rate requires **consistently decimating the data rate from an initial value to a final value**. In order to make sure that the initial sampling rate can easily be decimated to the output audio frequency, it is safe to define `samp_rate` as a multiple of the output audio frequency. For example, for a 48 kHz

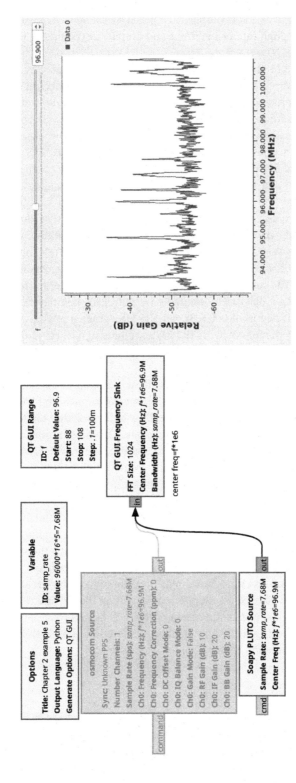

**Figure 2.10** Increasing the bandwidth thanks to the Adalm-Pluto source to monitor more stations: notice how the Adalm-Pluto hardware allowed for multiplying fivefold the sampling rate and hence the bandwidth.

output sampling rate, an input of $48 \times 32 = 1.536$ MHz complies with the RTL-SDR receiver sampling rate range. Similarly, for a 44.1 kHz output, an input at $44.1 \times 50 = 2.205$ MHz complies with the requirements of the sampling rate of the input and output, assuming the various processing blocks decimate by a factor of $50 = 5 \times 5 \times 2$.

The flowchart of Figure 2.11 is best implemented step by step, displaying in the frequency domain each processing output to check that the bandwidth and spectral content are well understood, depending on processing parameters. The low-pass filter right after the osmocom source selects a single FM station and rejects all other adjacent signals. Since after its decimation by $N$ (keeping 1 in $N$ samples), the output spectrum will lie in the $\pm samp\_rate/2/N$, the low-pass filter cutoff frequency must be equal to this threshold value $fs/2/N$ to avoid aliasing. The transition width defines the processing complexity and should be a tradeoff between how sharp the filter transition is and limiting the number of tap coefficients to as few as reasonable. The WBFM quadrature rate must be equal to the data rate at the output of the low-pass filter (i.e. `samp_rate/Decimation` of the low-pass filter). The WBFM output rate will be `samp_rate/LPF_decimation/WBFM_decimation`. If the low-pass filter decimation factor `LPF_decimation` and the WBFM Demodulator decimation factor `WBFM_decimation` have been wisely selected so that the resulting sampling rate is one of the acceptable values for the sound card, then connecting the output to the audio sink should allow listening to the broadcasting station content by connecting a headset to the audio output of the computer. As an example, if the osmocom source was selected as $48\,000 \times 32 = 1.536$ MS/s, then a low-pass filter decimation by eight feeds the WBFM Demodulator with 192 kS/s, well enough to cover the WBFM station spectral width, and a WBFM decimation factor of four leads to $192/4 = 48$ kS/s by definition of the initial sampling rate.

If erroneous decimation factors are selected and the final sampling rate is not one of the values compatible with the sound card configuration, then aU messages, meaning "Audio Underflow," are displayed in the GNU Radio Companion console, thus indicating that too few samples reach the sound card with respect to the expected sampling rate. On the opposite, aO means "Audio Overflow" and that too many samples reach the sound card. Since two clocks are present in the flowchart – the RTL-SDR source and the sound card – some synchronization loss is sometimes expected, but continuous messages of aU or aO mean the sampling rate is not consistent along the flowgraph.

Notice that by decimating the WBFM output, we have eliminated two of the three upper spectral components, namely the stereo sound component and RDS. The sound quality is still acceptable: indeed, commercial FM is broadcasting on the lower spectral components the sum of the left and right channels so that a mono receiver can still play the transmitted signal, and only if the second spectral component centered on 38 kHz, which holds the difference of the left and right channels, is analyzed can a stereo sound be played with different signals on the left and right ears. A 19 kHz continuous tone informs the receiver that a stereo program is being broadcast. Finally, the RDS signal is binary phase shift keying (BPSK) modulated, centered on a 57 kHz offset, and will be investigated in the phase modulation Section 2.5.

### 2.4.2 Stereo Sound and RDS

Three spectral components are frequency multiplexed at the output of the WBFM demodulator: left+right centered on 0 Hz, left-right centered on 38 kHz, and RDS centered on 57 kHz. However, demodulation requires that the useful signal is brought to baseband to be processed. We use again the Xlating FIR Filter introduced at the end of Chapter 1 for frequency transposition by multiplying

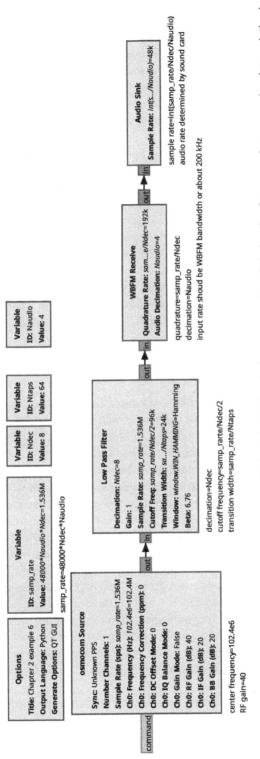

**Figure 2.11** Definition of the `samp_rate` variable as a multiple of the final audio frequency so that all decimation factors are integers and no interpolation is needed, and activating the audio output of the PC. All parameter definitions are given below each processing block.

with a numerically controlled oscillator followed by filtering to bring the signal of interest to baseband centered on 0 Hz, this time to bring the RDS signal to baseband.

Considering that the bitrate of the RDS digital information is $57\,000/48 = 1187.5$ bits/second, and the phase-modulation scheme uses a bandwidth equal to twice the bitrate, the Xlating FIR Filter low-pass filter taps are readily tuned by selecting such a cutoff frequency and a transition width wide enough to avoid excessive computational load from an excessively long filter while narrow enough to remove any leakage from the stereo channel. After selecting a local oscillator frequency of 57 kHz, the RDS signal is transposed to baseband and ready for phase demodulation in Section 2.5.3 (Figure 2.12).

The same processing technique for decoding Frequency Division Multiple Access (FMDA) resource sharing is applicable by transposing back to baseband the relevant signal using the Xlating FIR filter as long as all channels lie within the sampling bandwidth of the receiver, namely the 2.4 MHz sampling rate of the RTL-SDR in these examples. Thus, multiple FM station signals were recorded in the 2.4 MHz wide initial RTL-SDR recordings so that multiple FM stations can be demodulated simultaneously, with the number of stations being processed only limited by the available bandwidth and computational power. While listening to the audio broadcast of multiple stations simultaneously is questionable, this parallel processing of multiple channels recorded by a single signal source and each brought independently to baseband by its own frequency Xlating FIR was already demonstrated for ACARS (Figure 2.6) and is also valid, for example, for the pager protocol POCSAG ranging from 466.025 to 466.23125 MHz and decoded using `multimon-ng` (see Chapter 3).

### 2.4.3 FSK Modulation

Each analog modulation scheme (AM, FM, PM) has a corresponding digital implementation named ASK (amplitude shift keying), FSK (frequency shift keying), or PSK (phase shift keying). The extreme case of ASK is on-off-keying (OOK) or Morse, where a bit is broadcast by sending a carrier, and no bit is transmitted by stopping carrier emission. Most common though is FSK usage, where various frequency tones encode various bit states. In the simplest case, two-bit states (binary communication) are encoded with two frequencies, but the technique can be extended to multiple frequencies for transmitting multiple bits during each sampling period. As an example, the amateur M17 protocol uses 4-FSK, hence encoding two bits during each sampling period.

In this demonstration (Figure 2.13), the analog radio frequency modem XE1203, manufactured by Semtech, is used, encoding the RS232 compatible data as two possible frequencies, each encoding the two possible bit states. Hence, on the receiver side, an FM demodulator block converts back the frequency fluctuations as level fluctuations readily decoded as an RS232 stream with a known baudrate, a start bit followed by data bits and finishing with a stop bit, the data byte being encoded least significant bit first. In this implementation example, the communication protocol starts by sending a synchronization sequence for locking the receiver clock to the emitter clock (binary 10101010 or 0x55, also known as the ASCII symbole 'U'), followed by the letters "OK," followed by a unique identifier allowing for addressing the receiver. The payload message then follows this initialization sequence.

This demonstration aims at emphasizing how easy it is to decode messages transmitted over a radio frequency signal: the first layer of communication security, the hardware layer preventing monitoring of the transmitted signal, is breached with radio frequency communication since any listener can collect the transmitted signal and analyze its content. Many transceivers have been decoded this way, including wireless weather stations (e.g. http://wmrx00.sourceforge

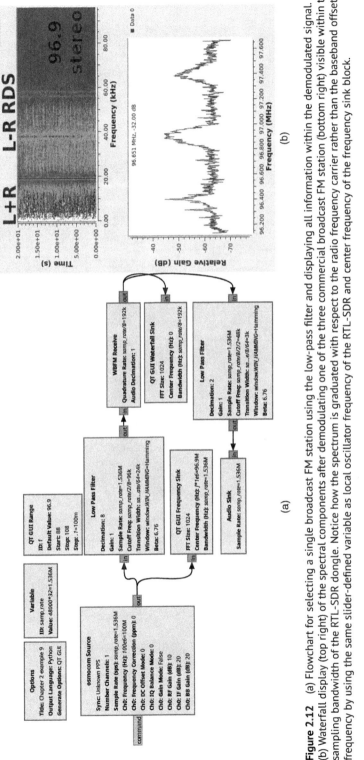

**Figure 2.12** (a) Flowchart for selecting a single broadcast FM station using the low-pass filter and displaying all information within the demodulated signal. (b) Waterfall display (top right) of the spectral components after demodulating one of the three commercial broadcast FM station (bottom right) visible within the sampling bandwidth of the RTL-SDR dongle. Notice how the spectrum is graduated with respect to the radio frequency carrier rather than the baseband offset frequency by using the same slider-defined variable as local oscillator frequency of the RTL-SDR and center frequency of the frequency sink block.

**52** | *2 Using GNU Radio with Signals Collected from SDR Hardware*

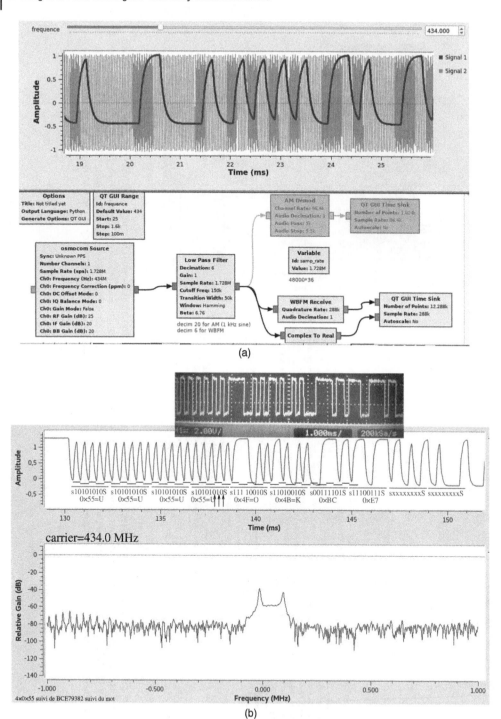

**Figure 2.13** Reception of a signal emitted by a Semtech XE1203F radiomodem programmed to send digital sentences on a 434 MHz carrier. (a) The time domain (thin line) FSK modulation signal transmitted by the modem, and (thick line) the output of the FM demodulator, with high-frequency sequences observed as rising voltage output and low-frequency modulation as lower voltage, as expected from an RS232 sentence. (b) The received sequence matches the transmitted oscilloscope trace (top), and the message payload can be decoded by manually following the RS232 protocol (middle). The two modulation frequencies are clearly seen on the spectrum (bottom).

.net/Arduino/OregonScientific-RF-Protocols.pdf and https://www.linux-magazine.com/Online/ Features/Reading-Weather-Data-with-Software-Defined-Radio) or meteorological balloons (radiosonde), as described at https://github.com/rs1729/RS.

## 2.5 Phase Modulation and Demodulation

We have introduced that historically AM, FM, and PM were successively developed due to increased complexity of the modulation and demodulation schemes. AM is incoherent in that it does not require recovering on the receiver side a copy of the emitted carrier compensating for the frequency offset between emitter and receiver local oscillators, as long as the frequency difference is smaller than the low-pass filter cutoff frequency following the rectifier of the envelope detector. Frequency demodulation involved tracking carrier frequency variations using a PLL, and any frequency offset would appear as a DC-component offset on the voltage output of the demodulator, requiring a high-pass filter when zero-crossing detection is used for bit state analysis. We shall now address PSK, the most recent and most complex modulation and demodulation scheme, since not only is the frequency difference between emitter and receiver local oscillators to be corrected, but also the phase must be tracked, which a PLL will not achieve. PSK, and its binary version BPSK, is most common and is used in such protocols as RDS and the GPS broadcasting on its L1 (1575.42 MHz) carrier at a data rate of 1.023 Mchips/s, well within the reception capability of the RTL-SDR dongle.

### 2.5.1 Phase Modulation

Although PSK is the most complex modulation scheme to demodulate, it is arguably the simplest modulation scheme to implement using hardware components. As an example, phase modulation is achieved by feeding a mixer (e.g. Minicircuits ZX05-1L-S+ when broadcasting in the 434 MHz European industrial, scientific, and medical band, Figure 2.14) on one side by a radio frequency carrier signal (LO port), and on the other hand by a square-wave signal with a mean value of 0 representing the signal (intermediate frequency – IF – port) to generate the signal sent to the antenna (RF port). Based on the internal schematic of the double-balanced mixer, the polarity of the modulating signal defines the side of the diode bridge through which the LO signal goes to reach the middle point of the transformer, and hence the phase (between 0 and $\pi$) is imprinted on the output signal.

---

**BPSK modulation of a carrier**

⚠ For BPSK modulating a carrier using a square wave digital signal, the following steps are used:

- Set digital data synthesizer as a **square wave** function generator with a **1 to 1.5 V amplitude** (IF input of the mixer), or any level high enough to saturate the diode bridge as advertised by the operating power of the mixer, namely +13 dBm for the Minicircuits ZX05 (+13 dBm or $13-30 = -17$ dBW on a 50 Ω load[0] is reached for a voltage of $\sqrt{10^{-17/10} \times 50} = 1.0$ V).

---

*(Continued)*

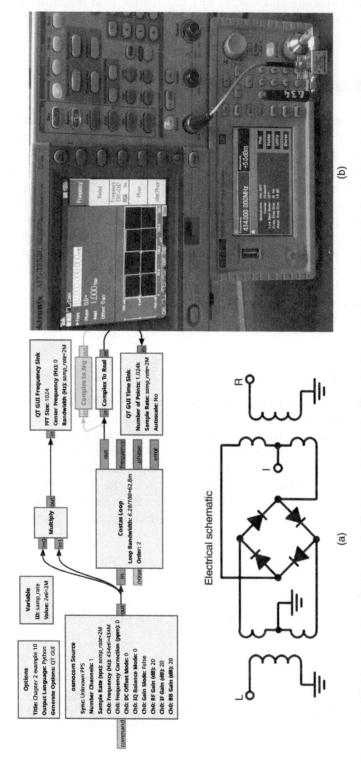

**Figure 2.14** (a) Schematic of the minicircuits double-balanced mixer, detailing the internal wiring, and GNU Radio Companion schematic demonstrating how to eliminate the modulation by squaring the BPSK modulated signal, as well as the Costas-loop carrier recovery. (b) Experimental setup. When running this experiment, make sure the IF signal does not include a DC offset, which might damage the diodes in the bridge.

- Make sure to remove **any DC component** on the IF digital input: if the digital signal is fed from a microcontroller general purpose input output (GPIO), a series capacitor will cancel the DC offset and avoid damaging the mixer diode bridge.
- Select the radio frequency carrier within an unlicensed industrial, scientific, and medical band compatible with the selected mixer, e.g. 434 MHz in Europe when using the ZX05-1L-S+ (LO input of the mixer).
- Connect the RF output to the antenna[0].

### 2.5.2 Phase Demodulation

The challenge of PSK demodulation lies in the recovery of the transmitted carrier in order to eliminate the frequency offset $\Delta f$ between the local oscillator and the incoming signal. Indeed, if this frequency difference is not cancelled, the phase of the signal $\Phi$ is, on the one hand, defined with a continuously changing contribution as a function of time $2\pi \cdot \Delta f \cdot t$ and, on the other hand, with the phase to be detected $\varphi \in [0, \pi]$ including the transmitted information so that the received signal is expressed as $s(t) = \exp(j(2\pi\Delta f + \varphi(t)))$. One way of estimating the carrier frequency offset is by eliminating the modulation and collapsing the energy into an image of the pure carrier, implemented in GNU Radio with the *Costas loop*, which provides the demodulated signal phase in addition to an indicator of the frequency offset. This result is achieved by noticing that an M-PSK signal transmits during every sampling period $\log_2(M)$ bits as a phase encoded as $n \cdot 2\pi/M$, $n \in [0 : M-1]$. With such a modulation, considering the $M$th power leads to an expression as

$$s^M(t) = \exp\left(j(2\pi\Delta f + \varphi(t))\right)^M = \exp(j(2\pi\Delta f \times M + \varphi(t) \times M))$$

and since $\varphi(t) = n \cdot 2\pi/M$ then $M\varphi(t) = 0$ and only a clean carrier at $M \times \Delta f$ is recovered. Care must be taken, though, that $M \times \Delta f$ lies within $\pm f_s/2$ and the loop bandwidth (Figure 2.15), reducing the accessible range of $\Delta f$ for increasing $M$. Once the carrier has been recovered, symbol synchronization is needed to lock the sampling period on the transmitted data rate, as will be discussed in detail in Section 8.14.

Computing the $M$-th power of the M-PSK signal is only the initial coarse frequency offset estimate, also called acquisition in GNSS terminology, while the Costas loop is actually a phase tracking (PLL) loop modulo $2\pi/M$. The Costas loop implementation is found in `gr-digital/lib/costas_loop_cc_impl.cc` of the GNU Radio source code.

### 2.5.3 Radio Data System (RDS)

In the case of the broadcast FM band, the two mono (left+right) and stereo (left–right) channels are located around the 19 kHz pilot signal of the radio frequency carrier, and on the other hand, the digital stream of the emitter (RDS) is located 57 kHz from the baseband. These two signals

---

0 The bel is defined as $\log_{10}(X)$ of the unitless quantity $X$, and the decibel is $10\log_{10}(X)$. Since $X$ must be unitless, it must be the ratio of two quantities with units, e.g. the ratio of two voltages in a Bode plot or the ratio of a measurement with its unit. If a power is given in W, then the logarithmic unit is given as $dBW$. If a power is given in mW, then the logarithmic unit is given as $dBm$. For a radiation pattern with respect to the isotropic antenna, the unit is given as $dBi$. Since 1 W = 1000 mW, then $dBm = dBW + 30$. Since the bel and the decibel are logarithmic representations of powers, the conversion of linear voltages to dB requires computing $20 \log_{10}(V_{out}/V_{in})$ since power and voltages are related by a square function. This is most significantly the case when measuring scattering coefficient, also known as $S$-parameters, which are voltage ratios.

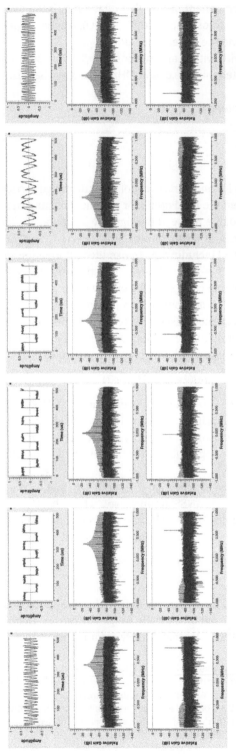

**Figure 2.15** In all charts, top is the demodulated signal's real part (cosine of the phase), middle is the spectrum of the received BPSK modulated signal, and bottom is the spectrum of the squared signal (clean carrier at twice the frequency offset). From left to right: local oscillator offset by −400 kHz with respect to the transmitted signal, −320, 0, +270, +310 and +370 kHz. The first, fifth, and sixth signals are out of the Costas loop bandwidth, and the carrier frequency correction for recovering the phase is not functional, as can be seen with the squared signal clean carrier peak reaching close to the Nyquist frequency of 1 MHz. In the second, third, and fourth cases, the Costas loop is able to compensate for the frequency offset, and the square wave of the 10 kHz BPSK modulation generated by the square wave applied to the IF port of the mixer is visible on the top time-domain chart.

are independently demodulated since all the information needed has been conserved during FM demodulation : the output of the WBFM block is provided at a rate of more than 115 ksamples/s. The Costas loop recovers the carrier frequency of an M-PSK modulated signal for $M \in [2, 4, 8]$ with $M$ called the loop `Order`. While the Costas loop takes care of identifying the carrier offset and compensating for the frequency offset to provide a copy of the emitted carrier, it does not solve the issue of selecting the sampling time: the symbol synchronization will be detailed in Section 5.3.1.3 to fully demodulate the phase-modulated signal and convert phase states (soft bits) to hard $\log_2(M)$ bits/sampling period.

As was mentioned earlier (Section 2.4.2), the bit rate of RDS is 1187.5 bits/s, but with a differential Manchester encoding, so that twice as many symbols are transmitted every second. This parameter is important for symbol synchronization estimation when the local sampling clock must be tuned to the emitter sampling clock: the `Symbol Synchronization` block adjusts the sampling rate to reach the `sps` value and outputs one unique sample per symbol after tracking the sampling clock. Figure 2.16(b), the constellation, which displays the imaginary part as a function of the real part, is initially rotating widely as the phase $\varphi$ is subject to the frequency offset $\delta f$ linear component over time $t$ through $\varphi = 2\pi \cdot \delta f \cdot t$, and once the Costas loop compensates for the frequency offset, the constellation stabilizes as a "horizontal" cloud along the real axis, with a cancelled imaginary component. However, samples are still collected during symbol transition, leading to two connected clouds centered on $\cos(0) = 1$ and $\cos(\pi) = -1$. By symbol synchronization, the samples are only collected far from the transitions, and the two clouds are disconnected, helping converting the soft bit (phase value) to a hard bit (0 or 1) decision by a threshold along the origin of the real axis.

Be aware that since the symbol synchronization is changing the data stream rate with a varying output rate with respect to the input rate (as tuned by its sampling clock rate), its output cannot be displayed on the same constellation sink since all inputs of a given processing block must be streamed at the same sampling rate.

Notice that in this demonstration, as well as in the analysis of the Meteor-M2N QPSK signal (Section 8.5), we first introduce the Costas loop to stop the constellation from rotating, and once the constellation is stable, we introduce the symbol synchronization for removing the transitions between stable soft-bit states. This order is opposite to the one demonstrated at the end of Section 5.4 but makes the sequential demonstration easier when addressing a new phase-shift keying modulation.

### 2.5.4   The Global Positioning System (GPS)

We conclude this section about PSK with the illustration of receiving GPS using an RTL-SDR receiver initially designed for the reception of television signals broadcast a few tens of kilometers away and yet able to recover the signal transmitted from 20 000 km away with the power of a light bulb (50 W) by atomic clocks on board the GPS satellite constellation.

GPS, or historically NAVSTAR (Navigation Satellite Timing and Ranging), was developed in the 1960s and deployed in the 1970s with the legacy channel GPS L1 broadcasting at 1575.42 MHz on a 2 MHz bandwidth, well within the capabilities of the RTL-SDR receiver. Fitting such a device with an active antenna polarized by a bias-T, the raw radio frequency signal is recorded, with individual satellite transmission differentiated by their unique pseudo-random (PRN) 1023-bit-long code (so-called Gold code) and communicating the NAVigation message by XORing each PRN code with the bit state being broadcast as BPSK modulation. While providing the full decoding sequence as implemented by `gnss-sdr` [Fernández-Prades et al., 2011] is beyond the scope of this presentation, we might wonder whether a relevant signal was recorded. Since the satellites are moving

**58** | *2 Using GNU Radio with Signals Collected from SDR Hardware*

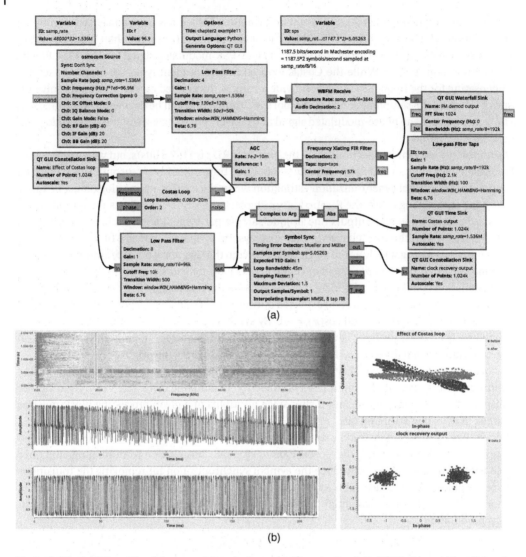

**Figure 2.16** (a) Schematic of the processing chain used to decode the digital RDS channel identifying the radio frequency emitter. (b) Top is the waterfall with the Fourier transform along the abscissa evolving over time along the ordinate, with the 19 kHz tone clearly visible and the 57 kHz-centered BPSK-modulated RDS stream. Middle: the phase of the RDS stream after transposition by the nominal 57 kHz exhibiting the slow drift from the inaccurate receiver local oscillator with respect to the emitter local oscillator. Bottom: the phase of the output of the Costas loop after correcting for the frequency offset between emitter and receiver local oscillator frequencies, leaving only the random bit state fluctuations between the two discrete states. Bottom right: constellation display before Costas loop, after Costas loop, and after symbol synchronization.

along their orbit, the received signal is plagued by Doppler shift in addition to spectrum spreading from the PRN code BPSK modulation.

The coarse frequency offset of a BPSK modulated signal is obtained by squaring the signal, collapsing the energy spread over a spectrum whose first null is determined by the data rate – here 1.023 MHz – into a clean carrier. Indeed, even if the receiver local oscillator was an atomic clock similar to those found on board the GPS space vehicles, the motion of the satellite would induce a Doppler frequency shift.

## 2.5 Phase Modulation and Demodulation

**Frequency shift induced by a satellite motion**

The motion of any moving radio frequency emitter or receiver induces a frequency shift due to the Doppler effect: the carrier frequency $f_0$ emitted or received by a vehicle moving at radial velocity $v$ is shifted by a frequency offset $df = f_0 \frac{v}{c}$, with $c$ being the speed of light in vacuum. The tangential velocity of a satellite in a circular orbit is solely determined by its distance to the center of the Earth: considering that geostationary satellites are located at an altitude of 36 000 km above the surface of the Earth, with radius $R_{Earth} = 40000/(2\pi) = 6370$ km, and orbit the planet at a rate of $T_{GEO} = 1$ rotation every day – hence appearing at a mostly fixed location in space – then Kepler's third law, stating that the ratio of the square of the orbital period to the third power of the radius of the orbit, leads to the GPS satellite period $T_{GPS}$ knowing their medium Earth orbit (MEO) of 20000 km:

$$T_{GPS} = T_{GEO}\sqrt{\frac{r_{GPS}^3}{r_{GEO}^3}} = 0.49 \text{ day}$$

or a 12-hour period. Considering the length of the orbit of $2\pi(20\,000 + 6370) = 165\,690$ km, the tangential speed of the GPS satellite is $165\,690/0.49/86\,400 = 3.92$ km/s. From this velocity and the carrier L1 frequency of 1575.42 MHz, the Doppler frequency shift could be expected to be as high as 20 kHz. However, since only the radial component toward the viewer $P$ assumed to be located on the surface of the Earth impacts the radio frequency link (Figure 2.17), a projection when the satellite rises over the horizon at $H$ or sets at $H'$ is needed. From basic geometrical considerations (Figure 4.5) where the viewer at the surface of the Earth draws a right triangle with the satellite rising over the horizon and the hypotenuse linking the center of the Earth with the satellite, the sine of angle $\vartheta$ between the tangential velocity and the vector component toward the receiver is related to $R_{Earth}/(R_{Earth} + r_{GPS})$, providing the projection factor whose numerical value is 0.242, and the maximum Doppler shift is $\pm 20 \times 0.242 = \pm 4.8$ kHz. Notice how large this Doppler shift is with respect to the navigation message bitrate of 50 bits/s or 20 Hz transmission rate, requiring fine-tuning of the receiver frequency with respect to the received signal as the satellite is flying along its orbit.

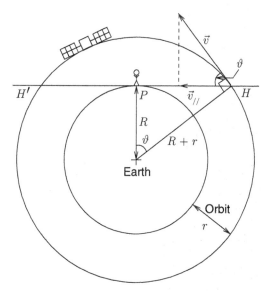

**Figure 2.17** Projection of the tangential velocity $\vec{v}$ toward the receiver $\vec{v}_{//}$.

**60** | *2 Using GNU Radio with Signals Collected from SDR Hardware*

Phase modulation requires fine cancellation of any source of phase fluctuation between the emitter and receiver so that only the random phase variation holding the transmitted information is left. From orbital considerations, the frequency shift from the nominal 1575.42 MHz might be expected to lie within the $\pm 5$ kHz offset range, but the poor quality of low-cost oscillators fitted on commercial, off the shelf (COTS) SDR receivers must be considered. With typical tolerances in the 100 ppm (parts per million[1]) range, the local oscillator driving the PLL generating the local oscillator of 1575.42 MHz is expected to introduce a variation in the $\pm 157$ kHz, which is more than 30 times the Doppler-induced frequency shift.

Collapsing the spread spectrum energy in the carrier, or codeless GPS decoding, uses squaring of the recorded IQ stream to estimate twice the frequency offset (due to squaring doubling the argument of the complex exponential) between emitter and receiver. The following charts are the results of processing signals recorded by an RTL-SDR dongle tuned to 1575.42 MHz and sampling at 2 MS/s. At first, the signal is squared and its Fourier transform on 32 768 bins is displayed with high averaging to lower the noise floor (Figure 2.18). The data were recorded using the command line `rtl_sdr` tool, which saves to file the raw unsigned 8-bit integers ("characters" in the C-language naming convention) interleaving I and Q coefficients. This information was recorded in the SigMF file with the data type "ci8" for "complex 8-bit integer." The GNU Radio flowchart reads the content of the file, converts unsigned characters to floating point numbers and orders the interleaved stream as complex numbers, squares the signal, and displays the time-domain data and spectrum of the squared signal. Notice that GPS power – defined with a minimum received power at ground of $-130$ dBm by the US Air Force standards – is well below the thermal noise in a 2 MHz bandwidth of $-174$ dBm/Hz$+10 \log_{10}(2 \cdot 10^6) = -111$ dBm, and that the spectrum will hardly display any signature of the presence of the signal before despreading by squaring or correlating with the known Gold codes.

While in continuous time signal processing, the frequency range of the Fourier transform spans from 0 to $f_s$ or, from the spectrum periodicity, from $-f_s/2$ to $+f_s/2$, it is important to notice that in discrete time, a Fourier transform computed on $N$ samples will lead to $N$ complex coefficients in the spectral domain at frequencies ranging from $-f_s/2$ to $+f_s/2 - f_s/N$. Using the Python language and its numpy library, the abscissa is expressed as `numpy.linspace(-fs/2,fs/2-fs/N,N)`.

Once the coarse frequency offset has been determined after codeless analysis of the signal, the Frequency Xlating finite impulse response (FIR) filter is used to bring the baseband signal close to 0 Hz by compensating for the local oscillator frequency offset, and additional filtering for decimating magnifies the abscissa of the spectra to emphasize the various satellite signals. This second flowchart (Figure 2.19) emphasizes the benefit of cascading FIR filters over one sharp filter with a large decimation factor that would induce an excessively long sequence of taps, inducing large computational power. By cascading filters, a first filter removes the excess bandwidth (low-pass filtering) and decimates before feeding a second filter whose transition width is defined with respect to the input decimated stream data rate and hence requires fewer taps with respect to a single filter on the initial stream.

Global Navigation Satellite Systems, including GPS, are time transfer systems before allowing for the location of the receiver. In this context, the time stamp of each sample is a relevant information when assessing the time difference between the GNSS timescale and the local clock, whose offset

---

1 The relative frequency offset of oscillators $\frac{\Delta f}{f_0}$ is defined as the frequency offset $\Delta f$ relative to its nominal frequency $f_0$. Since these values are in the range of a few $10^{-6}$, the quantity is traditionally multiplied by $10^6$ and expressed in ppm. This unit-less quantity allows to assess the frequency offset at any nominal frequency $f_0$ due to oscillator imperfection. Notice that this quantity might not be fixed but dependent on temperature, aging, stress, and other environmental factors.

## 2.5 Phase Modulation and Demodulation | 61

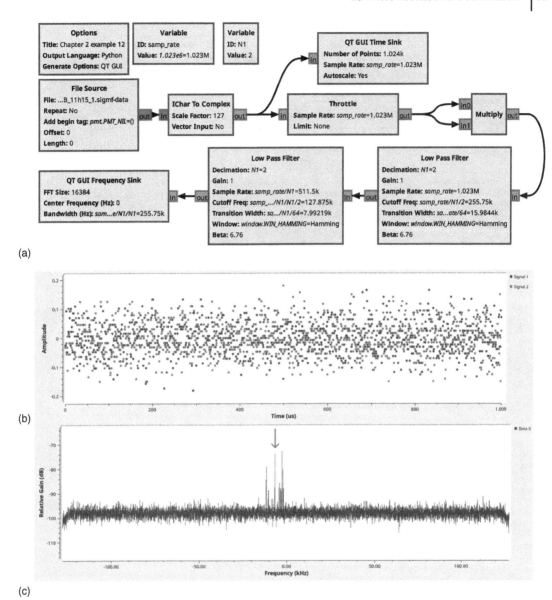

**Figure 2.18** (a) Flowchart for loading the GPS record encoded as 8-bit (`char`) interleaved IQ samples and displaying the time domain (b) and spectrum of the squared signal (c) since the energy of the BPSK-modulated signal collapses into a pure carrier at twice the frequency offset at baseband once squared. Each satellite is visible as a different frequency component representing the different Doppler shifts for each space vehicle in the spectrum of the squared signal (vertical arrow).

must be corrected. However, assuming a constant data flow with no interruption as needed for SDR processing, there is no need to individually time-stamp each and every sample. The data acquisition index $n$ is enough to assign a relative time stamp with respect to the beginning of the acquisition as $n/f_s$, and knowing how many samples have been collected since GNU Radio was launched implicitly provides the time stamp of each index. From this analysis, any post-processing delay after the

## 2 Using GNU Radio with Signals Collected from SDR Hardware

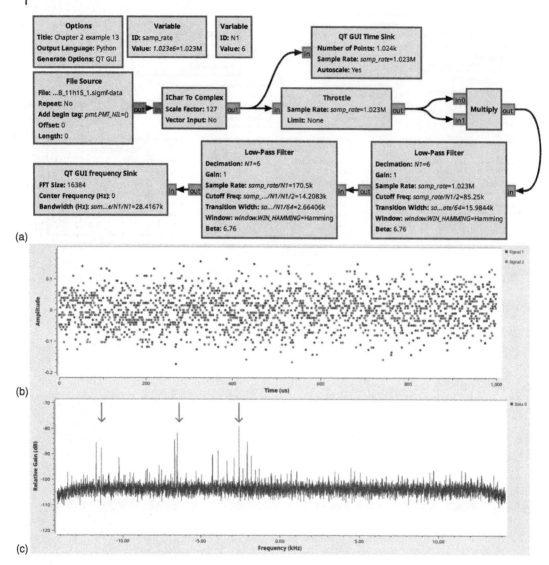

**Figure 2.19** (a) Improved flowchart from Figure 2.18 with coarse frequency transposition to compensate for local oscillator inaccuracy and zoom on the spectrum by decimating after low-pass filtering to avoid aliasing. (b) Time-domain display. (c) By decimating, the spectral components are better separated, and each individual satellite is well visible from widely spread carriers (vertical arrows).

ADC has sampled and stored a data is irrelevant to time stamping, and the asynchronous communication schemes (USB, Ethernet) or buffered communication will not impact the ability to assign a known time stamp to each sample. Similarly, any clock manipulation to control an oscillator on the data recorded by the SDR system, e.g. on the GPS timing signal, is only relevant on the clock signal driving the ADC, which usually also clocks an field programmable gate array (FPGA) in charge of data collection and preprocessing. The general-purpose processing unit (CPU) clock is irrelevant in this context since it is separated from the data collection layer by asynchronous communication methods with undetermined latencies. As a demonstration of this principle, the output of `gnss-sdr` includes the time offset between the local copy of the clock and GPS clock, which can be used to steer the SDR FPGA clock to match the spaceborne atomic clock reference [Rabus et al., 2021].

## 2.6 Spectral Occupation of the Various Modulation Schemes

The bandwidth of a communication channel defines the amount of information that can be transmitted. The modulation scheme induces a spectral occupation, further discussed in Chapter 5, and hence the distribution of the spectral components within the allocated bandwidth. Figure 2.20 illustrates the spectral occupation of three common modulation schemes discussed here – AM, FM, and PM – to encode the same input signal – a sine wave of fixed frequency and amplitude.

AM, despite its use in aircraft communication to avoid the FM capture effect, is notably inefficient due to its duplicate copying of the information in both upper and lower sidebands (even spectrum of a real signal) and the varying power broadcast depending on the signal amplitude. Alternatively, a single sideband can be broadcast, at the expense of a more complex modulation and demodulation scheme, but with improved spectral and power efficiency. Broadcasting only the upper side band (USB) or lower side band (LSB) is a classical modulation scheme in the ham radio community aiming at maximizing communication range (DX). An example of such a modulation scheme is demonstrated in Figure 2.21 where a broadcast signal (real signal source, top-left) is modulating only the upper or lower sideband using a complex (real and imaginary) modulation scheme.

## 2.7 Local Oscillator Leakage Issue

We have mentioned that most SDR receivers benefit from the improvements of mixers and adopt a direct conversion approach where the radio frequency incoming signal is frequency transposed by mixing with a local oscillator (I and 90° shifted component for $Q$), whose output feeds an ADC. Historically, the superheterodyne architecture was developed to compensate for poor mixer isolation: in the superheterodyne technique, the transposition of the radio frequency wave at $f_{RF}$ to baseband centered on 0 Hz is performed in two steps, from $f_{RF}$ to an intermediate frequency $f_{IF}$ much lower than $f_{RF}$, and a second low-frequency mixer transposes $f_{IF}$ to 0 Hz. Indeed, $\exp(j2\pi \cdot f_{RF}t) = \exp(j2\pi \cdot (f_{RF} - f_{IF})t) \cdot \exp(f_{IF} \cdot t)$, and the frequency transposition can be performed in two steps, with the advantage that the IF (lower-frequency) mixer exhibits better isolation than the the RF mixer and that image rejection filters can be included between both mixers. The local oscillator leakage (local oscillator [LO] leakage) issue arises when analyzing signals close to DC. Indeed, mixers with poor isolation will leak some energy toward the RF port connected to the antenna, where the impedance mismatch will reflect part of the energy back into the receiver amplifier and into the mixer. Because this leakage signal is, by definition, at the local oscillator frequency, the mixer output is at exactly 0 Hz and appears as a strong DC signal on the ADC input. Figure 2.22 demonstrates this issue when observing the GRAVES RADAR continuous wave signal emitted at 143.05 MHz: the RTL-SDR dongle exhibits poor LO leakage performance, and zooming close to 0 Hz (DC) leads to a strong peak in the spectrum. The GRAVES signal is actually offset by −7500 Hz due to the poor accuracy of the RTL-SDR local oscillator, which exhibits a −53 ppm offset. We benefit from this frequency offset considered in this analysis as an IF far away from the signal of interest so that no additional frequency offset is introduced on the local oscillator. By sampling at a wide enough sampling rate samp_rate, all signals are included in the baseband, and the LO leakage is shifted away from the 0 Hz centered band of interest in the digital domain (software processing) using an

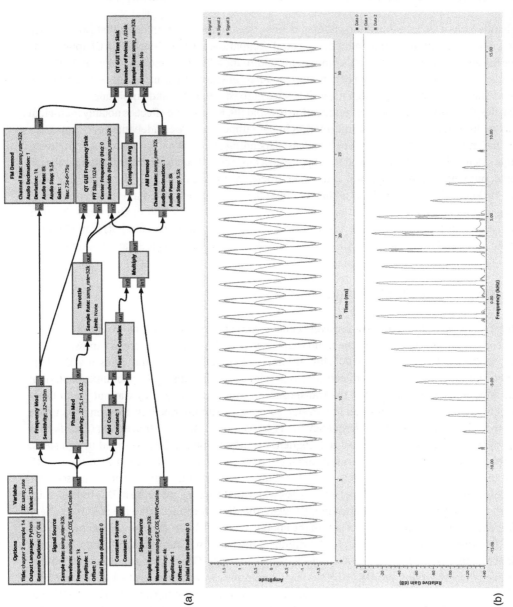

**Figure 2.20** (a) Flowchart for generating various modulation schemes of the same input signal, displaying their spectrum, and returning back to time domain by demodulating. (b) Frequency domain (spectral occupation) and time domain plots of the modulated and demodulated signals. The PM modulation parameters were selected to exactly overlap the FM modulation spectra, emphasizing the close relationship between the two modulation schemes.

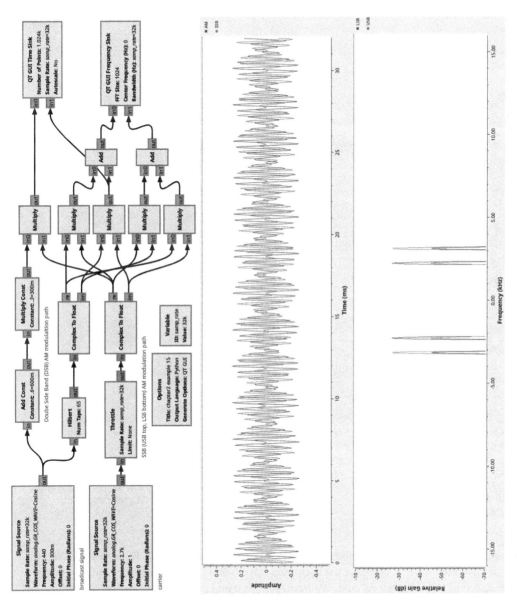

**Figure 2.21** (a) Flowchart for generating Upper Side Band (USB) or Lower Side Band (LSB) of an amplitude modulated signal, and comparison with the Double Side Band (DSB) AM modulation scheme (a). (b) Comparison of the spectra resulting from USB and LSB modulation, with the unwanted sideband attenuated by 20 dB with respect to the sideband carrying the information. The time domain chart illustrates the close relation between the AM and single side band (SSB) modulation waveforms.

**66** | *2 Using GNU Radio with Signals Collected from SDR Hardware*

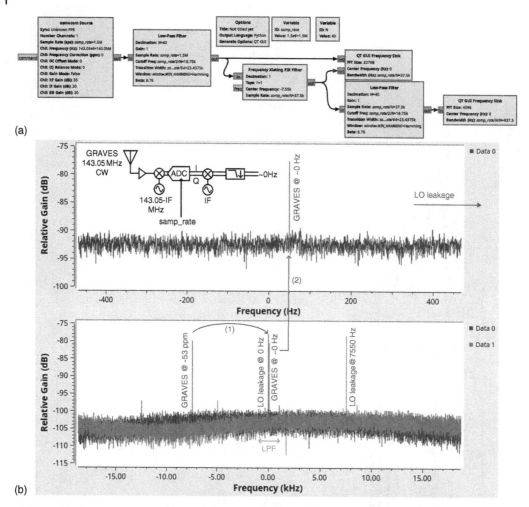

**Figure 2.22** (a) Flowchart for compensating for LO leakage by frequency transposition of the sampled signal to move the DC component to the side of the baseband and bring the signal of interest – here the continuous wave (CW) GRAVES RADAR signal – close to 0 Hz (1) before low-pass filtering and decimation to eliminate the LO leakage (2). (b) Demonstration on an actual signal recorded by a RTL-SDR. Notice that the LO leakage becomes visible only when the thermal noise power in a Fourier transform bin becomes lower than the LO leakage, meaning that a high number of bins (here 32 768-bin FFT) on a low sampling rate is needed to emphasize the LO leakage issue. The low-pass filter (LPF – bottom) indicates the baseband signal kept before decimating (middle).

`Xlating FIR Filter`. Doing so, the GRAVES signal is brought close to 0 Hz, a low-pass filter gets rid of unwanted spectral components including the LO leakage, and only the signal of interested broadcast by GRAVES is analyzed. We have already stated that the frequency offset induced by moving targets illuminated by GRAVES happens to be equal to the projected velocity in meters/second so that a frequency span of ±400 Hz is more than enough to tackle subsonic aircraft flying in view of the receiver, as shown in Figure 2.22(a).

We mention in the caption of Figure 2.22 that under most circumstances the LO leakage is not visible, and only with sufficient decimation and high number of bins in the Fourier transform does the issue become visible. Indeed, the background thermal noise is $k_B \cdot T = -174$ dBm/Hz with

$k_B = 1.38 \times 10^{-23}$ J/K being the Boltzmann constant and $T$ the temperature (in Kelvin) so that the background noise of the FFT with bin size $B$ is $k_B \cdot T \cdot B$. Since $B = samp\_rate/N$ with $samp\_rate$ the sampling rate and $N$ the number of bins in the FFT, under common conditions of high sampling rate and of low $N \simeq 1024$, the thermal noise is above the LO leakage power. Only by reducing $samp\_rate$ through a strong decimation and a high $N \simeq 32\,768$ as shown in Figure 2.22 does the LO leakage become significant, as is often the case when monitoring continuous wave (CW) RADAR or using SDR for physics experiments where an unmodulated carrier is investigated.

## 2.8 Conclusion

While Chapter 1 introduced the main concepts needed to tackle SDR processing using GNU Radio as a simulation tool analyzing synthetic signals, this chapter aimed at emphasizing how these flow-graphs could be used to process IQ samples collected from dedicated SDR receiver hardware. Doing so, we have seen how real signals could be demodulated in real-time or by post-processing records stored in files, demonstrating the most common analog modulation schemes encoding the information on amplitude, frequency, or phase, and their respective spectral occupation.

This chapter focused exclusively on reception since emission does involve regulation issues that the reader must meet before considering broadcasting, either by obtaining a license (ham radio license) or confining the emission to a dedicated environment, preventing interference with legal emitters (anechoic chamber). Broadcasting signals from SDR will be introduced in Chapter 5.

Finally, notice that while the sound card was designated as a GNU Radio **sink** to output sound when demodulating commercial broadcast FM or air traffic control signals, it can also be used as a **Source** when addressing very low frequency (VLF) signals below 96 kHz (half the sampling rate of 192 ksamples/s), as was the case when analyzing time of flight measurements of the German DCF77 timing signal broadcast on 77.5 kHz [Friedt et al., 2018].

## References

D. Bederov. Arithmetic based implementation of a quadrature FM demodulator SDR in GNU Radio. In *FOSDEM*, Brussels, Belgium, 2015.

J. Bloom. *Eccentric Orbits: The Iridium Story*. Open Road+ Grove/Atlantic, 2016.

C. Fernández-Prades, J. Arribas, P. Closas, C. Avilés, and L. Esteve. GNSS-SDR: An open source tool for researchers and developers. In *Proc. 24th Intl. Tech. Meeting Sat. Div. Inst. Navig.*, pages 780–794, Portland, Oregon, Sept. 2011. https://github.com/gnss-sdr/gnss-sdr accessed March 2024.

J.-M. Friedt and T. Laverenne. Bitstream clock synchronization in an ACARS receiver: Porting GR-ACARS to GNU radio 3.8. In *Software Defined Radio Academy (SDRA)*, 2020.

J.-M. Friedt, C. Eustache, É. Carry, and E. Rubiola. Software-defined radio decoding of DCF77: Time and frequency dissemination with a sound card. *Radio Science*, 53(1):48–61, 2018.

M. Frigo. A fast Fourier transform compiler. In *Proc. ACM SIGPLAN Conference on Programming Language Design and Implementation*, pages 169–180, Atlanta, Georgia, USA, 1999.

ICAO. Technical manual for Iridium aeronautical mobile satellite (route) service. Technical report, ICAO, 2006. https://www.icao.int/safety/acp/ACPWGF/ACP-WG-M-11/ACP-WGM11-WP04-Draft.

J. III. Mitola. *Software Radio Architecture: Object-Oriented Approaches to Wireless Systems Engineering*. Wiley, 2000.

D. Rabus, G. Goavec-Merou, G. Cabodevila, F. Meyer, and J.-M. Friedt. Generating a timing information (1-PPS) from a software defined radio decoding of GPS signals. In *2021 Joint Conference of the European Frequency and Time Forum and IEEE International Frequency Control Symposium (EFTF/IFCS)*, pages 1–2. IEEE, 2021.

T. Rondeau. GNU Radio for exploring signals. In *FOSDEM*, Brussels, Belgium, 2016. https://archive.fosdem.org/2016/schedule/event/gnuradio/.

Sec and Schneider. Iridium hacking. In *Chaos Communication Camp*, Zehdenick, Germany, 2015. https://www.youtube.com/watch?v=ahZOGhV8qnc.

Sec and Schneider. Iridium update, 2022. https://people.osmocom.org/laforge/OsmoDevCall/20220325-Iridium-Update-Public.pdf.

# 3

# Communicating with External Software (Python, Networking, ZeroMQ, MQTT)

While GNU Radio is well suited for processing continuous data streams, using external software might sometimes be efficient either because existing decoding tools have already been developed and are suited for processing the output of GNU Radio blocks, or for batch processing datasets when continuous streams are not needed (e.g. RAdio Detection And Ranging [RADAR] target signals), or when tuning the radio frequency frontend or processing parameters prior to implementing a dedicated GNU Radio block (out-of-tree [OOT] module). These three cases will be addressed in this section:

1) using first in first out (FIFO) file systems provided, for example, by Unix `mkfifo` to connect a pseudo-file output to the standard input, as will be demonstrated with the `multimon` software [Elias Önal, 2023] for decoding POCSAG sentences,
2) using Internet protocol (IP) datagram (user datagram protocol [UDP]) or connected (transmission control protocol [TCP]) communication [Fall and Richard Stevens, 2011] by running a dedicated thread next to the GNU Radio scheduler, allowing to access, for example, the dataflow control variable for the flowgraph,
3) using the XML-RPC (remote process control) to access the variables driving a GNU Radio flowchart and benefiting from a description of the remote function being called rather than a custom protocol as described earlier,
4) using the ZeroMQ and message queuing telemetry transport (MQTT) overlays to TCP/IP to wrap complex data structures such as vectors of complex numbers, well suited for a discontinuous communication with external processing software as will be demonstrated for Python – the native output of GNU Radio Companion – and Octave, the open-source version of the Matlab language.

As application examples we encountered, sentence decoding, once the header synchronization frame has been identified, is efficiently handled by external software identifying bit packets and parsing their content, as will be discussed later, for example, with weather satellite Meteor-M2N decoding according to a protocol compatible with the generic CCSDS[1] or for correlating signals reflected by moving targets, which are computationally intensive operations hardly achieved in real time with general-purpose processing hardware. Furthermore, communicating with external software allows for benefiting from the most efficient description language or using appropriate libraries, for example in a prototyping sequence where first recorded data are analyzed, and then interpreted language prototyping is used before switching to a compiled and hence faster language, before finally complying with the GNU Radio framework and encapsulating the processing in the

---

1 Consultative Committee for Space Data Systems.

*Communication Systems Engineering with GNU Radio: A Hands-on Approach*,
First Edition. Jean-Michel Friedt and Hervé Boeglen.
© 2025 John Wiley & Sons, Inc. Published 2025 by John Wiley & Sons, Inc.
Companion website: www.wiley.com/go/friedtcommunication

# 3.1 Connecting to an External Sentence Decoding Tool Using Named Pipes

C++ work function of an OOT block. With this development strategy in mind, we shall illustrate communication protocols between GNU Radio and Octave or Python.

## 3.1 Connecting to an External Sentence Decoding Tool Using Named Pipes

The Unix file system provides a mechanism for allowing various software to communicate seamlessly: pipes (the "|" symbol) and their file version, the named pipe. When creating a named pipe with mkfifo /tmp/myfifo, a pipe is created with any data entering the pipe from the software writing to the other end of the software reading. Hence, streaming the results of GNU Radio signal processing to a File Sink connected to the named pipe will allow for communicating with any software able to open the file and read its content.

To ensure proper functionality, launch the GNU Radio program as shown in Figure 3.1 and observe that nothing happens (no display on the Time Sink output oscilloscope) until the other end of the pipeline is connected. However, by running cat < /tmp/myfifo in a terminal, we can see that the scheduler starts generating data, the binary representation of which is displayed in the terminal. Furthermore, using Octave, we can perform the same operation with the following code:

```
f=fopen('/tmp/myfifo');while (1);d=fread(f,1000,'float');plot(d);refresh();end
```

which opens the file once and then continuously reads the last 1000 floating-point numbers (implicitly encoded as 4 bytes, i.e. 4000 bytes) in order to display their contents. By manipulating the slider bar that varies the frequency of the signal, we can observe increasing latency, as all the data cannot be consumed in real time. We attempted to close and reopen the file within the loop, but this did not change the fact that the pipeline does its job by storing all injected data until it is consumed.

Note that /tmp/myfifo must be created *before* launching GNU Radio; otherwise, a true file (without the p attribute in the first field of ls -la /tmp/myfifo) will be created in its absence and gradually filled without providing the expected result.

Because the named pipe is a FIFO, all data are buffered until processed so that after some time, a lag is observed between the data output of GNU Radio and the chart displayed by Octave as can be verified by moving the slider defining the sine wave frequency. This approach is hence suitable when no data must be lost and the data rate is low enough that the sink software can process all data coming from GNU Radio.

We will illustrate this use of the named pipe in a practical case in the following Sections 3.1.1 and 3.1.2. We first start with a simpler application of decoding pager messages, also known as POCSAG.

### 3.1.1 POCSAG Single Channel Decoding

A simple approach to analyze data streams from demodulated signals is to use an external tool reading from the Unix file input and written by GNU Radio to such a pseudo-file output created by the mkfifo command. Indeed, mkfifo creates a named pipe, exposed to the user as a file not storing the data on a mass storage device but to whatever software is reading the same file. We demonstrate this principle using a radio transmission decoding software named multimon and its latest revision multimon-ng. This software is able to decode a large number of protocols as listed with the -h flag, and we will here focus on the POCSAG protocol as historically used with pagers still in use in France with such users as fire fighters or computer system administrators who

3.1 Connecting to an External Sentence Decoding Tool Using Named Pipes | 71

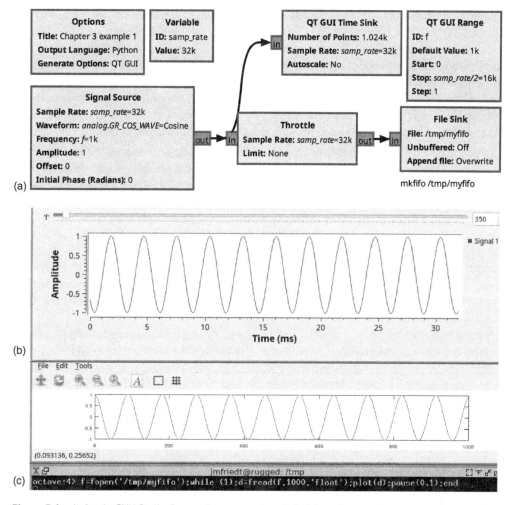

**Figure 3.1** A simple GNU Radio Companion processing chain (a) produces a data stream at a rate of approximately 32 000 samples per second (block `Throttle`) and feeds it into a file connected to a FIFO. At the other end, Octave opens this file, reads its contents, and displays it on a graph as fast as possible (c). The oscilloscope (b) allows validation of when the data are produced by GNU Radio during the execution of the processing chain.

do not wish to rely on existing cellular network infrastructures for critical messaging. Using the `raw` input interface type assumes that signed 16-bit integer data are streamed at 22 050 Hz and GNU Radio will be tuned to generate such a data stream.

When decoding POCSAG messages only, all other decoders supported by `multimon-ng` are eliminated with the `-c` flag, and graphical output as well as POCSAG messages transmitted at 1200 bauds are displayed with `multimon-ng -t raw -c -a POCSAG1200 -a SCOPE myfile`, where the pseudo-file FIFO `myfile` was created using the `mkfifo myfile` command under Unix.

We focus on signals emitted in France by pagers, also known at the time by their brand name Tam-tam or Tatoo and still used by the e*message company[2] with multiple channels broadcasting

---

2 https://www.emessage.fr/.

messages encoded according to this protocol at the six allocated frequencies of 466.{025; 05; 075; 175; 20265; 23125} MHz. Because POCSAG is frequency shift keying (FSK) modulated, GNU Radio is well suited for data acquisition, frequency demodulation, and decimation after filtering. Possible frequency transposition is needed when decoding multiple channels in parallel, whereas decoding the payload will be handled by the `multimon-ng` external tool fed through as many named pipes as there are decoded channels. The computational load will limit the number of channels decoded simultaneously since, on the one hand, a thread will be allocated to each GNU Radio processing block, and one process will be executed for each instance of `multimon` (Figure 3.2).

⚠ A stream through a named *pipe* only starts flowing when both ends of the pipe are connected. Launching the `gnuradio` software is not enough to run the acquisition, and the processing only starts *after* running all instances of `multimon-ng` connected to all the named pipes fed by GNU Radio.

The other configuration subtlety lies in complying with the expected data rate for `multimon-ng` to handle the data stream, namely a **rate of 22 050 Hz with 16-bit integers**. This data rate is most easily achieved if the initial radio frequency sampling block, e.g. RTL-SDR `Osmocom Source` block, sampling rate is set to a multiple of the targeted output data rate, with a radio frequency sampling rate broad enough to include all broadcast channels and meeting the hardware sampling rate range limited by the (USB2) digital communication bandwidth to 2.4 Msamples/s for the RTL-SDR.

Because a frequency demodulator block acts as a frequency-to-voltage converter, any frequency offset between emitter and local oscillator frequencies will appear on the receiver side as a voltage offset. Since `multimon-ng` defines state transition by zero-crossing, an excessive offset might prevent detecting the successive bit states: the high-pass filter included between the Frequency Modulation (FM) demodulator output and the FIFO input removes the continuous (DC) offset and is tuned to hardly affect each bit shape. This high-pass filter is also used to decimate the stream to the expected data rate.

### 3.1.2 POCSAG Multichannel Decoding

We use the opportunity of POCSAG decoding to emphasize the flexibility of SDR processing over hardware implementation. Having implemented one POCSAG channel decoding scheme, we wish to decode simultaneously all French POCSAG channels in the [466.025, 466.23125] MHz range, well within the 2.4 MHz bandwidth of the RTL-SDR receiver. So far, we have only decoded a single channel by tuning the hardware local oscillator to its carrier frequency. However, since the I/Q stream provides all the information carried by all the channels, we can decode the content of all communication channels in parallel (Figure 3.3).

From a signal processing perspective, we must band-pass each channel and process the associated information without being polluted by the spectral content of adjacent channels. Practically, we frequency-shift each channel close to baseband (zero-frequency) and introduce a low-pass filter to cancel other spectral contributions.

This processing scheme of frequency transposition followed by filtering and decimating is so classical that it is implemented as a single GNU Radio Companion block: the `Frequency Xlating FIR Filter`. This block, clocked by the input stream sample rate, includes the frequency defining the local oscillator with which the frequency must be shifted by mixing, and the low-pass filter taps. A low-pass filter is defined using GNU Radio Companion's `Low Pass Filter Taps` block to compute the taps: the block ID defines the variable name storing these coefficients and is used at `taps` argument of the low-pass filter.

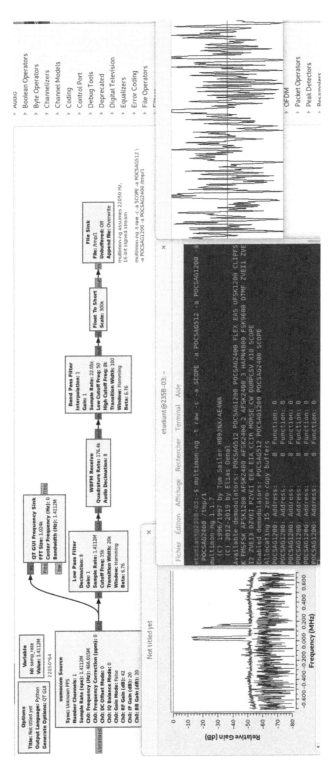

**Figure 3.2** Receiving a single POCSAG channel around 466 MHz and processing the demodulated signals streamed to a named FIFO.

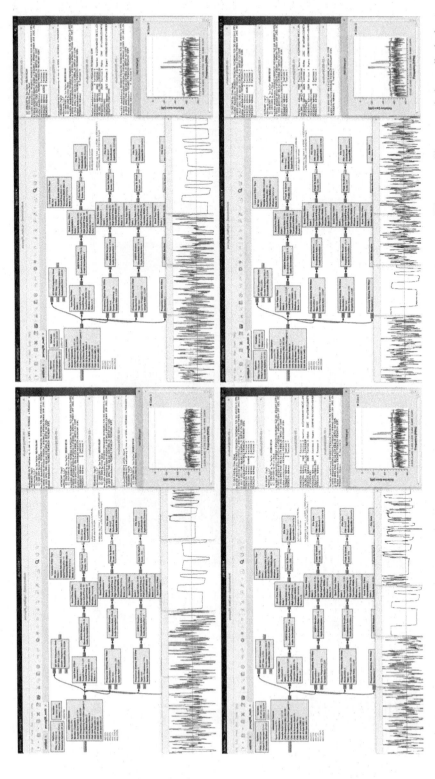

**Figure 3.3** Signal processing of the POCSAG signals using multimon-ng. Notice how different channels lead to different oscilloscope windows displaying a signal.

This processing strategy can be extended to any number of channels (Figure 3.4), as long as processing power is available. Practically, we have observed that it is unwise to process more channels than computer threads available on the processor.

Furthermore, this conclusion can be extended to any number or kind of modulation schemes transmitted within the analyzed bandpass, as will be shown next with the broadcast FM band.

## 3.2 TCP/IP Server Running in a Separate Thread

The named pipe is well suited for sharing data between GNU Radio and processing software running on the same host computer. In a distributed system, it might be desirable to remotely process the data or change the flowchart settings, whether to benefit from heterogeneous multiprocessing system architectures or transport data collected by an embedded system to a powerful processing framework. The Internet protocol (IP) was designed to homogenize decentralized computer networks and is the current de facto low-level standard for computer communication. A server is designed to wait for the connection of a remote system requesting its features called the client. In order to remotely change the settings of a GNU Radio flowchart, a server will be running next to GNU Radio, waiting for clients to connect and share commands to define the new settings. However, GNU Radio is already running an infinite loop scheduler so that running the server *next to* GNU Radio requires creating a new thread and sharing data structures between the spawning program and the thread (e.g. the variable setting callback functions). In this example, we will focus on a TCP/IP server, TCP meaning a connected mode in which each transaction between client and server is validated and ordered, written in Python since this language is the one used by GNU Radio Companion to connect blocks.

Let us start with the basics: the interoperability of computers connected on a network according to the principles of the internet, following a layer of abstraction formally known as the open systems interconnection (OSI) layers (Figure 3.5) according to the *open systems interconnection* standards.

OSI layering may seem arbitrary until one tries to implement it in a practical case (see Chapter 8) to discover that each layer implies different technical expertise and knowledge. Thus, the lowest layer – hardware – will be easily approached by an electronics specialist, while the highest layer – the application layer – involves many abstract computer concepts. In between, information needs to be assembled into packets, routed from one machine to another to allow for routing the information from the source to the destination, and the parties involved must agree on the representation of the information and the various services capable of processing the information (socket ports). The principle of the OSI hierarchy is that each higher layer assumes that the lower layers have been implemented and are functional. Therefore, no packet routing by TCP in connected mode, which guarantees the integrity of the exchanges, or UDP in which a server broadcasts information to clients that may or may not receive it, is possible without access control and conversion of physical address into a software address by address resolution protocol (ARP) (Figure 3.5).

Above the IP layer, which translates physical addresses into logical addresses, there are two modes of communication: TCP, which guarantees transactions (connected mode) and in which the server blocks its exchanges until they are acknowledged by the client, and UDP (datagram), in which the server sends data that may or may not be received by a client in an order that is not guaranteed depending on the packet routing along the route between the server and the client. In this second case, the server performs its operations independently of any connection from a client to receive or not receive the acquired data: this mode of communication is best suited when

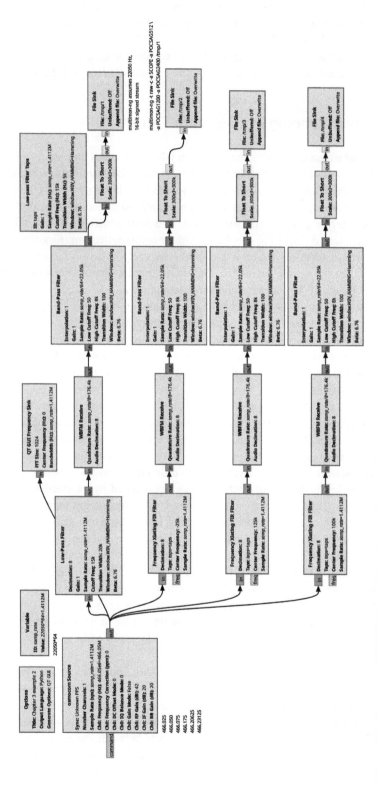

**Figure 3.4** Decoding 4 POCSAG channels simultaneously.

7 Application (HTTP, SMTP, NTP)
6 Presentation (ASCII, HTML, SSL, MQTT, XML)
5 Session (RPC, socket)
4 Transport (TCP,UDP)
3 Network (ARP, IP)
2 Link(Ethernet)
1 Physical (10BASE-T)

**Figure 3.5** OSI layer hierarchy describing the services needed for a communication through a computer network. We shall be interested here in the upper layers, although this layering will be referred to multiple times in this document when emphasizing how each abstraction layer involves different expertise and background, from the closest to physics at the bottom to computer science on the top. The OSI layers assume that each upper layer can rely on functional implementations of the lower layers.

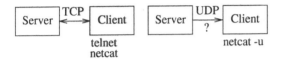

**Figure 3.6** Concept of a server – waiting to provide a service – and a client – requesting a service, exchanging data either in connected mode to guarantee exchanges (TCP) or in datagram mode without validation of transactions (UDP).

implementing a RADAR, for example, which can freely control a SDR receiver and move the antennas while a client receives data when conditions are favorable. Universal clients that make it easy to test servers are `telnet` and `netcat` for TCP, the latter with its `-u` option for receiving information in UDP. One might wonder, why not always use TCP, which guarantees data exchanges? A TCP exchange requires storing transiting packets that could be corrupted or whose order has been swapped due to changes in the routing rules on the network during communication: a TCP/IP stack is very complex to implement and resource-intensive, while a UDP exchange can be implemented in a few lines in the absence of any acknowledgments.

The concept of a socket is at the heart of a Unix system, which cannot function without it, as defined by the POSIX standard [POSIX Standards, 2023]. Just execute the `ss` (*socket statistics*) command to see the hundreds of communication pipes open on a GNU/Linux system, even when disconnected from the internet but exchanging information between its various services. Thus, a socket does not necessarily transport data from one computer to another over a computer network but can exchange information between processes running on the same computer: this is called *inter-process communications* or IPC. Remote control of processes is a special case of IPC called RPC for *remote process control*. The question then arises about the representation of the exchanged data and their encapsulation so that the parties involved can agree on their representation.

Indeed, a basic TCP server – remember that the **server** is constantly waiting to provide a service to **clients** (Figure 3.6) which connect occasionally to access this service – is written in Python as

```
import socket
import string
while True:
    sock = socket.socket(socket.AF_INET, socket.SOCK_STREAM)
    sock.setsockopt(socket.SOL_SOCKET, socket.SO_REUSEADDR, 1)
    sock.bind(('127.0.0.1', 4242))
    sock.listen(1)
    conn, addr = sock.accept()
    with conn:
```

```python
        print('connected from ', addr)
    while True:
        data=conn.recv(1)
        if data:
            data = data.decode()
            print(data)
            if 'q' in data:
                sock.shutdown(socket.SHUT_RDWR)
                sock.close()
                break
```

to connect a server to a socket (`bind`), wait for a client connection (`listen`), and exchange information (`recv`, `send`) following the same initialization sequence that would be found in C. These series of bytes have no structure and have meaning only because the two parties agreed in advance on their organization. These examples are still useful because, for example, in GNU Radio, the Python server proposed earlier is launched in an independent thread by

```python
import threading
import my_server
threading.Thread(target = my_server, args = (self,)).start()
```

in a `Python Snippet` executed during the initialization of the scheduler. Passing the argument `self` gives access to all the functions defined by the GNU Radio Companion flowchart, and in particular, the *setter* and *getter* associated with each variable declared in the processing chain: each GNU Radio Companion variable `var` is associated with a setter function `set_var` and a getter function `get_var`. Therefore, the thread can call `self.get_var()` and `self.set_var()` if the variable `var` has been defined in order to modify its content. We extensively use this mechanism when a client needs to sweep a parameter of a radio frequency link, such as the carrier frequency of the signal.

Thus, in Octave, a client of the form

```octave
sck=socket(AF_INET, SOCK_STREAM, 0);
server_info = struct("addr", "127.0.0.1", "port", 4242);
connect(sck, server_info);
send(sck, 's'); % start
```

will connect to port 4242, the same port to which the server has previously bound on the same computer running the server (127.0.0.1), to send a command, for example, the letter "s," which could indicate the start of a processing sequence. Here, a connected TCP connection, indicated by `SOCK_STREAM`, ensures that the transactions are guaranteed by acknowledgment of each exchange, unlike an unconnected transaction or *datagram* according to UDP in which information is transmitted without guaranteeing its reception. These two modes will be used depending on whether the information needs to be organized and acknowledged (TCP) or simply sent to clients that may or may not be listening and where the loss has little consequence (for example, a data stream coming from a radio frequency receiver).

Python and Octave are two interpreted languages that we often use together: Python for its flexibility in accessing hardware resources and its use in connecting together the processing blocks of GNU Radio produced by signal analysis chains designed in GNU Radio Companion, and Octave for the ease of its matrix implementation of linear algebra algorithms according to the language derived from Matlab. The programmer, more flexible than the author in Python, will have no difficulty in translating the algorithms proposed in Octave to NumPy without having to go through https://numpy.org/doc/stable/user/numpy-for-matlab-users.html. Thus, in this presentation, we will strive to exchange not only scalars but especially data vectors between Python and Octave.

### 3.2 TCP/IP Server Running in a Separate Thread | 79

> **From left to right or from right to left: byte order**
>
> Apart from certain Airbus Defence documentation for the European Space Agency [Flachs, 2015], it seems quite obvious in Western countries to place the most significant bits on the left and the least significant bits on the right, thus writing one thousand two hundred thirty-four as 0x4d2 or 0b10011010010 as indicated by dec2hex and dec2bin in Octave. The situation is less clear for the arrangement of bytes for a magnitude coded on 8 bits: historically, for the American DARPA and Western countries that dominated the development of the Internet, it seems natural to place the most significant bytes on the left and the least significant bytes on the right and thus write 0x4d2 as 0x04 as the byte at the address with the least significant weight ("to the left") and 0xd2 at the highest memory address ("to the right") so that the display of the memory contents from its lowest address to its highest address displays 0x04 0xd2. As historically the internet was developed [Hafner and Lyon, 1998] by BBN on Honeywell and IBM architectures and then by Sun Microsystems on SPARC and Motorola architectures, it was natural to adopt this order selected by these processor architectures, called *big endian* (most significant byte at the lowest address), to transmit information coded on multiple bytes over the network. However, Intel had the idea of placing the least significant byte at the lowest address, a choice that becomes logical when arithmetic operations are performed on a complex instruction set computer (CISC) architecture with variable length instructions: the arithmetic and logic unit reads the instruction (the opcode) and begins loading the arguments, and if the first argument read is the units, then the arithmetic operation can start while reading the next byte containing the tens and possibly the hundreds and thousands, thus propagating the carry during each suboperation. This organization of the least significant byte at the lowest address is called *little endian*. On the contrary, a big endian organization requires reading the integer by starting with the thousands before ending with the units to start the arithmetic operation.
>
> Even though today Intel/AMD little endian processors dominate the market for consumer personal computers, the internet, like Java and therefore Android, remains big endian. Exchanging information between the two architectures requires agreeing on the byte order. This is the purpose of the htons (or htonl for 4 bytes) instructions for organizing two bytes (a short in C) in the correct order, from host to network and vice versa at the other end. These macros are defined in /usr/include/netinet/in.h on GNU/Linux as an identity or byte swap depending on the architecture used (#if __BYTE_ORDER==__BIG_ENDIAN). While this operation must be performed explicitly in C, it will be implicit in the infrastructures we will see later (ZeroMQ or MQTT [Eurotech International Business Machines Corporation (IBM), 2010] speak of *network byte order* for the organization of their fields coded on more than one byte, but the content itself is only a packet of bytes that the developer must organize properly), or nonexistent for ASCII transactions (XML-RPC) in which the order of the arguments is that of the exchanged ASCII strings.
>
> Note that in the examples we will discuss later, the exchanges take place within the same computer (127.0.0.1), and since a processor is consistent with itself, any error in the byte order on transmission is corrected on receipt (two errors that compensate each other). In production, it is prudent to communicate with a machine of opposite endianness to identify potential sources of malfunction – Java is wonderful for this.

As opposed to TCP, which guarantees transactions, UDP just sends packets to whoever wants to hear them. Thus, Figure 3.7 presents a GNU Radio Companion processing chain that simply broadcasts single precision floating point numbers (float symbol in orange in GNU Radio Companion), while at the other end, Octave (left) or Python (right) executes

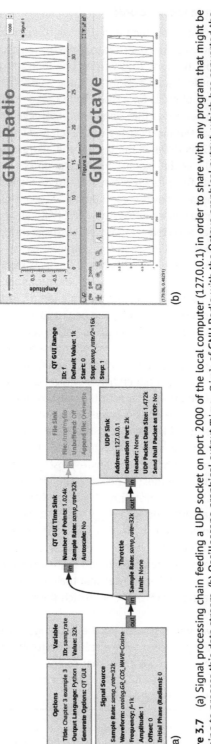

**Figure 3.7** (a) Signal processing chain feeding a UDP socket on port 2000 of the local computer (127.0.0.1) in order to share with any program that might be listening to process this data stream. (b) Oscilloscope graphical output Time Sink of GNU Radio with the Octave graphical output, which has converted the received array of bytes to float with typecast().

```
                                    import socket
pkg load instrument-control        import array
while (1)                           from matplotlib import pyplot as plt
  s = udpport("LocalHost",          s = socket.socket(socket.AF_INET,
  "127.0.0.1", "LocalPort", 2000);      socket.SOCK_DGRAM)
  val = read(s, 4000);             s.bind(("127.0.0.1", 2000))
  vector = typecast(val, "single"); while True:   # 4000 bytes=1000 float
  plot(vector); pause(0.1)           val, addr = s.recvfrom(4000)
  clear("s")                         vector = array.array('f', val)
end                                  plt.plot(vector)
                                     plt.show()
```

to open the UDP socket on port 2000 of the local computer (on which GNU Radio writes), display the acquired data after converting the byte packet into a vector of floating point numbers, and close the socket. This perpetual opening/closing of the socket may seem questionable, but it is the best way we have found to ensure that the processed data is the latest transmitted and not old data held in a buffer. In the case of UDP, even if some data is lost, it is not a problem since we guarantee obtaining a vector of the correct number of fresh data. The best way to synchronize the acquired data with a physical event such as the rotation of an antenna is to send a command to GNU Radio to perform the action, wait for the necessary time for the command to be completed or ideally an acknowledgment by a TCP communication in response to the request, then open the UDP socket and capture the desired number of data in this configuration, and repeat for all the envisaged configurations – for example, for a synthetic aperture radar, with all successive positions of the antennas.

When launching the GNU Radio program transmitting the data on port 2000 of the local socket (127.0.0.1) in UDP mode, we can validate the emission of data using `netcat` with `nc -l -p 2000 -u 127.0.0.1` with `-l` to listen and `-u` for UDP.

Similarly, we can use GNU Radio to provide a processing chain for signals acquired by an interface broadcasting its information via UDP and utilizing the UDP Sink as proposed in Figure 3.8, this time exchanging 4-byte (32-bit) encoded integers as indicated by the green icon in the GNU Radio Companion processing chain. In this example, we send a ramp in Octave (`val=int32([k:k+1024]);v=typecast(val,'uint8');` on the left) or in Python (`numpy.arange(0+k,1024+k,dtype=numpy.int32)` on the right) but we could, of course, send any sequence of measurements, for example, acquired by RS232 from an instrument.

Based on these lower layers of the OSI hierarchy, we will now explore some mechanisms, organizing transactions and facilitating client access to the interfaces exposed by the servers.

## 3.3 XML-RPC

A TCP/IP or UDP/IP client–server requires agreeing on the protocol for exchanging information between the client and server. In order to organize these transactions by encapsulating them in a format that can be easily processed automatically, it may seem logical to announce the nature of the required service and the associated variable. In the previous example, only the designer knows that the variable "var" exists with its associated read and define functions, and an external client may not be aware of the available function list. It seems natural to encapsulate the data in messages and thus to encapsulate them in a syntax easily decodable by a computer, such as JSON or XML, at the expense, of course, of increased message size and dependency on message decoding libraries.

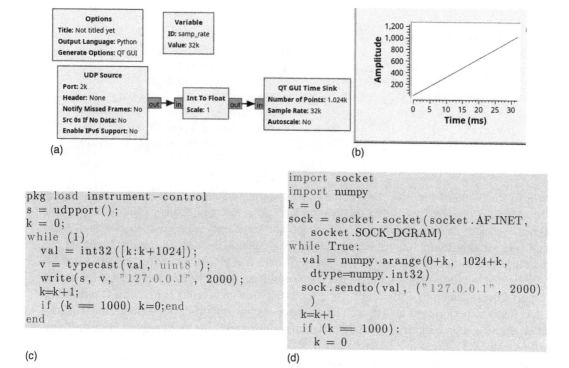

**Figure 3.8** (a, b) GNU Radio signal processing chain for fetching the data stream coming on port 2000 of the computer executing the script and displaying its content. Octave (c) and Python (d) scripts generating ramps and sending the data on a UDP port to be interpreted as 32-bit integer values.

In the implementation of XML-RPC, calls to remote functions (remote procedure call) are encapsulated in XML messages as described in Winer [2003]. In this way, a client only needs to inform the server of which service (function) it wants to use in order to modify its content (variable value). Just run "apt-cache search xmlrpc" under Debian/GNU Linux to see the multitude of languages implementing this protocol, especially Python. In this language, a client can be written in just a few lines

```
from xmlrpc.client import ServerProxy
s = ServerProxy('http://localhost:8080')
s.set_freq(5000)
```

to modify the server's configuration that we defined in GNU Radio using the processing chain proposed in Figure 3.9. In this processing chain, the frequency of the signal from the Signal Source block is a variable freq and we observe the effect of the remote command to the freq by the change in the period of the sine wave on the time domain graphical output.

In order to test the proper functioning of the server, it is not even necessary to write a line of code since the shell command xmlrpc is provided in the package libxmlrpc-core-c3-dev of Debian/GNU Linux. This command can be used with the command xmlrpc localhost:8080 set_freq i/1664 to modify the variable freq and assign the integer value (i) of 1664. Alternatively, a string can be transmitted by prefixing the argument with "s" or a floating point number with "d" (the types are described in man xmlrpc).

## 3.3 XML-RPC

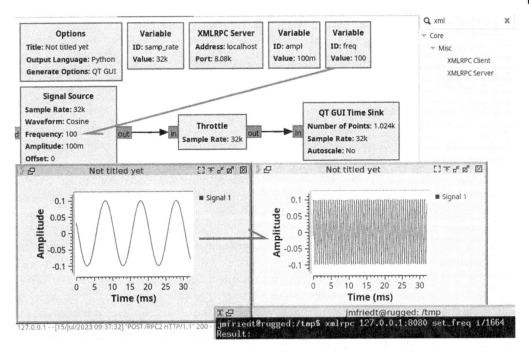

**Figure 3.9** Reconfiguration of a parameter of a GNU Radio processing chain – therefore Python – through a command sent from the shell. The variable `freq` defines the frequency of the sine wave signal source and therefore exposes the `set_freq` method, which we call from the shell through `xmlrpc` (bottom right). The Python server acknowledges the reception of the command (bottom left) in the GNU Radio Companion console, and the frequency is effectively modified (arrow) upon sending the command.

According to the XML-RPC specifications [Winer, 2003], we can construct the message in XML format to send the value 1664 as a 4-byte integer to set the variable `freq` using its setter `set_freq`:

```
curl -X POST -H 'Content-Type: text/xml' -d '<methodCall>\n<methodName>
set_freq</methodName>\\n<params><param><value><i4>1664</i4></value></param>
</params>\n</methodCall>'\'http://localhost:8080/RPC2'
```

Thus, any language that does not support XML-RPC but can communicate via HTTP using the POST method can communicate with the server.

At the moment, we are not able to expose the variables used in a GNU Radio processing chain, as the request to the `listMethods` method of the client with

```python
import xmlrpc.client
proxy = xmlrpc.client.ServerProxy("http://localhost:8080/")

for method_name in proxy.system.listMethods(): # liste des fonctions
    if (method_name.find("set_")>=0):           # expos'ees par le
    serveur
        print(method_name)

try:
    setampl = proxy.set_ampl(0.2)       # echec (par de variable ampl)
except xmlrpc.client.Fault as err:
```

```
    print("Unsupported function")
try:
    setfreq = proxy.set_freq(200)     # succes, freq redefinie
except xmlrpc.client.Fault as err:
    print("Unsupported function")
```

refuses to provide the list of variables. Therefore, only trying the different functions and intercepting errors (`try: ... except:`) allows to test if a variable exists or not. In the example mentioned earlier, `set_freq` is successful, but `set_ampl` fails because the amplitude is not a defined variable in the processing chain.

In order to expose the list of methods, https://docs.python.org/3/library/xmlrpc.server.html explains that we need to enable this functionality in the server. To do this, we add a Python snippet in GNU Radio containing the command `self.xmlrpc_server_0.register_introspection_functions()` to activate the ability of the server `xmlrpc_server_0` (associated block ID) to provide all services and consequently only call the variables that are actually defined (Figure 3.10).

We now have two approaches to define server parameters from the client side: a TCP/IP server or an XML-RPC server. These connected connections are appropriate to ensure that the message sent by the client is well understood and acknowledged by the server. However, it is common for the server to continue its acquisition and signal processing activities even if no client is listening to its services. This non-blocking connection is supported by UDP, which once again does not encapsulate information but only groups bytes to communicate to potential clients listening. If no one is listening, the information is simply lost, and if the routing changes during communication, neither the order nor the integrity of the data stream is guaranteed. In order to facilitate the organization of transmitted information, we will complete the previous demonstration of UDP exchanges by using a more abstract library called ZeroMQ (0MQ or ZMQ).

## 3.4 Zero MQ (0MQ) Streaming

TCP/IP client–server allows, in a connected mode, to make sure that all data transmitted by the server is received by the client and ordered. However, not only is the TCP/IP stack complex and computationally intensive, but some conditions require that data are streamed from the server to any client listening to the stream and willing to process the collected data. In the context of SDR, we might want for data to be streamed continuously from the server, and if no client is listening, then the data are lost, and we leave to any client the freedom to connect and catch up with the ongoing broadcast at will.

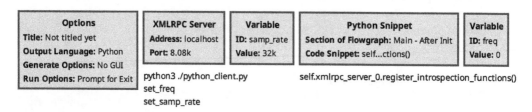

**Figure 3.10** Activation of the `register_introspection_functions()` method in a Python snippet in order to allow XML-RPC clients to fetch the list of variables known by the server.

## 3.4 Zero MQ (0MQ) Streaming

This datagram communication scheme is classically implemented using UDP packets over IP. However, a popular application layer alternative to UDP/IP and TCP/IP is the ZeroMQ (0MQ) framework [Dugan, 2021; Hintjens, 2013; Akgul, 2013] providing a higher-level abstraction framework for achieving similar results. We will illustrate continuously streaming data and collecting parts of these information using the ZeroMQ publish–subscribe reminiscent of UDP in its capability to continuously transmit from the server, irrelevant of any client listening. Under such conditions, GNU Radio implementing the ZeroMQ publish is collecting data from the SDR source, preprocessing, and streaming to a server, while on the other hand of the network (possibly on the same host if needed), a subscribe client in any language – GNU Radio, but possibly Python, Octave, or C(++), depending on the usage of the collected data – implements the discontinuous data stream processing. Many languages support ZeroMQ as shown with:

```
$ apt-cache search mq | grep z
python3-zmq - Python3 bindings for ZeroMQ library
libczmq4 - High-level C binding for ZeroMQ
libgnuradio-zeromq3.10.5 - gnuradio zeromq functions
octave-zeromq - ZeroMQ binding for Octave
```

to name just a few.

As a demonstration of this principle, the flowgraph of Figure 3.11 uses a ZeroMQ publish sink to stream the collected data, possibly preprocessed by GNU Radio prior to streaming. If a known number of elements are desired in each packet, the conversion from a continuous stream to a vector of known length allows to control the minimum size of the ZeroMQ output packet.

Like TCP and UDP, ZeroMQ offers a connected mode that guarantees transactions – request-reply in ZeroMQ's terminology – but with a heavy bidirectional protocol for acknowledgment, and a mode for broadcasting information without guarantee of reception, similar to the UDP datagram – named publish–subscribe at ZeroMQ (Figure 3.12). An example of a Python client receiving arrays of values (vectors) and process using NumPy is

```
import numpy as np    # pkg load signal;
import zmq            # pkg load zeromq;
import array
from matplotlib import pyplot as plt
```

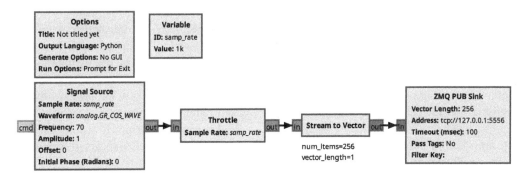

**Figure 3.11** GNU Radio Companion flowgraph illustrating the use of the ZeroMQ publish sink for streaming data to clients if connected: the flowgraph will always be running continuously even if no client is connected.

**Figure 3.12** ZeroMQ (or 0MQ) provides an application layer above TCP/IP and UDP/IP to abstract the transmitted information and encapsulate it in packets with metadata. Like TCP or UDP, ZeroMQ provides a connected mode (REQ–REP) that guarantees blocking transactions and a nonconnected mode (PUB–SUB) that we will favor for our applications of transferring data acquired by software-defined radio receiver for remote processing.

```
Nt = 256
context = zmq.Context()

sock1 = context.socket(zmq.SUB)   # sock1=zmq_socket(ZMQ_SUB);
sock1.connect("tcp://127.0.0.1:5556");
sock1.setsockopt(zmq.SUBSCRIBE, b"")
vector1 = []
while (len(vector1)<Nt):
  raw_recv = sock1.recv()
  recv = array.array('f',raw_recv) # f->l for integers
  print(recv)         # vector1tmp=recv[0::2]
  plt.plot(recv)      # vector2tmp=recv[1::2] for interleaved
  plt.show()
```

while a similar Octave client, if the `octave-zeromq` package was installed, is

```
pkg load zeromq;
Nt = 1024
sock1 = zmq_socket(ZMQ_SUB);
zmq_connect(sock1, "tcp://127.0.0.1:5556");
zmq_setsockopt(sock1, ZMQ_SUBSCRIBE, "");
recv = zmq_recv(sock1, Nt*8, 0);
% vector=typecast(recv,"single complex");   % if we received complex
    numbers
vector = typecast(recv,"int64")             % if we received long
    long integers
```

Although the server keeps on running even if no client is recording the transmitted data, for example, when the client is displaying a chart and waits for the user to close the graphical window, all data are buffered, and the next packet received by the client might be obsolete by the time it is being processed. Hence, if the client does not disconnect its SUB socket, the sequence remains contiguous and the packets are stored until they are processed. The only way we have identified to ensure that the data stream is the most recent and not remnants of a previous message – for example, when processing radio frequency signals acquired by a mobile antenna to ensure that the data are acquired at the new antenna position – is to close and reopen the SUB socket to eliminate the queue of pending data.

Similar to collecting and preprocessing data from GNU Radio and sending the result to an external post-processing framework, ZeroMQ can be used for sending data to GNU Radio running a subscribe client. In this case, the Python server for sending arrays of floating point number is

```python
import numpy as np    # pkg load signal;
import zmq            # pkg load zeromq;
import time
port = "5556"

context = zmq.Context()
sock = context.socket(zmq.PUB)
sock.bind("tcp://*:"+str(port))  # broadcast
k = 0
while True:
    payload = np.arange(0+k, 1024+k)
    k = k+1
    sock.send(payload)
    time.sleep(1)
```

On the other hand, in a REP–REQ scenario, a ping-pong game requires the REQuest client to ask the REPly server for new data (Figure 3.13) and not to resend a packet without being invited to do so, otherwise risking receiving an error of the type " Operation cannot be accomplished in current state."

REQ client

```python
import socket
import zmq
import array
context = zmq.Context()
sock1 = context.socket(zmq.REQ)
sock1.connect("tcp
    ://127.0.0.1:5556");
while True:
 noerror = 1
 while noerror:
  sock1.send(b"Hello")
  rcv = sock1.recv()
  # print(rcv.decode('ascii')) si
    str
  r = array.array('i', rcv)
  print(f"{len(r)} {r[0]} {r[1]}
    {r[-1]}")
```

REP server

```python
import numpy as np    # pkg load
    signal;
import zmq            # pkg load
    zeromq;
import time
port = "5556"
payload = "Hello"
context = zmq.Context()
k = 0
while True:
  sock = context.socket(zmq.REP)
  sock.bind("tcp://*:"+str(port))
    # broadcast
  while True:
    message = sock.recv()
    print(message)
    payload = np.arange(0+k,
    1024+k, dtype = np.int32)
    sock.send(payload)
    print(k)
    k = k+1
    time.sleep(1)
```

Producing the data flow from GNU Radio makes prototyping more fun by continuously generating data, potentially from a physical radio frequency receiver but here from synthetic signals, which therefore require a `Throttle` block to limit GNU Radio scheduler's data production to samp_rate samples per second. In Figure 3.14, we observe that the data produced in Python by the code generated by GNU Radio Companion are properly read in C (bottom left) or Python

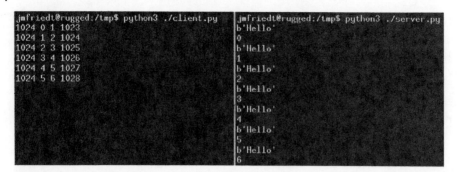

**Figure 3.13** Exchanges between client and server in which each vector is required by the client REQ to be provided by the server REP, ensuring transaction sequencing and absence of data loss.

(bottom right), which allows displaying the waveform using `matplotlib` and verifying its adequacy with the sawtooth pattern produced but with a variable number of transmitted data as imposed by the GNU Radio scheduler. Indeed in this context, we cannot make assumptions about the length of the transmitted vectors and will either have to wait to accumulate enough points to process the data (for example, convolution or Fourier transform, which requires a known number of values in a vector to calculate the integral) or only process the useful subset and store the other values.

Similarly, the communication between GNU Radio (using Python) and Octave is demonstrated in Figure 3.15, which highlights the necessity of explicitly converting the byte packet produced by the publisher in order to interpret it correctly, either as floating point numbers (`single`) or as complex numbers for an in-phase/quadrature (IQ) data stream or as integers (`int32` for 4 bytes/integer) – the list of arguments supported by Octave's `typecast` is provided by `help typecast`. This conversion of the byte packet to the appropriate type is also valid for Python with the `array` argument as documented at https://docs.python.org/3/library/array.html. Therefore, it is the responsibility of the developer to ensure that the exchanged data types are consistent or to implement a protocol that guarantees transaction coherence if the nature of the exchanged data is likely to vary.

ZeroMQ allows for easy exchanging information either through connected or broadcasted protocols, with the concept of topics that clients can subscribe to in order to only keep a subset of the transmitted information. However, it should be noted that the information is broadcast in clear text, which guarantees efficiency but also comes with the risk of easy manipulation of IP packets nowadays (Figure 3.16). In the output of `tcpdump`, we can easily recognize the 32 bytes of the IPv4 header starting with 0x40 [Fall and Richard Stevens, 2011], along with a TCP transaction (the "06" in 0x4006, with the first nibble being the maximum packet time-to-live [TTL] set to 64), the source IP and the destination IP (0x7f000001 or 127.0.0.1 in decimal), the communication port 0x15b4, which equals 5556 in decimal, and so on.

Indeed, in the context of broadcasting data from a radio frequency receiver, eavesdropping on the transmitted data is probably of little importance, but injecting erroneous data could be catastrophic. It is not so much the obfuscation of the data that is important but rather their integrity, which could justify an additional layer of authentication that is clearly missing when we display the transmitted data to the SUB client in C using `tcpdump`.

Enabling ZeroMQ input as the means to feed processing functions with data is a very powerful approach since it allows using any hardware supported by GNU Radio with the associated processing software. As an example of this usage, `gr-iridium` does not support the Pluto+ SDR

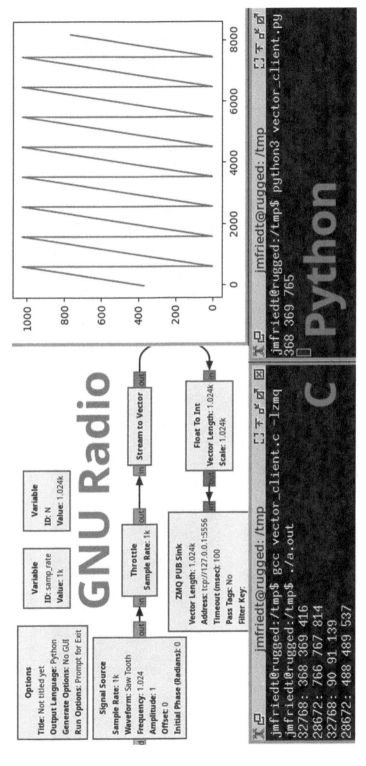

**Figure 3.14** Generation of a sawtooth signal by GNU Radio (*Signal Source* of shape *Saw Tooth*, amplitude 1, and frequency 1.024) produced at a rate of 1000 samples/s and unit amplitude multiplied by 1024 when converting from floating point number – orange symbol – to 32-bit integer – green symbol – transmitted in ZeroMQ PUB and received in Python at the bottom right by a ZeroMQ SUB for display using matplotlib, validating the consistency of the transaction, and in C at the bottom left, confirming that multiple clients can simultaneously receive the published data vectors.

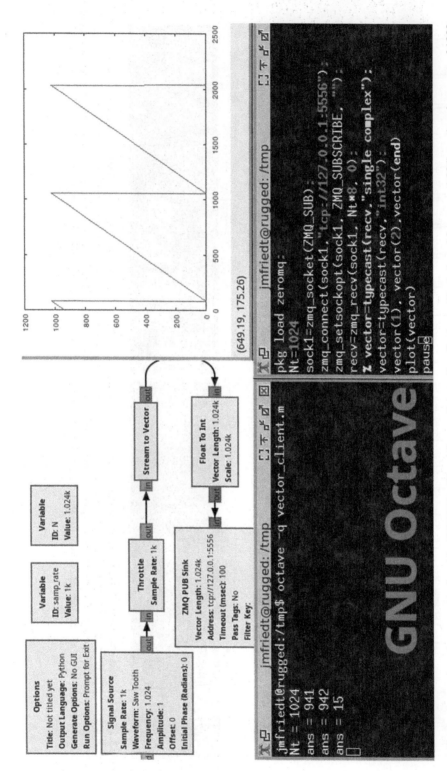

**Figure 3.15** Generation of a sawtooth signal by GNU Radio (*Signal Source* of shape *Saw Tooth*, amplitude 1, and frequency 1.024), produced at a rate of 1000 points/s and unit amplitude multiplied by 1024 when passing from floating point number – orange symbol – to 32-bit integer – green symbol – transmitted through ZeroMQ PUB and received by GNU Radio through ZeroMQ SUB for graphical display, validating the coherence of the transaction.

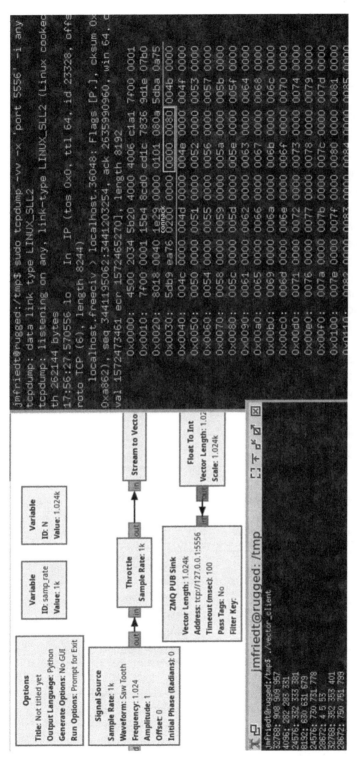

**Figure 3.16** Observation using tcpdump -vv -x 'port 5556' -i lo (more selective than any) of the data transmitted by ZeroMQ: after the 32 bytes of the IP header, we can see that the overhead of the ZeroMQ protocol is minimal, with some information about the nature of the transaction (CONNACK for a server-to-client connection), the payload size (rectangle highlight) 0x8000 = 32768 (remember that the order is little endian on the network), followed by the clear text data (underlined), in accordance with the information provided by the server in C (first transaction).

hardware as an input stream but does support a ZeroMQ Sub input. Hence, a basic GNU Radio flowchart recording from the Pluto+ using the `FMComms 2/3/4 Source` to a `ZMQ PUB Sink` will stream the data to `gr-iridium` running as `./apps/iridium-extractor -D 4 --multi-frame ./examples/zeromq-sub.conf`, with care taken to match the `Pass Tags` settings in both configurations. Using this approach, the decoding software might not even run on the recording hardware, allowing to share processing power between multiple computers. This approach is generalized in https://github.com/muaddib1984/stillsuit where ZeroMQ streams are combined with XML-RPC servers to allow remote control of various hardware supported by GNU Radio and streaming the collected data over a ZeroMQ socket.

> **ZeroMQ buffer size and impact on closed loop control bandwidth**
>
> ZeroMQ provides the capability of recording SDR signals with a receiver located at one site, streaming the collected signal over a network and generating a control signal on an remote site. Because of the nature of Ethernet communication assembling data into packets, the emitter accumulates enough samples before generating a data packet, which is sent over the network. On the receiver side, assuming a continuous sampling, the time stamp of each recorded sample can be computed despite receiving one large batch of data, and the control signal is generated continuously from the processed data. If the communication bandwidth is larger than the data acquisition rate, the data stream can appear continuous, but a latency has been introduced to fill the data queue: this latency is given by the packet size multiplied by the sampling rate.
>
> The packet size is determined by the GNU Radio scheduler and on assumption should be made on its value since it will be dependent on sampling rate and processing load. However, we can experimentally assess the packet size in a given experimental condition. The following flowgraphs
>
>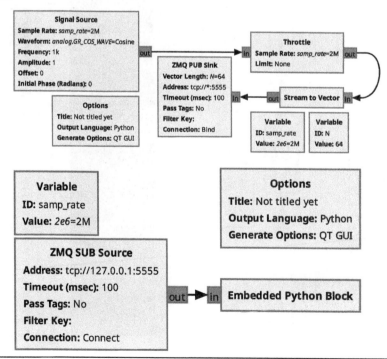

generate a data stream on the PUBlish side (left) and receive the data stream on the SUB-scribe side (right) to feed a custom sink block (see Chapter 9 about custom blocks) limited to displaying the packet size:

```
def __init__(self):
    gr.sync_block.__init__( self, name='ZMQ size',
        in_sig=[np.complex64], out_sig=[] )
    self.n=0

def work(self, input_items, output_items):
    if self.n<100:
        print(len(input_items[0]))
    self.n=self.n+1
    return 0
```

where the prototype with no output (out_sig=[]) indicates the block is a sink block, and whose work function running continuously on the input stream only displays the packet length for the first 100 packets (the GNU Radio Companion console is slow and will slow to a crawl if displaying too many information).

Running these flowcharts indicates that the packet size is in the 8192 sample range, meaning that with a sampling rate of 2 Msamples/s, a latency of 8192/2 = 4096 µs has been introduced. This latency might not be relevant in an open-loop configuration where the data are only streamed from emitter to receiver but will impact the closed-loop bandwidth if the control signal is computed based on the received datasets: in this example, the closed-loop bandwidth is reduced to a fraction of a kilohertz despite the 2 MHz sampling rate.

This conclusion can be generalized to closed-loop systems controlled by SDR: while general-purpose processors exhibit tremendous computation speed, digital processing **introduces latencies which are multiples of the sampling rate**, and every finite impulse response (FIR) filter with $N$ taps will generate a mean delay of $N/2$ sampling periods. Hence, despite samplings rate in the Msamples/s range, closed-loop bandwidths in the tens of kilohertz only are not uncommon, limited by all the digital processing block delays [Adapted from Matusko et al., 2023].

## 3.5 MQTT

MQTT (*message queuing telemetry transport*) is emerging as a protocol for the "internet of things" (IoT), where embedded systems are designed to communicate. The need for a TCP/IP stack to implement MQTT is surprising in this context: IP, internet control message protocol (ICMP), and UDP took up only a few kilobytes of RAM and flash, much less memory than is needed to store TCP packets that must accumulate in case of packet loss or rerouting that would change their order. It seems that the main benefit of MQTT, beyond its centralization on a single server (referred to as a *broker* in MQTT terminology), is the secure sockets layer (SSL) encryption of the exchanged packets (once again at the expense of the client's computing power seeking to transmit its information).

Similar to ZeroMQ, MQTT operates in a publish–subscribe context, but this time not in a point-to-point connection but through a single *broker* that centralizes the transactions and therefore appears

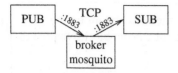

**Figure 3.17** MQTT relies on a data hub – the *broker* – which centralizes the exchanges, and even though we remain in a publish–subscribe model with published topics that a client may or may not subscribe to, this time all participants are called clients that connect to the server which is the *broker* that communicates on port 1883.

as a weak point in the network (Figure 3.17). Thus, every service in MQTT is implemented as a client, whether a *publisher* or *subscriber*, with a single server that is the *broker*. Each publisher can offer services through filters, and each subscriber can filter the information it wants to process. This mechanism appears inefficient in the context of exchanging radio frequency data streams that aim for efficiency, but https://opensource.com/article/18/6/mqtt provides a concrete example of the use of these functionalities in the analysis of energy production in the state of New York, where a very rich dataset is slowly made available by electricity producers and accessed via MQTT with appropriate filters according to a hierarchy resembling a file system structure.

### 3.5.1 MQTT for Python, Bash and Octave

An implementation of the MQTT *broker* is called `mosquito` and this library is the one we will use once installed with `sudo apt install mosquitto mosquitto-clients` on Debian/GNU Linux. On the client side, an implementation of MQTT, also promoted by the Eclipse Foundation [Eclipse Foundation, 2023], is called Paho and provides compatibility for a multitude of languages including C and Python. On Debian/GNU Linux, we therefore install `sudo apt install libpaho-mqtt-dev` for C, identified by searching for which package in the distribution provides the header describing the content of libraries `MQTTClient.h` and `python3-paho-mqtt` for Python.

First, we make sure that a *broker* is running on the Linux operating system to allow these developments, either by `ps aux | grep mosq`, which should indicate `/usr/sbin/mosquitto -c /etc/mosquitto/mosquitto.conf`, or by reading the content of `/var/log/mosquitto/mosquitto.log` as an administrator. The communication between this server and MQTT clients will take place through port 1883, which must be open through any network transaction protection mechanism (firewall). We validate the proper functioning of the *broker* from the command line through some simple publish–subscribe exchanges:

```
$ mosquitto_pub -t "mycomputer" -m "Hello"
$ mosquitto_pub -t "mycomputer" -m "World"
$ mosquitto_sub -t "mycomputer"
Hello
World
```

Convinced of the broker's good operating conditions, we start by implementing an exchange of data vectors, always ensuring the coherence of the exchanges since MQTT simply transmits byte arrays without encoding the data organization. The Python code for communicating over MQTT is as short as

```
import paho.mqtt.client as mqtt
import numpy
```

```
client = mqtt.Client()
client.connect("127.0.0.1")
data = numpy.arange(0, 1024, dtype=numpy.int32)
client.publish("float_vect", data.tobytes(), 0)
```

and is validated by verifying that a transaction is executed by publishing data received by the shell command `mosquitto_sub -t "float_vect"` with a subscribe service that subscribes to the `float_vect` stream as we have selected when creating the publisher. However, since `mosquitto_sub` does not know the nature of the data structure being transmitted, it only displays bytes as ASCII characters that have no meaning. Therefore, we need to write clients – *subscribe* – able to decode the transmitted information, for example, by taking inspiration from

```
import paho.mqtt.client as mqtt
import numpy
import time

def callback_func(client, userdata, message):
    print("rcv ", numpy.frombuffer(message.payload, dtype=numpy.int32))

client = mqtt.Client()
client.connect("127.0.0.1")
client.subscribe("float_vect")
client.on_message = callback_func
client.loop_start()
time.sleep(40)
```

which highlight an interesting mechanism of automatically invoking a callback function when receiving a packet without the need to explicitly use a separate thread, as the C convention would require all reads to be blocking.

Although MQTT is not provided as a Debian/GNU Linux package for Octave, it can be installed without issues from the available source archives at https://sourceforge.net/p/octave-mqtt. This new package installed through the Octave `pkg install` command provides access to the functions necessary to connect to the server (*broker*), and thus all transactions are seen from the perspective of a client: for a subscribe connection that receives the messages

```
pkg load mqtt
client = mqttclient("tcp://127.0.0.1");
subs = subscribe(client, "float_vect");
vector = []
do
  recv = read(client, "float_vect");
  if (isempty(recv)==0)
  % vector = typecast(recv, "single complex");
    vector = typecast(recv.Data, "int32")
  end
until (isempty(vector)==0)
```

## 3 Communicating with External Software (Python, Networking, ZeroMQ, MQTT)

and for a publish connection that sends:

```
pkg load mqtt
client = mqttclient("tcp://127.0.0.1")
vector = [1:1024]
data = typecast(vector, "char")
write(client, "float_vect", data, "QualityOfService", 1);
```

However, Octave does not allow strong typing of the exchanged variables, and we observe that the emitted array is in the form of floating point numbers expressed in double precision (thus 8 bytes per data) and that from Python's point of view, this array is read by modifying the callback function using

```
def callback_func(client, userdata, message):
    print(numpy.frombuffer(message.payload, dtype=numpy.float64))
```

hence a conversion of the byte array to double-precision floating point, thanks to `dtype=numpy.float64`.

Finally, we conclude this overview of MQTT by highlighting the simplicity of integrating a Python library into GNU Radio: indeed, https://github.com/crasu/gr-mqtt offers an interface between GNU Radio and MQTT by simply encapsulating the functions that we have just explained in the `work` method of a dedicated Python block compatible with GNU Radio calls.

## 3.6 Conclusion

We have endeavored to demonstrate how to communicate from GNU Radio to different languages in order to distribute processing either by making the most of each language or by sharing resources across separate computers. To do this, we explored XML-RPC, ZeroMQ, and MQTT for socket-based communication or Unix pipes for communicating through the file system.

These frameworks will be used in the Chapter 4 to distribute processing, either by selecting the most appropriate language to each task or by sharing computational load between one computer dedicated to data acquisition and other processing units for signal processing and display.

## References

F. Akgul. *ZeroMQ*. Packt Publishing, 2013.

B. Dugan. KV4FV. Understanding ZMQ-blocks, 2021. https://www.youtube.com/watch?v=LPjZaOmNfxc.

Eclipse Foundation. Eclipse paho downloads at https://eclipse.dev/paho/index.php?page=downloads.php, 2023.

Elias Önal. https://github.com/EliasOenal/multimon-ng, 2023.

Eurotech International Business Machines Corporation (IBM). MQTT V3.1 protocol specification, 2010. https://public.dhe.ibm.com/software/dw/webservices/ws-mqtt/mqtt-v3r1.html.

K. R. Fall and W. Richard Stevens. *TCP/IP Illustrated: The Protocols, volume 1*. Addison-Wesley, 2011.

H. Flachs. Sentinel-1 SAR space packet protocol data unit. Technical report, Airbus Defence & Space, 2015.

K. Hafner and M. Lyon. *Where Wizards Stay Up Late: The Origins of the Internet*. Simon and Schuster, 1998.

P. Hintjens. *ZeroMQ: Messaging for Many Applications*. O'Reilly Media, 2013.

M. Matusko, I. Ryger, G. Goavec-Merou, J. Millo, C. Lacroûte, É. Carry, J.-M. Friedt, and M. Delehaye. Fully digital platform for local ultra-stable optical frequency distribution. *Review of Scientific Instruments*, 94(3):034716, 2023.

POSIX Standards. Section sockets at https://pubs.opengroup.org/onlinepubs/9699919799/functions/V2_chap02.htmltag_15_10, 2023.

D. Winer. XML-RPC specification, 2003. http://xmlrpc.com/spec.md.

# 4

# Correlating: Passive and Active Software-Defined Radio (SDR)–RADAR

## 4.1  SDR–RADAR Requirements and Design

Software-defined radio (SDR) is characterized by giving access to the raw in-phase/quadrature (IQ) radio frequency stream to the user, prior to any processing other than transposing from the radio frequency band to baseband by mixing with the two copies of the local oscillator, in phase and in quadrature. Hence, SDR is perfectly suited for investigating properties of the radio frequency signal, whether direction of arrival (DoA), as was demonstrated for Global Positioning System (GPS) spoofing detection when analyzing the same signal received by multiple antennas [Feng et al., 2021], or radio-frequency detection and ranging (RADAR). More generally, the concepts introduced in this chapter, aimed at illustrating how to detect time-delayed copies of a known transmitted signal (target range) through the *correlation estimator*, are useful whenever a known pattern is searched in a noisy signal, whether a header identifying the beginning of a sentence in a stream of bits or whether the code assigned to a given emitted in a code division multiple access (CDMA) scheme. We shall demonstrate how correlation is efficiently calculated in the Fourier domain and hence allow the reader to understand existing software such as `gnss-sdr`, the Global Navigation Satellite System (GNSS) decoder relying on GNU Radio for scheduling and acquiring GNSS signals before decoding to provide position, velocity, and time solutions.

The signals associated with this chapter are available at https://iqengine.org in the `GNU Radio SigMF Repo` → `passive radar`, with two sets of measurements including the reference channel and the surveillance channel sampled at a rate of 2.048 MS/s, interleaving real and imaginary parts as 8-bit integers.

A RADAR system is designed to detect static or moving targets, with range being identified with a time of flight and velocity by Doppler shift. A RADAR system requires that targets are illuminated by a radio frequency source: this radio frequency emitter can be cooperative when the user generates a dedicated radio frequency signal to be used for probing the environment or noncooperative in the so-called passive RADAR approach. The latter strategy is well suited for amateurs since it benefits from existing radio frequency sources and analyzes the time of flight between a reference channel fitted with an antenna directed toward the noncooperative source and a second surveillance channel fitted with an antenna directed toward the target(s). The dual-channel coherent design is a core aspect of the following discussion. Since general users will not be allowed to generate powerful, broadband signals, the passive RADAR approach is attractive for becoming familiar with the concepts of RADAR detection and processing, but limitations in their practical use often arise from the noncooperative source location or radiation pattern.

---

*Communication Systems Engineering with GNU Radio: A Hands-on Approach*,
First Edition. Jean-Michel Friedt and Hervé Boeglen.
© 2025 John Wiley & Sons, Inc. Published 2025 by John Wiley & Sons, Inc.
Companion website: www.wiley.com/go/friedtcommunication

## Continuous wave RADAR for measuring target speeds

The GRAVES RADAR (*Grand Réseau Adapté à la Veille Spatiale*) is a continuous wave bistatic RADAR with an emitter located 30 km from Besançon, France, where one of the author's laboratories is located. This strong source illuminates all moving targets and is initially designed to detect satellites flying over Metropolitan France, with a receiver located south of France. This strong source is ideally suited for a distributed passive RADAR system where the lack of range from the continuous wave is compensated for with DoA measurements from spatially distributed receivers. In this measurement, multiple Ettus Research X310 SDR receivers are time and frequency synchronized over a White Rabbit network. The following figure illustrates the need for a stable local oscillator – here a hydrogen maser, but the reader can reproduce with a good-quality oven-controlled crystal oscillator (OCXO, e.g. Hewlett Packard HP10811-60111) or a rubidium clock (e.g. Symmetricom X72) available as second-hand items – when addressing fine frequency measurements as introduced by Doppler shift. Notice that in a RADAR system, the Doppler shift is $\delta f = 2 \cdot f_c \frac{v}{c}$ with $f_c$ being the emitted carrier frequency, $v$ the projected target velocity toward the receiver, and $c$ the speed of light. In the case of GRAVES with $f_c = 143.05$ MHz, the Doppler frequency shift $\delta f$ happens to be close to the projected speed of the target in m/s so that subsonic aircrafts induce frequency variations in the $\pm 300$ Hz range as seen here.

Below (top waterfall chart) is the temperature-controlled crystal oscillator (TCXO) fitted on the X310 SDR receiver fluctuation over time ($y$-axis), leading to an inaccurate measurement of the plane velocity observed as the Doppler frequency shift ($x$-axis). Middle-bottom waterfalls: the X310 are clocked by hydrogen-disciplined White Rabbit time and frequency dissemination systems, providing high stability references. Thanks to these features, the Doppler shift is measured with high accuracy, allowing to estimate the target velocity as seen from the stable direct coupling signal frequency measurement at the middle of each chart. Bottom: one of the spectra illustrates two targets, one incoming (positive Doppler) and one outgoing (negative Doppler shift). All these measurements are completed in the second Nyquist zone since the GRAVES RADAR emits at continuous wave a 143.050 MHz, while the 200 MHz sampling rate allows for sampling up to 100 MHz carrier frequencies. The signal aliased from 56.95 MHz is recorded by setting the baseband local oscillator to this frequency.

4.1 SDR–RADAR Requirements and Design | 101

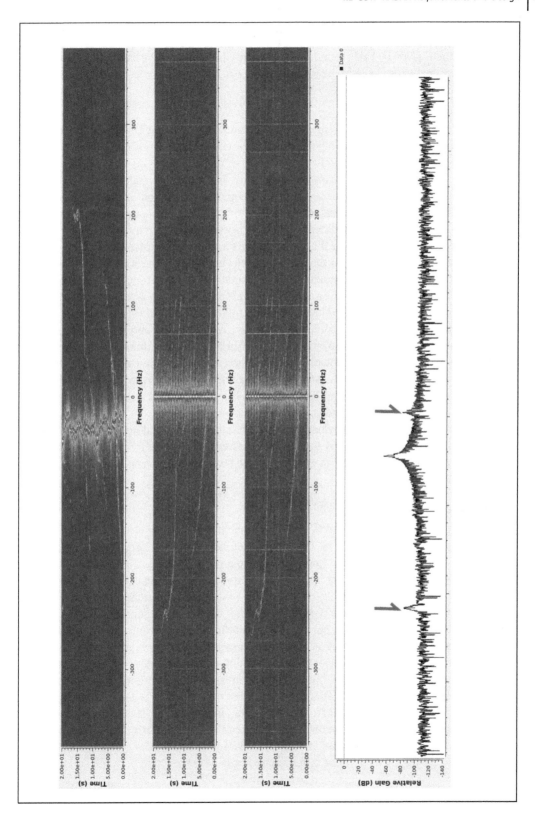

4 *Correlating: Passive and Active Software-Defined Radio (SDR)–RADAR*

The approach discussed in this chapter aims at bringing the passive-RADAR concept to the active-RADAR system by using the so-called noise-RADAR strategy. The range resolution $\Delta R$ of a RADAR system is given by half (from the two-way trip) the inverse of the bandwidth $B$ times the speed $c$ of electromagnetic waves in the medium being considered, or $c = 300$ m/µs in air:

$$\Delta R = \frac{1}{2}\frac{c}{B}$$

Once the basic concept is stated, the means of reaching $B$ compatible with the desired $\Delta R$ vary with technology, requirements, and regulations. The pulsed RADAR emits a short but powerful pulse and records the time of flight at the expense of poor filtering capability from the short pulse duration but most significantly requiring dedicated hardware hardly compatible with the generic SDR approach. Alternatively, a frequency domain sweep characterization of the medium is performed, and the inverse Fourier transform of the recorded reflection coefficient in the range $B$ provides the time-domain echoes and hence range to targets: this is the frequency stepped continuous wave (FSCW) approach. Rather than probing the scene under investigation by sequentially emitting each spectral component, emitting all frequencies at the same time is achieved using noise, where all spectral components are simultaneously transmitted. While emitting from a noise generator would not be legal for the unlicensed amateur, digital communication systems use waveshapes closely matching the requirements of broadband spectral occupation. In this demonstration, we shall use a commercial, off-the-shelf WiFi transmitter as a covert emitter, and the characteristics of the recorded signal will allow for recovering the range (and velocity) of illuminated targets. Since the WiFi-transmitted signal is unknown, as is the emission of a noncooperative passive RADAR source, it must be recorded during transmission by the reference channel, while the surveillance channel monitors the signals reflected by dielectric and metallic targets to observe features that digital communication designers would call multipath components.

In this demonstration, the WiFi payload will be a random sequence and no meaningful content will be transmitted, although the dual use of the radio frequency source for target detection and communication is a popular application of multipurpose broadcasting systems [Saddik et al., 2007].

---

**RADAR source requirements**

RADAR measurement requires powerful, broadband signal sources, usually not compatible with most amateur or university activities. The range resolution drives the need for bandwidth as discussed in the text. The need for power comes from the RADAR equation stating that the received power decays as the *fourth* power of the distance from a monostatic RADAR system and the target acting as a point-like source. Indeed free space propagation loss or energy conservation requires that the energy from the emitter decays as the square of the distance, and if the target is considered as a point-like source reflection, the fraction of emitted power that reached its location acts itself as a source whose reflected power again decays as the square of the distance to the receiver. The RADAR link budget is hence less favorable than a one-way communication system, but increasing emission power is only relevant as long as the coupling between the emitter and receiver does not increase the receiver background direct signal when increasing the emitted power: isolation between emitter and receiver is the core characteristics of the system defining the range to the furthest static target before direct coupling is stronger than the returned power. Moving targets benefit from the frequency offset introduced by the Doppler shift to be differentiated from direct coupling between emitter and receiver.

**Figure 4.1** Architecture of the hardware for implementing the various RADAR systems described in this chapter, whether passive or active. The emitter is assumed to transmit a signal within the recording bandwidth of the receiver. The core requirement is for the receiver to provide two coherent (clocked by the same local oscillator) channels, one called the reference and the other the surveillance. In the case of an active RADAR, the transmitter output is coupled (CPL) to the reference channel input with the strong enough attenuation to avoid wasting transmitted power and in order not to saturate the reference channel. Coupling of −20 to −40 dB is typical in such a setup.

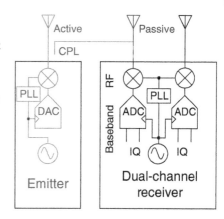

Two SDR–RADAR architectures will be considered, sharing the same receiving signal processing scheme but using different hardware with different financial impact. The first approach (Figure 4.1), easiest to implement but most expensive, uses a dual-channel coherent receiver such as the Ettus Research B210. The AD9361 radio frequency front end fitted on this receiver provides two coherent input channels – RXA and RXB – clocked by the same local oscillator and providing simultaneous sampling of both analog-to-digital converters (ADC). The second approach uses two different RTL-SDR receivers clocked by a common radio frequency source: although the local oscillators and analog-to-digital converters will be running at the same rate, the nondeterministic USB bus will introduce a random latency between channels. This latency will be characterized, and processing assumes that the data flow never stops during an acquisition. Since both RTL-SDRs will be streaming `samp_rate` samples/s with the clock generated by the same oscillator, only the random delay introduced when sequentially starting both data flows must be determined. This requirement justifies the use of ZeroMQ publish streaming of the data to a dedicated processing external software: with this approach, both RTL-SDRs keep on streaming data continuously when the processing software only regularly grabs packets of the reference and surveillance channels for post-processing, the data being otherwise lost if no user is listening on the subscribe socket.

## 4.2 Correlation: GNU Radio Implementation

Searching for a known pattern in a noisy signal is performed through the cross-correlation function between time-dependent $t$ pattern $p(t)$ and time-dependent signal $s(t)$, expressed for a delay $\tau$ as

$$xcorr(p, s)(\tau) = \int p(t) s^*(t + \tau) dt$$

with * being the complex conjugate. When the pattern $p$ matches signal $s$ shifted by a time delay $\tau$, then the negative signal in $p$ is multiplied by the same negative signal in $s$, as do the positive parts of the pattern and signal, and energy accumulates in the integral, leading to a correlation peak at the time delay $\tau$ representative of a target range $r = c \times \tau/2$ with $c = 300$ m/μs, the speed of the electromagnetic wave in vacuum.

Although GNU Radio provides the means for correlating a known sequence ("Access Code") in the received signal to detect the beginning of a frame, the cross-correlation is not available as a native processing block. Displaying the cross-correlation in real time during data acquisition would be welcome to make sure all peripherals are properly connected and that the RADAR system is operating properly, so we aim at computing a cross-correlation using native GNU Radio block. This result is achieved by starting from the convolution theorem and remembering that if we write $FT(x(t))$ as the Fourier transform of time-dependent function $x(t)$ defined as $FT(x)(f) = \int x(t) \exp(j2\pi ft)dt$ then the convolution between functions $s$ and $p$ is expressed as

$$conv(p,s)(\tau) = \int p(t)s(\tau - t)dt$$

which can be expressed in the Fourier domain as

$$FT(conv(p,s)) = FT(p) \times FT(s)$$

so that the Fourier transform of the convolution is the product of the Fourier transforms of each term of the function. In discrete time, the shortest vector between $s$ and $p$ is zero-padded to match the length of the longest vector, allowing for the product of the elements of each vector after computing the Fourier transform. This relation is important because it allows to benefit from the fast Fourier transform (FFT) using the symmetries of the trigonometric function to compute the result through a binary tree, reducing the complexity of a Fourier transform on $N$ elements from $N^2$ for the classical Fourier transform ($N$ multiplications for each of the $N$ terms) to an $N\log_2(N)$ complexity algorithm. The demonstration of the convolution theorem is as follows:

$$\left. \begin{array}{l} conv(x,y)(\tau) = \int x(t) \cdot y(\tau - t)dt \\ FT(x)(f) = \int x(t)e^{j2\pi ft}dt \end{array} \right\} \Rightarrow FT(conv(x,y)) = \iint x(t)y(\tau - t) \cdot dt \exp(j2\pi f\tau) \cdot d\tau$$

We define $u = \tau - t \Rightarrow du = d\tau$ and $FT(conv(x,y)) = \iint x(t)y(s)dt \cdot \exp(j2\pi f(t+u))du \Leftrightarrow FT(conv(x,y)) = \int x(t)\exp(j2\pi ft)dt \times \int y(u)\exp(j2\pi fu)du$ leading to the solution that $FT(conv(x,y)) = FT(x) \cdot FT(y)$.

Convolution and correlation are closely related since only the time flips in the second argument, and the convolution theorem is readily adapted to the cross-correlation by noting that time is flipped by considering the complex conjugate of a function. Indeed, the complex conjugate of $\exp(j\omega \cdot t)$ is $\exp(-j\omega \cdot t)$, so that

$$FT(xcorr(s,p)) = FT(s) \cdot FT^*(p)$$

emphasizing the need to consider the complex conjugate of the second argument when expressing the time domain cross-correlation.

Under GNU Radio, this expression is implemented through the FFT block. However, GNU Radio handles continuous streams of data, while the FFT block requires a vector of known length. Hence, a stream to vector block defines the FFT length, which matches for the RADAR analysis the pulse repetition interval (PRI), which defines the unambiguous range measurement. Indeed, if two pulses are emitted at time interval $T_{PRI}$, then the RADAR is unable to differentiate a target located at range $r = \tau/2 \cdot c$ when an echo is detected after a delay $\tau$ or a target at $r = (\tau + T_{PRI})/2 \cdot c$. Similarly, when selecting the FFT length, the PRI becomes $T_{PRI} = N/f_s$, with $N$ being the number of elements in the FFT and $f_s$ the sampling frequency (Figure 4.2).

Because the Fourier transform is linear, any number of targets returning time-delayed copies of the emitted pattern are detected in the inverse Fourier transform as many correlation peaks as targets have been detected. Figure 4.3 illustrates the case of two targets and the resulting two correlation peaks, where the Fourier transform of the emitted pattern was computed once, and the returned signal includes two time-delayed copies of the emitted pattern. Furthermore, in this case, the inverse Fourier transform is not explicitly computed, but the spectral analysis of GNU Radio through the frequency sink is used to display the time domain correlation peaks.

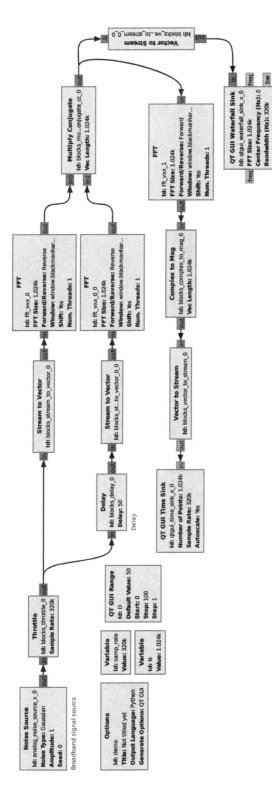

**Figure 4.2** Implementation of the cross-correlation between a signal and its time-delayed copy, with the time delay introduced manually with the delay block controlled by a slider, measured as the position of the cross-correlation peak.

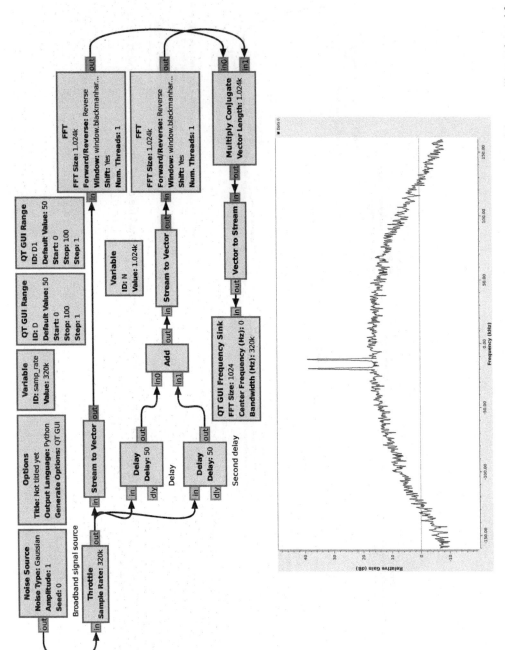

**Figure 4.3** Illustration of the detection of two targets whose time-delayed copies are summed in a signal representative of the surveillance channel for correlation with the top signal path representative of the reference channel. Since the Fourier transform is linear, each delayed component is identified and visible as a separate correlation peak after the inverse Fourier transform. Each delay is manually defined by the sliders D and D1.

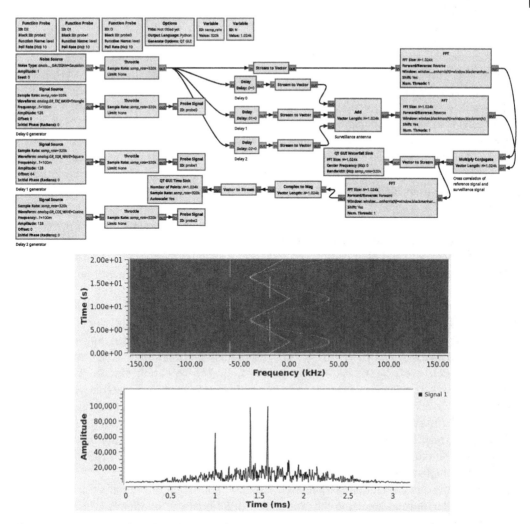

**Figure 4.4** Demonstration of the linearity of the correlation with three targets following a predetermined pattern driven by signal source blocks, using the probe signal and function probe to automate the delay definition along easily recognizable patterns of sine, triangle, and square waves.

This principle in generalized to multiple targets in Figure 4.4, where each target is assumed to follow a pattern with time delays defined as sine wave, triangle wave, or square wave, rather than manually introducing the time delay using sliders. This example is the opportunity to introduce the function probe capability of GNU Radio, where a signal source output feeds a probe signal block, creating a dynamic variable equal to the source output value, and this variable defines the delay through the function probe block. The result is shown in Figure 4.4 where the poor timing accuracy of the throttle block is visible with the distorted patterns of the sine, triangle, and square waves, but nevertheless, the path of each target is clearly visible on the correlation evolution along time on the ordinate axis in this waterfall depiction.

### Correlation for CDMA decoding: the case of GPS

CDMA allows for encoding each transmitted bit in a code unique to each emitter, to multiplex the channel between multiple speakers emitting on the same carrier frequency. CDMA is in particular used by GPS, in which all satellites broadcast the legacy L1 signal modulated as binary phase shift keying (BPSK) on the same 1575.42 MHz nominal frequency, each space vehicle being assigned a unique Gold code, from the name of its inventor R. Gold [1967]. The requirement for identifying each transmitter individually is that the cross-correlation of one code with another must lead to vanishingly small integrals (noise), whereas the autocorrelation accumulates energy in the signal. Hence, correlating the received signal with each known code sequence will provide identification and timing of the received messages from each individual emitter.

The challenge of the GPS acquisition phase, in which a cold-start receiver aims at identifying which satellite is visible and what Doppler shift is induced by the satellite motion, is that the correlation energy accumulation is plagued by the additional frequency offset introduced by the Doppler shift and the local oscillator frequency differences. Nevertheless, a CDMA signal with a repeating code can always be identified by autocorrelating, which is immune to frequency offset $\delta f$ since $xcorr(s\exp(j\omega t), s\exp(j\omega t))$ with $\omega = 2\pi\delta f$ is expressed as $\int s(t)\exp(j\omega t)\cdot(s(t+\tau)\exp(j\omega(t+\tau)))^*dt$ and since $\exp(j\omega\tau)$ is constant and not dependent on time, we obtain $\exp(-j\omega\tau)\int s(t)\exp(j\omega t)\cdot s^*(t+\tau)\exp(-j\omega t)dt = \exp(-j\omega\tau)\int s(t)\cdot s^*(t+\tau)dt = \exp(-j\omega\tau)xcorr(s,s)$ and the magnitude results in $|xcorr(s\cdot\exp(j\omega t), s\cdot\exp(j\omega t))| = |xcorr(s,s)|$. The repeating pattern is detected using the flowchart given (see page 110) as a comb of correlation peaks separated by the code period. As a demonstration, the autocorrelation of GPS signals (see page 111; top), and for comparison of the signal-to-noise ratio, the squared signal are shown (see page 111; bottom) emphasizing how squaring the signal rises the noise level while correlating improves the signal-to-noise ratio, thanks to the coherent accumulation of energy during the integration acting as averaging.

However, the autocorrelation computation lost the information of the nature of the code and the emitter. Beyond detecting the presence of GPS signals, correlation is used when the Gold code sequence is known to identify which satellite is broadcasting with which frequency offset, induced either by Doppler shift or local oscillator difference. The brute force approach of the acquisition phase of GPS in which all possible codes are tested for all possible frequency offsets is demonstrated in the flowchart (see page 112). In the flowchart, the loop on possible frequency offsets is manually handled by a slider in order to compensate for the Doppler shift of each satellite, here represented by the Gold code sequences number 9 and 23 (bottom-left "File Source" blocks). The Doppler shift and hence frequency tuning are different for each space vehicle, and the appropriate frequency offset must be identified for each satellite.

The question of the frequency offset range and step of the frequency correction is addressed as follows:

- From celestial mechanics, it is known that the period of GPS satellites orbiting at a medium Earth orbit of 20 000 km is 12 hours, so their tangential velocity is $v = 14\,100$ km/h= 3900 m/s.
- By projecting the tangential velocity (Figure 2.17) toward the ground-based receiver when the satellite rises over the horizon or sets below the horizon, it is demonstrated that the projected velocity is $v\frac{R_E}{R_E+h}$ with $R_E \simeq 6400$ km the radius of the Earth and $h = 20\,000$ km the altitude of GPS satellites, so that the projected velocity is in the $v_\parallel \simeq \pm 1000$ m/s, and the associated Doppler shift $f_c\frac{v_\parallel}{c}$ with $f_c = 1575.42$ MHz is in the range of $\pm 5$ kHz.
- Since the trigonometric function $\exp(j\delta ft)$ with mean value 0 will lead to a null correlation if the received signal is significantly frequency-shifted by $\delta f$ with respect to the local (unshifted) copy of the code, the residual frequency offset $\delta f$ after mixing with a software numerically controlled oscillator must be much smaller than the chip rate of 1 ms in the case of GPS-L1. Thus, the frequency correction step must be much smaller than 1 kHz, typically in the range of 10–100 Hz.

Hence, the slider for defining the frequency correction on the received signal and correcting for local oscillator offset and Doppler shift must range at least from $\pm 5$ kHz with 10–100 Hz steps, and as much as $\pm 150$ kHz with 10–100 Hz steps when analyzing signals recorded with RTL-SDR dongles whose local oscillators can be off by as much as 100 ppm or 150 kHz at $f_c$.

*(Continued)*

**110** *4 Correlating: Passive and Active Software-Defined Radio (SDR)–RADAR*

(Continued)

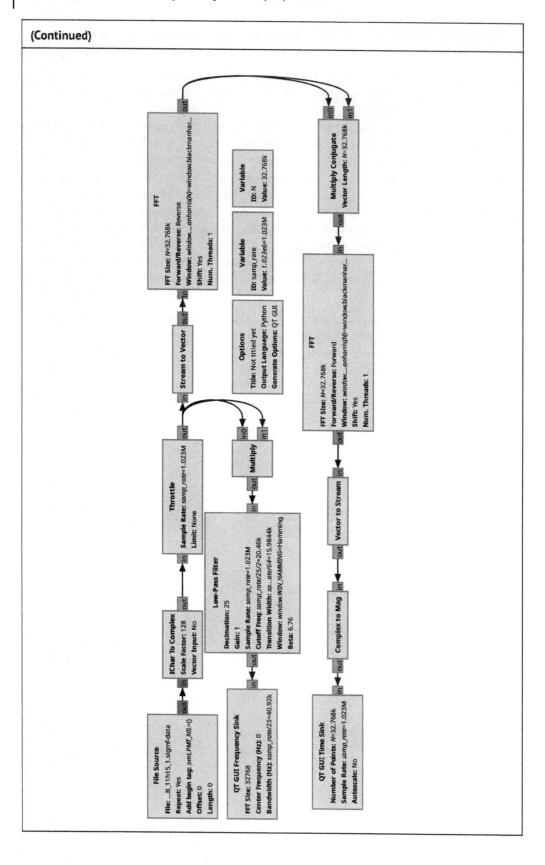

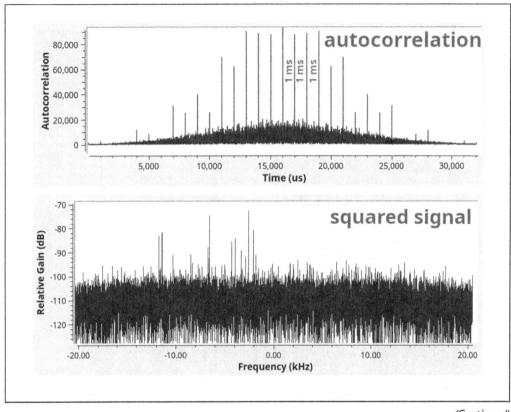

*(Continued)*

**112** | *4 Correlating: Passive and Active Software-Defined Radio (SDR)–RADAR*

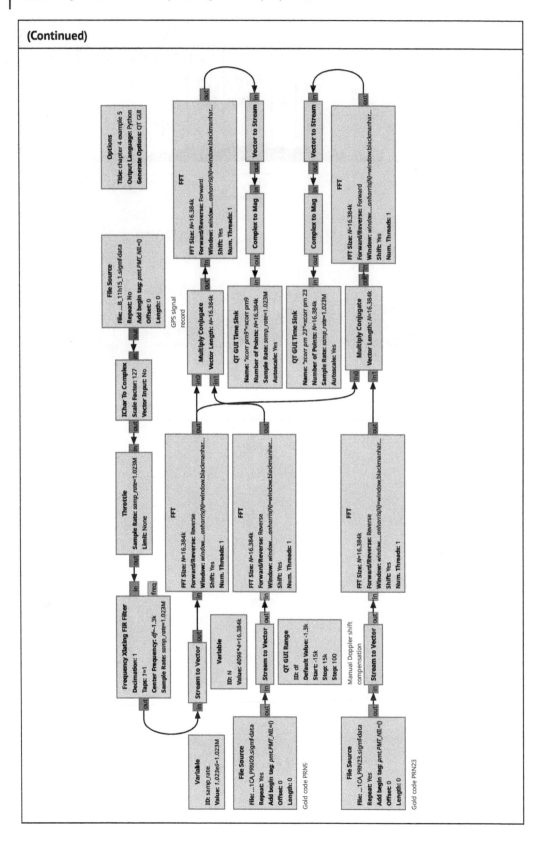

(Continued)

## 4.3 Passive RADAR Principle and Implementation

A passive RADAR benefits from existing radio frequency sources: historical demonstrations include mobile phone base stations, commercial broadcast FM stations, or analog television stations, but these sources have either become obsolete or are too narrowband to yield useful range resolution. Digital television terrestrial-broadcast (DVB-T) provides one widely available signal source matching both requirements of power and bandwidth, each channel being 8 MHz wide and multiple channels being analyzed sequentially for improved range resolution if the target remains at the same location during the complete measurement. Passive RADAR measurement was also demonstrated with spaceborne sources, including digital video broadcast-satellite (DVB-S) and the European spaceborne RADAR Sentinel-1, although these techniques will not be developed here due to the weaker power and intermittent availability of the signal in the latter case.

The relevance of a signal for passive RADAR application is determined by the ambiguity function defined as the autocorrelation of the signal for all possible frequency offsets representative of possible Doppler shifts

$$amb(p)(\tau, f) = \int p(t) \cdot p^*(t + \tau) \cdot \exp(j2\pi ft)dt$$

Indeed, any energy accumulation, as would be achieved by the correlation of the received noisy signal $s(t)$ after Doppler compensation by multiplying with $\exp(j2\pi ft)$, with the transmitted pattern $p(t)$, is represented by this ambiguity function when $s = p$: energy peaks would be intrinsic to the pattern structure and would mislead for a target at a range representative of pattern repetition with a frequency offset introduced in the pattern channelization (e.g. OFDM).

## 4.4 Active RADAR Principle and Implementation

Our objective in the following presentation is to develop the necessary framework for ranging, using electromagnetic waves only, a building estimated to be a few tens of meters away from the location of the experimental setup (Figure 4.5). The outline of the development will be as follows:

1. spectrum spreading for range resolution and using correlations for time of flight measurement
2. frequency stacking to concatenate measurements at multiple frequency ranges and hence compensate for the limited sampling rate of SDR hardware
3. complement the single-location measurement with spatial diversity of antenna location to add azimuth resolution and separate targets not only based on time of flight differences but also DoA. This approach of synthetic aperture RADAR (SAR) is best suited for SDR applications with respect to multiple input single output (MISO) or multiple input multiple output (MIMO) RADAR since in the latter case, random phase offset sources must be compensated for, while moving a single hardware setup in the SAR approach **keeps the phase offset constant** from one measurement to the next, allowing for azimuth compression with no additional phase calibration. The MISO/MIMO phase calibration is a challenging preliminary requirement for azimuth compression, which is naturally solved in the SAR approach at the expense of longer measurement duration since the antennas need to be physically moved. Nevertheless, the basic principle of SAR where a single setup is moved is also applicable to the MIMO approach where a single radio frequency hardware is switched from one antenna to another or even broadcast to all emitting antennas simultaneously if the transmitted signals are orthogonal – their cross-correlations cancel, and only correlating the given emitting signal with the received signal leads to energy accumulation and a non-null correlation.

**Figure 4.5** Geographical context of the measurement setup. From the balcony of the left building, the French Geographic Institute (IGN) website claims a house is located 50 m away: a RADAR setup is assembled (right) to assess this claim. Source: Géoportail.

## 4.5 Measurement Principle

Active RADAR systems are often associated with a pulse generator and a time-of-flight measurement by a receiving system recording the pulses reflected by targets. While the principle of such a system is easy to grasp conceptually as solely described in the time domain, the implementation is complex and leads to a poor signal-to-noise ratio with the inability to cancel noise sources. A strong pulse generator is complex to design and hardly compatible with radio frequency emission regulations, if only because spectral occupation control is challenging. Indeed, any change in the environment of the emission stage, including aging and warming or dielectric environment of the emitting antenna, might lead to a change in the shape of the transmitted pulse and hence the frequency occupation.

The classical RADAR range resolution analysis states that the only parameter defining the resolution for detecting a target is the occupied bandwidth $B$. In a pulsed RADAR, the bandwidth is the inverse of the pulse duration, but many approaches other than a short pulse are available for spectrum spreading: one such approach uses frequency steps from a start to a stop frequency, as would be performed by a network analyzer in a monostatic ($S_{11}$) or bistatic ($S_{21}$) measurement configuration. Another technique uses continuous frequency sweep in the frequency modulated continuous wave (FMCW) approach, where the transmitted and received signals are mixed with each other, leading to a beatnote whose audio frequency is proportional to the range, but the preferred approach in the subsequent presentation is noise RADAR [Bourret, 1957; Horton, 1959; Lukin and Narayanan, 2011; Sun et al., 2003], where a known (pseudo)-random pattern is transmitted and the time-delayed copies are identified by correlation. Hence, spreading a signal on a bandwidth $B$ leads to a range resolution $\Delta R$ through the time of flight, assuming an electromagnetic wave propagating at $c = 300$ m/µs in vacuum, halved for the two-way trip:

$$\Delta R = c/(2B)$$

In a pulsed RADAR, the shorter the pulse, the broader $B$ and hence the better $\Delta R$, within technological and regulatory constraints limiting $B$. We will benefit from the fact that most modern communication interfaces, including WiFi, appear at the hardware layer as random, although some periodic features might lead to unwanted repeating pattern as will be assessed by the *ambiguity function*. Indeed while correlation is the matched filter used for detecting a known pattern in a noisy received signal, a moving target induces a Doppler frequency shift $f_D$ determined by

$$f_D = 2f_c \frac{v_{\parallel}}{c}$$

with $f_c$ being the center carrier frequency of the transmitted signal, $v_{\parallel}$ the velocity component of the moving target toward the monostatic RADAR emitter and receiver, and $c$ the speed of light. If $f_D$ is greater than the inverse of the integration duration, then energy will not accumulate during the correlation integral since the mean value of the trigonometric functions is null. Hence, the correlation which was used for detecting a static target now becomes

$$c(\tau, f) = \int s(t)p(t + \tau) \exp(j2\pi ft)dt$$

between a known pattern $p$ and a received signal $s$ frequency shifted by $f$. The resulting 2D map indexed by $\tau$ and $f$ is the {delay, frequency shift} map or {range, velocity} map, a computationally intensive processing. As was discussed earlier in the context of passive RADAR and using existing waveshapes, mistakenly identifying some repeating structure in the noncooperative transmitter signal is avoided by assessing the ambiguity function

$$a(\tau, f) = \int p(t)p(t + \tau) \exp(j2\pi ft)dt$$

in the {delay, frequency shift} plane and making sure any structure is not mistakenly attributed to a moving target.

While broadcasting a radio frequency signal is regulated and transmitting a powerful, broadband signal is not allowed, the two approaches we consider for experimenting with RADAR signal processing are, on the one hand, using commercial, off-the-shelf transmitters such as those provided for digital communication, e.g. WiFi transmitters, and on the other hand, transmitting below noise levels. Indeed since the broadcast signal is known and detected using the matched filter of correlating the received signal with the transmitted pattern, the signal-to-noise ratio is improved by the correlation length or transmitted signal duration, similar to the sliding average used to reduce noise. The signal-to-noise ratio improvement is quantified through the *pulse compression ratio* (PCR) defined as the product of the signal bandwidth $B$ by its duration $T$. This unitless quantity defines how appropriate a wave shape is to RADAR processing. The higher the $PCR = B \times T$ product, the better the signal-to-noise ratio improvement while keeping the sharp timing capability needed for ranging with a correlation peak width equal to $1/B$ leading to the range resolution $c/(2B)$. In a digital communication scheme, the message bandwidth is determined by the bitrate, and the message duration is determined by the number of bits in the message times the bitrate, so that the PCR becomes the number of bits transmitted in a pseudo-random pattern: the longer the non-repeating pattern, the better the signal-to-noise ratio improvement after correlation. By knowing the transmitted pattern, even a weak signal transmitted below noise level can be detected after correlating the noisy received signal, assuming the illuminated scene remains static during the data collection. For moving targets, the velocity inducing a Doppler shift must remain constant during the acquisition duration for energy to accumulate coherently after Doppler shift removal when mapping the received signal in the {delay, frequency offset}, also called {range, velocity} plane.

## 4.6 From Theory to Experiment: Ranging by Frequency Stacking

SDR hardware is plagued by two limitations for RADAR applications: bandwidth limited by ADC speed and communication bandwidth. The latter issue is alleviated in RADAR applications with respect to digital communication since only burst of reference and surveillance data are needed rather than continuous streams, and buffers can be used for storing the data prior to communicating with the signal processing hardware. Under the assumption of a static illuminated scene, $N$ successive frequency bands separated by $f_s$ (the sampling rate) can be measured and the collected spectra stacked to generate a signal with global bandwidth $B = N \times f_s$, only limited by the transmitted signal bandwidth and measurement duration to meet the static target assumption.

We have discussed earlier how correlation is best computed in the frequency domain as $iFT(FT(s) \times FT^*(p))$ through the FT of the surveillance signal $s$ and the reference pattern $p$. Since the Fourier transform of the received quantities is computed, frequency stacking is naturally performed by measuring $s$ and $p$ in successive frequency bands determined by programming the hardware local oscillator of the SDR receiver front end and keeping $f_s$ constant, and concatenating the spectral components in an array indexed by the appropriate frequency. Considering $N$ samples of the $s$ and $p$ channels are collected at each local oscillator setting $f_{LO} = f_0 + q \times f_s$, $q \in [0 : Q-1] \in \mathbb{N}$, and then the array $a$ is indexed as $a(q \times N : (q+1) \times N)$, and the final spectrum frequency resolution is $f_s/N$ over a bandwidth $Q \times f_s$ leading to a range resolution of $\frac{c}{2(Q \times f_s)}$ and a maximum target range of $\frac{N \times c}{2 \times f_s}$. These considerations determine $Q$ and $N$, the two experimental parameters, once $f_s$ has been set by the SDR hardware and $f_0$ by the transmitter characteristics and available antenna bandwidth. When using a custom SDR transmitter for generating the pseudo random pattern $p$, both emitter and receiver local oscillators are reprogrammed to follow the same sequence $f_0 + q \times f_s$, while using a Commercial Off The Shelf (COTS) transmitter such as a WiFi transmitter requires selecting $f_s$ equal to the channel spacing as determined by the communication standards. For example, for 2.4 GHz WiFi, the 802.11b direct-sequence spread spectrum (DSSS) standard defines 22 MHz wide, 22 MHz spaced channels, which are readily sampled by selecting $f_s = 5.5$ MHz and sampling each transmitted signal as four successive receiver sub-bands by tuning the receiver local oscillator accordingly. Similarly, in the 5 GHz band, 15 MHz wide channels are separated by 20 MHz from channels 96–144, covering a full span from 5477.5 to 5722.5 MHz for a range resolution of 60 cm, although some unavoidable gaps lie in channels 98, 114 and 130, leading to some unwanted artifacts later when computing the RADARgrams (Figure 4.6). The range resolution benefit is nevertheless dramatic over the one that would be achieved with a single channel monitoring since the 5 MS/s sampling rate limits the range resolution to 30 m.

The impact of frequency stacking on the correlation is illustrated in Figure 4.7 where a PlutoSDR is used as a noise generator by phase-modulating using a pseudo-random sequence with a fixed amplitude carrier to emit a controlled broadband spectrum, replacing the WiFi dongle of the previous description over a transmitting antenna. A B210 dual-channel SDR monitors the reference

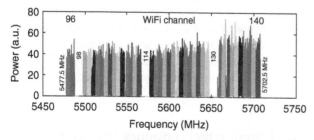

**Figure 4.6** Frequency stacking the 5.8 GHz 802.11ac WiFi signal emitted by a Alfa Network AWUS036ACS transceiver and received with a B210 dual-channel SDR receiver at a sampling rate of 5 MS/s, with each 15 MHz wide WiFi signal being sampled twice, in its lower and upper half, before switching to the next channel centered on the adjacent 10 MHz.

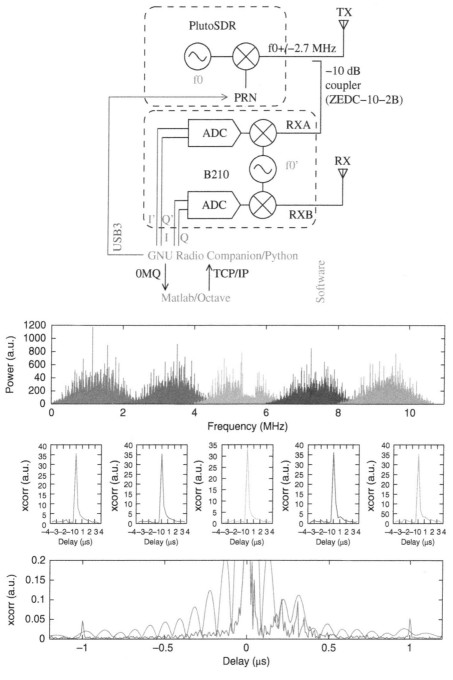

**Figure 4.7** Stacking measurements acquired in successive spectral bands in order to improve the time and hence range resolution. Top: Each spectrum is acquired on a 2.7 MHz bandwidth and the center frequency of the broadcast signal incremented with 2 MHz steps. Middle: The correlation of signals collected in each individual spectral band only exhibits a large peak at 0 delay and an asymmetry hinting at some clutter in the positive time delay, but no discrete target can be resolved. Bottom: Stacking the spectra improves the time resolution, in light by stacking 10 MHz bandwidth hinting at some individual target response between 200 and 300 ns and stacking 150 MHz bandwidth (dark) clearly separates these two reflectors.

signal through a coupler connected between the PlutoSDR output and the antenna and records the signal received from an antenna in a bistatic configuration. Each invidual spectrum leads to a poor time (range) resolution when correlating narrow frequency bands. Frequency stacking, leading to wide band signals, allows through constructive interferences at delays where targets are located and destructive interference otherwise, to sufficiently improve range resolution that targets at different ranges are separated. The separate emission source, here provided by the PlutoSDR, aims at reducing the cross-talk between emission and reception channels that would inevitably occur if a single component is used for both transmitting and receiving, limiting the RADAR range since the gain must be reduced to avoid saturation. Even if a single B210 solution was considered, the transmitted signal must anyway be recorded by the B210 receiver since the Analog Devices AD936x transceivers fitting most of the consumer grade SDR platforms use different local oscillators for transmitting and receiving, hence inducing an uncontrolled frequency offset between emitted and received signals with one channel of the dual-channel receiver dedicated to receiving the direct signal for offset monitoring.

We notice that each spectrum amplitude varies on both the reference $r$ and the surveillance $s$ signals, and the amplitude variation effect is magnified when correlating through the product of the Fourier transforms. An alternative way of flipping the time in the second argument of the product is that instead of multiplying with the complex conjugate, the correlation is calculated as

$$FT(xcorr)(s, r) \simeq FT(s)/FT(r)$$

Since both amplitudes of $s$ and $r$ are observed to fluctuate identically due to the noncooperative wave shape, the ratio cancels the artifacts induced by amplitude variations. Notice though that this approximation is only valid for broadband signals and numerically diverges for narrowband signals when $FT(r)$ includes many terms with a small amplitude.

The experimental setup leading to this result (Figure 4.7, left) justifies the previous lengthy explanation on communicating between GNU Radio and external processing software. Indeed, the local oscillator of the receiver must be programmed for each new spectrum acquisition and tuned according to the transmitted signal center frequency, whether emitted from the PlutoSDR acting as pseudo-random sequence phase noise generator produced by a GNU Radio flowchart or by a WiFi emitter requiring changing the channel broadcasting the information. These results are achieved with a TCP server running next to the GNU Radio flowchart controlling the B210 receiver, and the collected IQ stream is transmitted through a publish server through 0MQ. A client grabs the few samples representative of the electromagnetic wave interaction with its surroundings at the frequency band under investigation, and the measurement repeats for a new channel until the full bandwidth has been covered. Assuming a few tens of thousands of samples are collected in each band – at 60 cm range resolution, $N = 1000$ samples allow for detecting targets at a maximum range of 3 km since the correlation will provide time offsets from $-N/2$ to $+N/2$, and averaging $M$ times to improve the signal-to-noise ratio by $\sqrt{M}$ requires multiple independent measurements. Since the transmitted data are pseudo-random, the averaging process must take place at the correlation *output*, after the time domain samples have been Fourier-transformed and correlated by multiplying the spectra.

Figure 4.7 illustrates these concepts on real signals collected by broadcasting signals toward a building located, from aerial image analysis, about 40 m from the bistatic RADAR (with separate emitting and receiving antennas) system with both emitter and receiving antenna co-located. Since the speed of light is 300 m/µs, the two-way trip to a target located about 40 m away is expected to be in the $40/300 \times 2 = 270$ ns range. The slow accumulation of resolution as spectra are stacked by constructive interference of the complex quantities at delay where targets are located is visible

4.6 From Theory to Experiment: Ranging by Frequency Stacking | **119**

on the bottom chart even when each individual spectrum lacks the resolution to identify discrete targets and only exhibits clutter in the positive delay. The real-time plot of the correlation remains a useful indicator of a sound experimental setup during acquisition. Each local oscillator center frequency tuning range is set by GNU Radio, and interleaved IQ samples of both channels are streamed through 0MQ to a subscribe client. For *Nfreq* successive measurements separated by *finc* sampled at *fs*, assuming matrices of the surveillance *s* and reference *r* signals have been assembled with time series as vectors stacked for each successive measured center frequency, then the frequency stacking is implemented in Octave as

```octave
Nfreq = 50                % swept frequencies [no unit]
fs = 2.7;                 % sampling freq [MHz]
finc = 1.0;               % frequency increment (consistent with Python) [MHz]

span = floor(finc/fs*length(r)); % bin index increment every time LO
    increases by 1 MHz
s1 = zeros(floor((finc*Nfreq+3*fs)/fs*length(r)), 1); % extended spectral
    range
s2 = zeros(floor((finc*Nfreq+3*fs)/fs*length(s)), 1); % ... resulting from
    spectra concatenation
w = hanning(length(r)); % windowing function
for f = 0:Nfreq-1                              % center of
    FFT in the middle
    s1(f*span+1:f*span+length(x)) = s1(f*span+1:f*span+length(r))+w.*fftshift
    (fft(r(:, f+1)));
    s2(f*span+1:f*span+length(m)) = s2(f*span+1:f*span+length(s))+w.*fftshift
    (fft(s(:, f+1)));
end
x=fftshift(ifft(conj(s2).*(s1)));
fs2 = finc*Nfreq+3*fs;
N = 200;
tplot = [-N:N]/fs2;
resfin(:, m) = abs(x(floor(length(x)/2)-N:floor(length(x)/2)+N));
plot(tplot, resfin(:, m)/max(resfin(:, m)));
```

One should take care here of the two different conventions when computing a discrete Fourier transform, of storing in the first array index either the 0 frequency (DC) component and in the last index the $f_s - fs/N$ component as selected by Matlab, Octave, or NumPy, with $N$ being the Fourier transform length, which differs from the physical spectrum shape where the first index represents $-f_s/2$, the negative Nyquist frequency component, the DC component is located at half the vector position, and the last sample represents the $f_s/2 - f_s/N$ spectral component. Although these two representations are identical for each individual spectrum through the periodicity of the discrete Fourier transform spectra, they differ during the frequency stacking when the second representation matches the spectral distribution: the `fftshift` function will convert the output of the former computation to the latter representation and must not be omitted. The time axis is finally computed as the inverse of the stacked spectra bandwidth so that no degree of freedom remains when plotting the time delay to the targets.

## 4.7 Results

During this measurement, the two analyzed bandwidths are 50 and 150 MHz, i.e. by stepping 1 MHz through the bands 650–700 MHz and 650–800 MHz, respectively, or the upper part of the bands allocated to digital terrestrial television, for which we can expect antennas to be tuned. The resolution gain is significant (Figure 4.8) and allows to clearly identify a reflector closer than expected by about 50 m: the echo at around 200 ns must be located at a distance of about $150 \times 0.2 = 30$ m (150 being half of the 300 m/µs light speed, with the half accounting for the round trip).

We systematize this series of measurements in Figure 4.9, which complements the previous view with the baseline noise level when the antennas are pointing toward the sky and a measurement at 250 MHz bandwidth. We extended the correlation visualization to ±1 µs to show the two artifacts at these dates, which correspond to the inverse of the spectrum acquisition step. Indeed, we clearly observe during the accumulation of spectra that the windowing creates a frequency comb pattern spaced by 1 MHz, whose Fourier transform is a time comb spaced by 1 µs. These correlation peaks have nothing to do with reflectors, as evidenced by their symmetry in negative delays. In the inset, a zoom on positive times helps identifying the distances of the reflectors associated with the strongest echoes.

Let us attempt a preliminary analysis in the geographical context of measuring the source of these reflectors with only the range information (Figure 4.10). While it is almost unquestionable

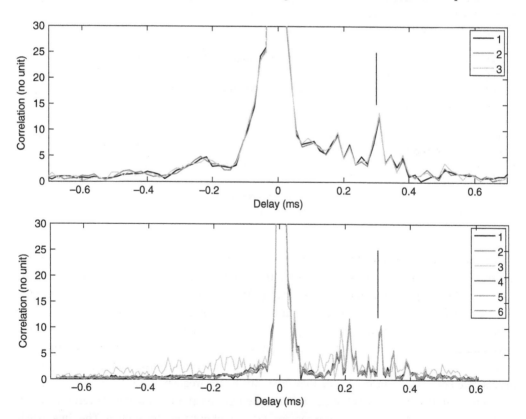

**Figure 4.8** Result of the measurement, top with a bandwidth of 50 MHz (650–700 MHz) or 3 m range resolution, and bottom with a bandwidth of 150 MHz (650–800 MHz) or 1 m range resolution. In both cases, the black vertical line is at 300 ns or a target at 45 m.

*4.7 Results* | **121**

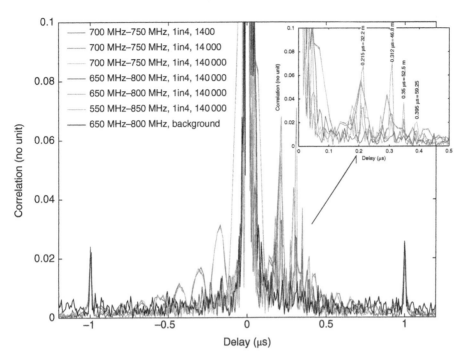

**Figure 4.9** Result of the measurement, with bandwidths ranging from 50 MHz (3 m resolution in distance) to 250 MHz (60 cm resolution in distance). The analysis of the echoes in the inset in the top right includes the conversion of the round trip time to a distance to the reflector.

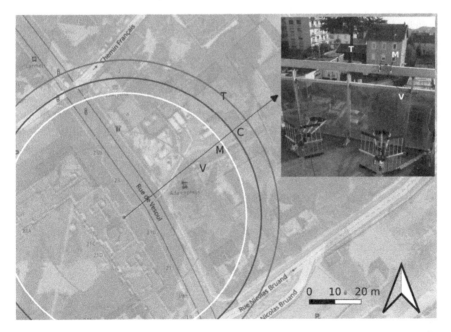

**Figure 4.10** Analysis of the source of echoes in a geographical context: the arrow starting point is the measurement location also shown in the inset; the circles correspond to the distances discussed in the text. The map background overlays the vector limits of OpenStreetMaps with the transparency aerial photographs from Google Earth, including the vehicle parked about 30 m from the RADAR source. The arrow indicates the direction in which the antennas are aimed.

that the echo at 46.8 m is the wall of the house (M), the other echoes are subject to interpretation. A car (V) parked in the illuminated field by the transmitting antenna would correspond to the echo at 32.2 m, while the metal chimney (C) on the roof of the house could be attributed to the reflector at 52 m. More questionable, the reflector at 59.3 m is attributed to a tree trunk (T) that is not indicated on OpenStreetMaps and whose foliage obscures visibility on Google Maps but could just as well be due to one of the doors of the garages lined up a few meters behind this tree.

## 4.8 Conclusion on Range Measurement

Thanks to two software radio platforms, one for transmission and a dual channel for reception, we were able to explore the principle of noise RADAR and experimentally test the relationship between distance resolution and signal bandwidth, even concatenating adjacent spectra acquired successively. These basic principles open up many perspectives:

- Complement the single receiving antenna with a network of equidistant antennas to regain angular information of azimuth of each target. Assuming that the targets are "far" from the antennas (far-field condition with a plane wave illuminating the network), the arrival direction is obtained by Fourier transform along the antenna axis in the matrix of temporal samples acquired by each antenna.
- Reduce the communication latency on the USB bus by moving the generation of the pseudo-random sequence to the Zynq of the PlutoSDR and performing the correlation in the FPGA of the B210. This requires mastering Very High Speed Integrated Circuit (VHDL) and RF Network-on-Chip (RFNoC), which is not easy, but with 56 MHz bandwidth between the AD936x and the logic gate arrays, the scanning speed gain (at least 20) is significant.
- We afforded the luxury of the B210 while R820T2 digital terrestrial television receivers could do the job. In S. Scholl [2017], the author teaches us at 12 minutes and 30 seconds of his presentation that it is possible to reprogram the frequency of the radio frequency (RF) oscillator of the R820T2 without losing samples on the transmitted flow from the receiver to GNU Radio, thus ensuring the maintenance of coherence between the two (or more) receivers. The library implementing these modifications is available at https://github.com/DC9ST/librtlsdr-2freq.

Finally, using a laptop is not very satisfactory when the Raspberry Pi 4 (Figure 4.11) is equipped with two USB2 ports and two USB3 ports. All necessary tools have been ported to Buildroot – fully exploiting the 64-bit processor, at least for the acquisition phase: a completely embedded version of the experiment is thus assembled using the packages provided by G. Goavec-Merou at https://github.com/oscimp/PlutoSDR/tree/for_next. It should be noted that all transactions between Octave and Python are done through sockets, so it is trivial to control the acquisition from Python on the Raspberry Pi using Octave executed on a PC.

The complete codes, of which excerpts are presented in this chapter, are available at github.com/jmfriedt/active_radar to encourage the reader to replicate the experiment or perform signal processing on the proposed signals.

## 4.9 Azimuth Resolution Through Spatial Diversity: Synthetic Aperture RADAR

We have identified targets as time-delayed echoes backscattered by a scene illuminated by a noise RADAR broadcasting wideband signals. However, the time domain plot

### 4.9 Azimuth Resolution Through Spatial Diversity: Synthetic Aperture RADAR

**Figure 4.11** A Raspberry Pi 4 controls the experiment and acquires data from the B210 through its USB3 port while sending pseudo-random sequences to the PlutoSDR through the USB2 port.

- lacks DoA information that would help identify the location of the targets through their azimuth information since only the range information has been calculated,
- multiple targets at a same range cannot be separated.

Converting the one-dimensional time domain A-scan to a two-dimensional azimuth-range B-scan requires repeating the measurement with different antenna orientations so the beam is targeted at different targets with different azimuth. However, the beam width $\vartheta$ is determined by the antenna aperture $D$ as

$$\vartheta \propto \frac{\lambda}{D} \text{ rad}$$

when radiating an electromagnetic wave with wavelength $\lambda = c/f_c$ at frequency $f_c$. Under such conditions, differentiating two targets 2 m apart at a range of 50 m would require an antenna diameter of $\lambda/(4 \times 10^{-2})$ or at 5 GHz where $\lambda = 6$ cm, $D \geq 1.5$ m. Not only would such an antenna be bulky and sensitive to wind pressure, it would be impractical to move accurately to point in different azimuth directions. The alternative is to *simulate* such an antenna aperture by moving a small antenna with large beamwidth over distance $D$ and by phase analysis of the signal reflected by each target illuminated by each antenna position and reconstruct the DoA of each signal. This approach is called synthetic aperture RADAR, and when implemented on a ground-based moving system, it is called ground-based synthetic aperture RADAR (GB-SAR).

While range resolution was solely dependent on signal bandwidth and the carrier frequency was not involved in the calculation, the selection of the carrier frequency becomes relevant when implementing a SAR system. Indeed, the antenna dimensions shrink with rising frequency, but also a shorter wavelength will lead to improved azimuthal resolution. Among the unlicensed industrial, scientific, and medical (ISM) bands, the 2.4 GHz band spans from 2.4 to 2.5 GHz, although 802.11b/g/n WiFi can only access the first 80 MHz, while the 5 GHz band spans from 5.725 to 5.875 GHz, still accessible to consumer-grade SDR transceivers. Benefitting from noncooperative sources in a passive RADAR implementation, DVB-T(2) signals are broadcast in Western Europe from 470 to 694 MHz but lead to bulky Yagi-Uda antenna setups impractical for fine motion, whereas the microwave signals are efficiently emitted and received from smaller patch antennas with gains in the 6–10 dB range.

The most intuitive SAR geometry is the uniformly spaced linear array (ULA): in this geometry, one or both antennas in the bistatic RADAR configuration are moved along a linear path with steps $\vec{d}$. The phase introduced by a plane wave incoming with wavevector $\vec{k}$ is $\vec{d} \cdot \vec{k} = \frac{2\pi}{\lambda} d \cos(\vartheta)$ where $\vartheta$ is the azimuth angle of the incoming wave. Thus, each antenna indexed with $n \in \mathbb{N}$ of the ULA separated by distance $d$ receives a signal expressed as $A \exp(j\frac{2\pi}{\lambda} n \cdot d \cos(\vartheta))$ so that $nd$ and $\cos(\vartheta)$ can be considered as two dual quantities in the expression of a discrete Fourier transform. In other words, taking the inverse Fourier transform along the antenna index when collecting matrices shaped with the stacked spectra leading to correlations (ranges) along the columns and the antenna position along the lines will lead to azimuth compression where the abscissa of the accumulated energy indicates the angle of arrival and the ordinate indicates the range.

Similar to Shannon's sampling theorem stating that the sampling frequency must be greater than half the highest spectral component of the sampled signal, the spatial equivalent states that if both transmitter and receiver antennas are moving from one position to the next, the phase variation of the two-way trip as argument of the complex exponential to a target located at range $R$ is expressed as $2\sqrt{(d - R\sin\vartheta)^2 + (R\cos\vartheta)^2}$ whose first-order Taylor series if $d \ll R$ is $2(R - d\sin\vartheta_0)$. Considering only the azimuth-dependent term, $2\vec{k}\vec{d} \simeq 2\frac{2\pi}{\lambda} \times 2d \leq 2\pi$ to avoid full phase rotations from one antenna step to the next, and hence $d \leq \lambda/4$. The sampling theorem allowing to recover unambiguously the DoA of each target signal states that the monostatic RADAR displacement step must be smaller than the quarter wavelength.

In case one of the antennas of the monostatic array remains fixed, then a factor of 2 is removed in the phase expression, and the moving step $d \leq \lambda/2$, but the subsequent processing step of inverse Fourier transform to compress energy in the azimuth direction requires inserting half the physical displacement step since one static and one moving antenna act as a virtual array of both antennas moving by half the distance of the physical ULA as classically considered when addressing MIMO RADAR geometries.[1]

Since the azimuth resolution is given by the length $D$ of the ULA, the only limitation on the angular resolution is the displacement length and the data acquisition duration of the $4D/\lambda$ positions along the ULA. At first, the antenna(s) are moved manually along a graduated linear path before automating the measurement with a motorized rail. Similar to frequency stacking by reprogramming the local oscillator from GNU Radio, the TCP server started as a separate thread next to the GNU Radio scheduler infinite loop accepts new commands controlling the rail position, and the client only fetches IQ samples through the 0-MQ subscribe socket when the antenna position has stabilized.

The previous measurements demonstrated in the ultra high frequency (UHF) band are reproduced here in the super high frequency (SHF) industrial, scientific and medical (ISM) band (Figure 4.12) obtained by scanning a frequency band of $2450 \pm 50$ MHz. Plotting circles of constant delay on an aerial photograph centered on the location of the RADAR allows to identify the echo at 50 m as *probably* reflected by the house and specifically its roof, and the echo at 25 m as *probably*

---

1 See https://hforsten.com/mimo-radar-antenna-arrays.html for a detailed explanation about virtual arrays in 1D and 2D-MIMO RADAR architectures.

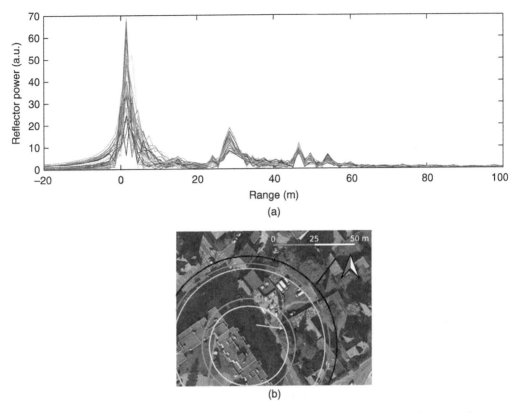

**Figure 4.12** (a) Echoes as a function of time for 30 successive measurements, and their mean value displayed as a thick line, with striking features at 24, 28.5, 46.5, 49.5, and 54 m. (b) Circles representing these ranges to try and identify the targets. The circles at about 50 m from the source, lead to reflectors attributed to the house we aim at mapping. The closer sources remain uncertain.

associated with the parking area and the changes in slope on the roofs of the adjacent boxes. Azimuth measurement should help resolve this uncertainty.

We emphasized the importance of separating tasks and assigning them the most appropriate language: GNU Radio (Companion) for continuous data acquisition, Python for controlling the acquisition parameters (local oscillator frequency, eventually antenna position), and Octave or possibly numpy for signal processing. In doing so, we modified the Python code generated by GNU Radio Companion, preventing any backtracking to graphically modify the processing chain: once the Python code was modified (to add the TCP server in particular), we could only manually add elements to the processing chain. This approach is now obsolete with the addition of two functionalities in GNU Radio Companion: the `Python Module` and `Python Snippet` blocks [Müller, 2020]. The `Python Module` block allows defining Python code that will be accessible from the block generated by GNU Radio Companion: as a result, integrating the TCP server in such a block does not prevent later graphical modification of the processing chain, as the Python code is still accessible. However, the TCP server thus defined must be instantiated at some point when starting the thread that waits for a connection on the socket. This is the role of the `Python Snippet` block, which inserts a piece of code at the desired location during the initialization of the processing chain – in our case, just after the initialization of the data structures and before launching the GNU Radio scheduler in its *Top Block*. With these two blocks, we can now integrate our own command-line programmed functionalities into a graphically defined processing flow. The example we propose here consists of changing the frequency of the local oscillator, but also, if GNU Radio is executed on a Raspberry Pi, GPIOs can be used, for example, to control a motor that moves an antenna (Figure 4.13).

The only issue we encountered in this development is sharing local methods and variables from the calling class for their use and modification in the thread. This issue is resolved by passing the variable `self` as an argument when creating the thread. This variable is accepted as an argument of the server (here named `tt`), and therefore all variables and methods of the calling class can be modified from the thread.

### 4.9.1 OFDM RADAR (WiFi)

WiFi, standardized by the IEEE 802.11 standard, is based on a modulation using division of the communication channel into orthogonal subcarriers (OFDM for orthogonal frequency-division multiplexing) for the 2.45 GHz band that interests us.

The principle of OFDM, which divides a wide spectral band into small sub-bands, is based on the observation that in an environment with reflectors – typically an urban environment – certain spectral components are canceled out due to destructive interference between the reflected signal and the direct signal between the transmitter and the receiver [Braun, 2015]. Instead of simply losing the connection, OFDM shares the spectrum into sub-bands and spreads the information among these bands. Due to redundancy and error correction codes, the loss of a communication channel does not result in the loss of the entire message.

IEEE 802.11 has been implemented in GNU Radio by Bastian Bloessl at https://github.com/bastibl/gr-ieee802-11. The most important point for us in his repository is the directory `utils/-packetspammer/` which provides tools to switch a USB-WiFi adapter to *monitor* mode, which accepts low-level commands such as continuously emitting packets without complying with the WiFi information exchange standard. Similar to the first approach, the WiFi transmission is coupled ($-10$ dB) through 20 dB of attenuation to the first reference channel of the B210, while the second channel acquires the signal from the receiving antenna. The message transmitted is irrelevant since only the structure of the physical layer induces the necessary spectral spreading. However, care should be taken to select a USB-WiFi adapter that supports this mode of operation; not all chipsets do. All experiments were conducted with a Alfa AWUS036NEH WiFi adapter capable of emitting 20 dBm. When we wanted to emit less power (to avoid excessive direct coupling), we inserted a 20 dB attenuator at the output of the WiFi transmitter to achieve an emitted power of 0 dBm.

The configuration of github.com/bastibl/gr-ieee802-11/tree/maint-3.10/utils/packetspammer is a tradeoff between the highest possible repetition rate with the longest possible packets without attempting to send a command to the WiFi interface while it is busy. We used `packetspammer wlan0 -n 1000000 -r 400 -s 1350` not forgetting to use `iwconfig wlan0 mode monitor` first and then `iwconfig wlan0 channel 3` to select the channel to communicate on. This way, the carrier frequency can be scanned by defining the channel index from the Octave script using `system(['/sbin/iwconfig wlan0 channel ',num2str(1+(frequency-1)`

`*finc/5)]);`, which, of course, must be adapted to the name of the WiFi interface (here `wlan0`). Since the transmission is not continuous, we keep only the sequences during which the reference channel observed a transmission from the WiFi card by testing the acquired vector from the ZeroMQ interface and looping until the required number of relevant points is obtained:

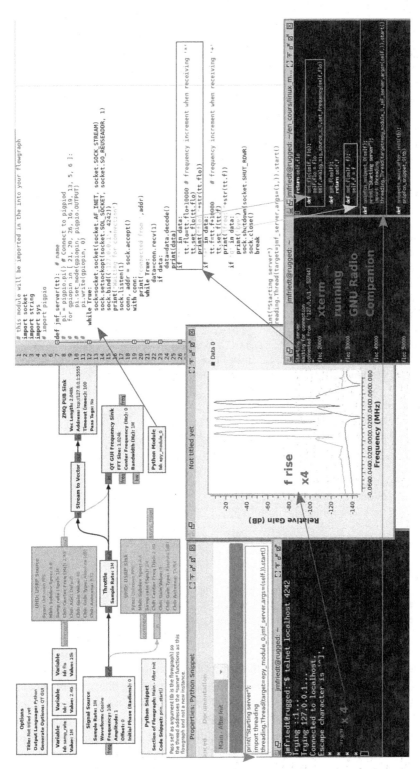

**Figure 4.13** Example of integrating a TCP/IP server into GNU Radio Companion. On the right, the server code, launched in its own thread, is defined in a Python Module. The thread is launched after the initialization of GNU Radio structures by the Python Snippet. Note that the server is tested by telnet on the port listened by the socket (here 4242) and that sending a command (here "*") does have an effect on the GNU Radio processing chain (modification of the oscillator frequency here).

## 4 Correlating: Passive and Active Software-Defined Radio (SDR)–RADAR

```
pkg load zeromq
pkg load signal
total_length = 70000*2;
error_threshold = 0.05;
error_number = 100;
error_vector = ones(error_number, 1);
for frequence = 1:Nfreq
 system([ '/sbin/iwconfig wlan0 channel ', num2str(1+(frequence-1)*finc/5)]);
 tmpmes1 = [];
 tmpmes2 = [];
 while (length(tmpmes1)<(total_length))
   sock1 = zmq_socket(ZMQ_SUB); % socket-connect-opt-close = 130 us
   zmq_connect(sock1, "tcp://127.0.0.1:5555");
   zmq_setsockopt(sock1, ZMQ_SUBSCRIBE, "");
   recv = zmq_recv(sock1, total_length*8*2, 0); % *2: interleaved channels
   value = typecast(recv, "single complex"); % char -> float
   tmpv1 = value(1:2:length(value));
   tmpv2 = value(2:2:length(value));
   zmq_close (sock1);
   toolow = findstr(abs(tmpv2)<error_threshold*max(abs(tmpv2)),
   error_vector');
   if (!isempty(toolow))
     if ((toolow(1)>100) && (max(abs(tmpv2))>0.05))
        tmpmes1 = [tmpmes1 tmpv1(1:toolow(1)-99)];
        tmpmes2 = [tmpmes2 tmpv2(1:toolow(1)-99)];
     end
     if ((toolow(end)<length(tmpv2)-100) && (max(abs(tmpv2))>0.05))
        tmpmes1 = [tmpmes1 tmpv1(toolow(end)+99:end)];
        tmpmes2 = [tmpmes2 tmpv2(toolow(end)+99:end)];
     end
   else
     if (max(abs(tmpv2))>0.05)
        tmpmes1 = [tmpmes1 tmpv1];
        tmpmes2 = [tmpmes2 tmpv2];
       else
          printf("*");   % transmitter shutdown
       end
   end
 end
 mes1(:, frequence) = tmpmes1(1:total_length);
 mes2(:, frequence) = tmpmes2(1:total_length);
end
```

This approach is also used if the noncooperative source is not controlled by the user but emits opportunistically (as in the case of passive RADAR). The principle is to define a vector of consecutive values below which we consider the emission to have ceased (error_vector is a series of error_number values at error_threshold times the maximum of the acquired sequence) and to search for this condition in the vector of data acquired on the reference channel (tmpv2) using the findstr function. If no gap is observed, we still verify that the emission has not simply ceased (condition on abs(tmpv2) greater than an arbitrary threshold of 0.05, considering that

GNU Radio works between −1 and +1; otherwise display a "*" to indicate that the WiFi emission has ceased). In case of temporary interruption of the emission, we still keep the relevant fraction, either at the beginning or at the end, and concatenate these measurements in tmpmes until reaching the number of data necessary for the subsequent correlation calculation. In the end, the matrices mes are formed by the temporal sequences for each channel frequency.

Based on this analysis, the emission is almost continuous and particularly desirable for a RADAR application, and in any case, we only keep the points during which the transmitter has been active. We now need to search for the occurrence of reflections from targets in the signal acquired on the measurement channel by correlating it with the pattern acquired on the reference channel. However, WiFi is a structured emission designed to transmit digital information, not a noise whose correlation is a single Dirac peak.

This leads us to study the physical layer. Each WiFi channel is allocated 20 MHz. The channels are separated by 5 MHz, so the 80 MHz wide band can accommodate 14 channels, of which 11 are practically accessible with the WiFi transmitter used in Europe. The central frequency of channel $N \geq 0$ is $2412 + N \times 5$ MHz. Within each channel, 64 subcarriers are defined with a spacing of $20/64 = 0.3125$ MHz. Some channels on the edges of each band ($-32$ to $-27$ and 27 to 32) are not used to avoid spilling over to the adjacent channel, but more importantly for us, channel 0 is not used to avoid amplifying power around DC after baseband transposition. Therefore, considering the reduced bandwidth of the USB receiving the signals from the B210, which even with USB3 is limited to about 6–6.5 Msamples/s on both channels, we have chosen to acquire only a fraction of each WiFi band with a width less than 10 MHz in order to reject the hole at the central frequency (Figure 4.14).

During the acquisitions, we observed periodic correlation peaks that hide any echo reflected by the targets reflecting the emitted signal. It appeared that the cause of this periodic structure is the OFDM modulation, which power modulates the spectrum within each transmitted channel. When we concatenate the spectra acquired on the adjacent WiFi channels, this structure is maintained and is found in the time domain, with the Fourier transform of a comb being a comb. Therefore, we need to eliminate this frequency-dependent power dependency. To do this, we will replace the *xcorr* correlation estimator expressed as

$$xcorr(ref, mes) = iFT(FT(ref) \cdot FT^*(mes))$$

with *FT* being the Fourier transform of the reference channel *ref* and measurement *mes*, *iFT* the inverse Fourier transform, and * the complex conjugate, with

$$xcorr(ref, mes) = iFT(FT(ref)/FT(mes))$$

From the phase point of view, the product by the complex conjugate or the quotient comes down to the same thing, but this time we are taking the ratio of the magnitudes instead of their product. By doing this, we eliminate the fluctuations introduced by OFDM, and we end up with a flat spectrum during correlation. Only the echoes appear as correlation peaks during the inverse Fourier transform (Figure 4.15).

## 4.10 Acquisition for Azimuth Measurement

We have decided to resolve the azimuth uncertainty by synthesizing a large aperture antenna. We have seen that in order to satisfy the spatial transposition of the well-known sampling theorem in its temporal version, the spacing between the antennas must be at most $\lambda/2$, when one of the

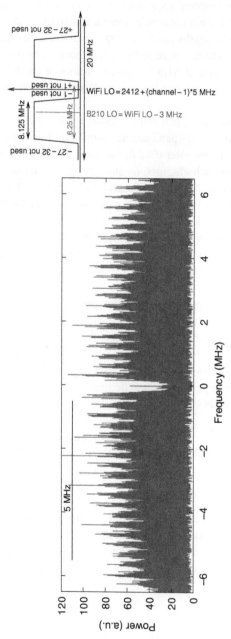

**Figure 4.14** (a) Spectrum of a WiFi channel acquired on a 13 MHz band, highlighting the absence of subcarrier 0 and the structure induced by OFDM modulation (power fluctuations with spectra centered every 312.5 kHz). (b) Measurement method to eliminate the unused channel 0 in order to obtain a flat stacked spectrum: the reception frequency is shifted by 3 MHz compared to the transmitted channel frequency, and a 6.25 MHz band is sampled to keep only 5 MHz when concatenating the spectra.

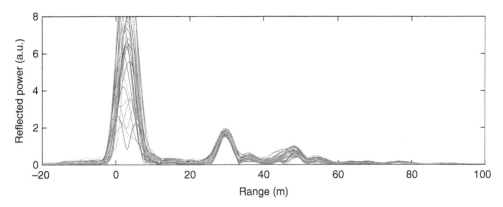

**Figure 4.15** Echoes acquired by a WiFi transmitter scanning 11 channels and whose emitted signal is recorded by a channel of the B210 receiver, the second channel recording the signal received by the receiving antenna. The loss of range resolution, due to the division by two of the analyzed bandwidth, compared to 4.12, is obvious, but the correspondence of the targets is otherwise excellent, validating the relevance of the measurement by an OFDM source.

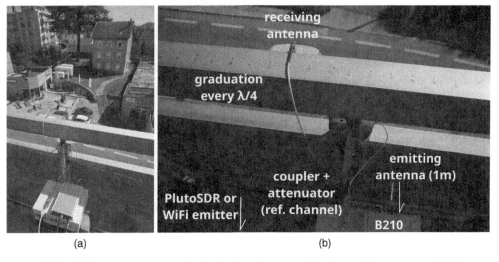

**Figure 4.16** (a) Zoom on the graduation indicating where to place the receiving antenna, with the receiving antenna placed on the balcony floor to maximize the distance between the receiving and transmitting antennas, thereby limiting direct coupling and increasing reception gain. (b) Overall view of the setup equipped with a WiFi key as a source of wideband signals.

two antennas remains fixed, in order to remove any ambiguity about the DoA of the signal. This means that in order to cover the 2.3 m wide balcony of an antenna array operating at 2.45 GHz, 38 antennas will be needed. This will be expensive in terms of B210 for the receivers or will require RF switches with more channels than we can obtain.

The solution to scan the receiving antenna position while keeping the transmitting antenna fixed will allow us to meet the specifications of only using available equipment: two Huber–Suhner WiFi antennas, reference 1324.19.0007 (2.4 GHz circularly polarized patch antenna, nominal gain of 8.5 dBi), are indeed available to us. Thus, we replace the network of antennas with a single receiving antenna that we will manually move along the balcony railing (Figure 4.16), while moving by $\lambda/4 \simeq 3.1 \pm 0.1$ cm.

## 4.11 Suppressing Direct Coupling Interference

We have encountered a loss of sensitivity during our developments, depending on certain geometries, which ended up being attributed to the antenna emitter being too close to the receiving antenna during its scanning. It is important to maintain this distance as large as possible – in our case, by placing the transmitter on the balcony floor while the receiver moves on the railing 1 m higher – in order to avoid saturating the receiver's amplification stage. Under this condition, the power leakage from the transmitter to the receiver is not particularly limiting if the spread spectrum is properly done, without periodicity that would result in delayed echoes by the repetition rate of the pattern. The leakage will result in an intense correlation peak close to zero delays, but since our interest focuses on targets several tens of meters away, we will simply omit analyzing the closest targets ($< 3$ m or 2 pixels in a 100 MHz bandwidth measurement), which are coupling artifacts.

## 4.12 Signal Processing

We have established that we are capable, through pulse compression along range, of correlating the emitted signal with the received signal to coherently accumulate energy at various delays $\tau$ of the pulses reflected by a target at distance $d$, with $\tau = 2d/c$ ($c$ being the speed of light, 300 m/μs). Recall that correlation is an inverse Fourier transform of a product of Fourier transforms (of the reference signal and the complex conjugate of the measurement signal to reverse time based on the convolution theorem) – this relationship will be useful to us.

---

**Fourier transform along the antenna scan direction for azimuth compression**

An intuitive way to understand why a Fourier transform on the scanning direction of the receiving antenna induces azimuth compression is to analyze the expression of the phase shift induced by the arrival angle of a signal with orientation $\vartheta$ relative to the normal of a uniformly distributed linear antenna array (ULA) separated by a distance $d$. The projection of the wave vector $k = 2\pi/\lambda$ onto the antenna array base $d$ indicates that the phase shift is $\vec{k} \cdot \vec{d} = k \cdot d \sin(\theta)$. For the $n$th antenna, the phase shift is $\varphi_n = nkd\sin(\theta)$, which is the product of two dual quantities: the antenna position $nd$ and the wave vector $k = 2\pi/\lambda = 2\pi f/c$. The series of measurements at frequencies $f$ induces compression in distance (inverse resolution of the measurement bandwidth, thus of the excursion of $f$), while the scanning in position $nd$ induces azimuth compression:

$$S_t(nd) \cdot \exp\left(j\frac{2\pi ndf}{c}\right)$$

is the Fourier transform of $S_t$ with a phase that can be interpreted as translation. We find the two dual quantities which, through a bidimensional Fourier transform, allow us to deduce the distance $R_0$ and azimuth $\theta_0$ of the target, and therefore its position is $(R_0 \cos(\theta_0), R_0 \sin(\theta_0))$.

**Figure 4.17** Organization of the acquired data: each column represents a new position of the receiving antenna, while time elapses vertically within each column. Therefore, there are, of course, many more rows (typically hundreds of thousands of samples to take advantage of averaging, which cancels out the noise during correlation) than columns (a few tens of antenna positions along the rail).

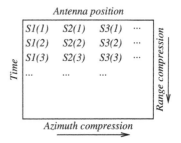

By scanning the position of the receiving antenna, we introduce a new variable that should result in azimuthal resolution ("right" or "left") in addition to the range resolution (Figure 4.17). Ideally [Forstén, 2019], in order to find a known condition, we would obtain an expression relating the measured quantities – delay at each frequency $f_q$ for each receiving antenna position $x_r$ – to the desired quantities, which are the position $(x_0, y_0)$ of each target. If we had a relationship of the form $\exp(j2\pi f_x x_0) \cdot \exp(j2\pi f_y y_0)$ with $f_x$ and $f_y$ as dual quantities to $x_0$ and $y_0$ related to $f_q$ and $x_r$, then we would be able to determine the position of each target through a bidimensional Fourier transform, where the solution is a Dirac delta function at $(x_0, y_0)$. The various transformations we will discuss aim to achieve such an expression in order to convert the complex matrix containing the measurements at the $f_q$ frequencies and $x_r$ positions into a matrix containing the power reflected by each target at position $(x_0, y_0)$ through a Fourier transform – in one direction for range compression and in the other direction for azimuth compression.

Let us assume there is a target at position $(x_0, y_0)$, which can be expressed in polar coordinates as $(R_0 \cos(\theta_0), R_0 \sin(\theta_0))$, with $R_0$ as the distance to the target and $\theta_0$ as its azimuth angle. The signal received by the antenna at position $x_r$ is of the form

$$s(q, r) \propto RCS \exp\left( j\frac{4\pi}{c} f_q \cdot \underbrace{R_r(r_0, \vartheta_0)}_{\text{target}} \right)$$

while noting that the phase induced is the product of the wave vector $k = 2\pi/\lambda = 2\pi f_q/c$ and the distance traveled $R_r = \sqrt{(x_r - r_0 \sin \vartheta_0)^2 + (r_0 \cos \vartheta_0)^2}$, where the target is located at $x_0 = r_0 \sin \vartheta_0$ and $y_0 = r_0 \cos \vartheta_0$.

When the receiving antenna moves in the x-coordinate, with its y-coordinate assumed to be fixed at an arbitrary origin $y = 0$, then the distance traveled between the RADAR and the target is approximately (using a first-order Taylor expansion approximation for small displacements) $R_0(x_r) \simeq R_0 - x_r \sin \theta_0$.

With this approximation, the received signal is of the form

$$s(q, r) \propto \exp\left( j\frac{4\pi}{c} f_q \cdot (r_0 - x_r \sin \vartheta_0) \right) \simeq \exp\left( j2\pi( \underbrace{2f_q \cdot r_0/c}_{\text{range } \alpha} - \underbrace{2x_r \sin \vartheta_0 / \lambda_c}_{\text{azimuth } \beta} ) \right)$$

assuming $f_q/c = 1/\lambda \simeq 1/\lambda_c$ is the wavelength at the center frequency of the emitted band, if the bandwidth is narrow enough to approximate $\frac{1}{x} = \sum_n (-1)^n \cdot (x-1)^n \simeq 1$ around $x \simeq 1$, keeping only the $n = 0$ term. Thus, the expression $\exp(j4\pi f_q(r_0 - x_r \sin \theta_0)/c)$ has successfully separated $r_0$ and $\theta_0$, with $r_0 = \alpha \times c/(2f_q)$ and $\sin \theta_0 = \beta/(2\lambda_c)$ in polar coordinates or $x_0 = r_0 \sin \vartheta_0 = \alpha \beta \cdot c \cdot$

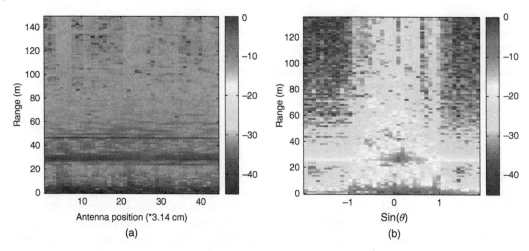

**Figure 4.18** (a) Raw measured signals resulting from correlation as a function of antenna position. (b) Result of azimuth compression by inverse Fourier transform along the antenna scanning direction (abscissa). Since correlation is the inverse Fourier transform of the product of temporal measurement Fourier transforms, this latter operation can also be considered as a bidimensional inverse Fourier transform.

$\lambda_c/4$ and $y_0 = r_0 \cos \vartheta_0 = c\alpha/2 \cos(\mathrm{asin}(\lambda_c \beta/2))$ in Cartesian coordinates. This allows to establish the dual quantities between Fourier space and real space, connecting $(f_q, x_r)$ to $(x_0, y_0)$ using the two-dimensional Fourier transform.

This expression of $\alpha$ and $\beta$ is important as it establishes the quantitative relationship between the measurement parameters $(f_q, x_r)$ and the position of each target $(x_0, y_0)$. There will be no degree of freedom to project the measured echoes at distance $r_0$ and azimuth $\vartheta_0$ onto a map to identify the nature of the reflectors.

This azimuth compression is illustrated in Figure 4.18, which provides a graph in the form of distance to the target on the $y$-axis and sine of the azimuth on the $x$-axis, which is incompatible with a projection on a map.

The conversion from these polar coordinates to Cartesian coordinates is achieved through the proposed mapping with the code

```
Nf = 100   % number of frequencies
Na = 73    % number of ant. pos,
c = 3e8    % speed of light
f = 2.45e9 % center frequency
df = 1e6;  % freq. sweep step
lambda = c/f
dx = lambda/4/2;

fs_r = 1/df;
r = (0:Nf-1)*fs_r/Nf*c/2;

fs_a = 1/dx;
alpha = (0:Na-1)*fs_a/Na-fs_a/2;
sin_thta=alpha*lambda/2;
```

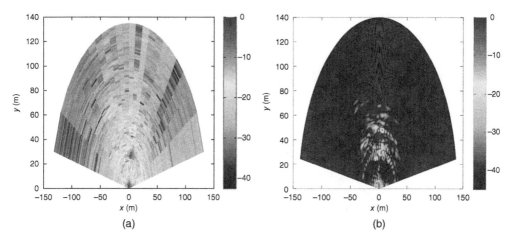

**Figure 4.19** Mapping conversion from {distance, azimuth} to Cartesian coordinates compatible with map projection. (a) Raw data obtained from the two-dimensional inverse Fourier transform for azimuth compression, and (b) A thresholding process to eliminate background noise and highlight the targets producing the strongest echoes.

```
[R,ST]=meshgrid(r, sin_thta(abs(sin_thta) <= 1));
X = R.*ST;Y = R.*sqrt(1-ST.^2);
Z=Img_focus(:, (abs(sin_thta) <= 1));
pcolor(X.', Y.', 10*log10(Z));
```

which concludes with Figure 4.19

The illustration in Figure 4.19 clearly suffers from energy leakage along the azimuth, an artifact of the inverse Fourier transform: as a reminder, the Fourier transform of a rectangle (point target) is a sinc function with lateral lobes that reduce the resolution after the transform.

## 4.13 Result Analysis

We have seen that the conversion between the measurement parameters $(\alpha, \beta)$ and the physical distances $(r_0, \vartheta_0)$ in polar coordinates or $(x_0, y_0)$ is bijective: there is no degree of freedom to adjust the position of the echoes by scaling. Only the orientation and position of the origin are manually positioned in QGIS. Figure 4.20 illustrates the result for the same set of data displayed at multiple scales.

Figure 4.21 compares the result on a larger scale, using a lower power threshold for display, between a measurement by noise RADAR and a RADAR realized by a WiFi transmitter. Once again, the comparison is quite convincing.

So far, the range resolution measurement is limited by the bandwidth, which is either limited by the antenna bandwidth, emitter bandwidth, or receiver bandwidth. However, we have only been using the magnitude of the cross-correlation. In the following section, we investigate the use of the phase for sub-wavelength range variation measurement yet with half-wavelength ambiguity.

**Figure 4.20** Overlay of aerial photographs from Google Maps with our resolved images in azimuth and distance of the reflectors visible by the RADAR located at the top of the cone. Note the excellent correspondence between the targets visible as intense echoes and the roof of the house at a 50 m range and its metal joint at the top, acting as a reflector (inset), or closer, the echoes between 20 and 30 m corresponding to the joints between the parking spaces and altitude breaks on the roof, also covered with metallic joints that act as radio frequency corner reflectors. Parked cars also act as reflectors due to their rounded shape, which always presents a reflective surface normal to the incident electromagnetic wave vector.

(a)           (b)

**Figure 4.21** (a) Distribution of reflectors on a Google Maps background for a dataset acquired by noise RADAR. (b) Distribution of reflectors on a Google Maps background for a dataset acquired by WiFi transmitter RADAR. The observation at a longer range – up to 150 m in this case – of targets illuminated by an incident wave of 1 mW (0 dBm) has reduced the threshold, which drowns nearby targets in clutter, even though the roofs of houses and parking garages can still be distinguished as the most powerful reflectors. Source: Google LLC.

## 4.14 Interferometric Measurement

We have seen that by spreading the spectrum of a signal over a bandwidth $B$, we can compress the pulse through cross-correlation between the transmitted and received signals to identify the range to the reflector with a resolution of $c/(2B)$, where $c = 300$ m/μs is the speed of light in a vacuum. By accumulating the spectra acquired over multiple adjacent bands (frequency stacking [Prager et al., 2019]), we can overcome the reception bandwidth limitation imposed by the reduced bandwidth of low-cost software-defined radios.

We have seen that by moving the receiving antenna along a straight axis with constant steps (ULA), a relatively simple formalism resembling an inverse Fourier transform allows to determine the azimuth of the target by considering that the phase of the received signal varies linearly with the target-antenna distance. The choice of a constant step simplifies the problem in the same way that a Fourier transform naturally provides a discrete and equidistant series of frequencies. The equivalent of spectral aliasing, but applied in space rather than time, implies that the step size must be smaller than half the wavelength of the transmitted signal to avoid ambiguity in the DoA.

We will now combine these two pieces of information to exploit the phase of the received signal, not just its magnitude as we have done so far, by observing this quantity at the most powerful reflectors during successive measurements. If the target moves less than $c/(2B)$, then the echo remains at the same location in the {distance, azimuth} map but the associated phase varies. This measurement technique is called interferometric SAR since, similar to optical interferometry in a Michelson or Mach–Zehnder setup, it is the phase between a reference beam (initial image in the case of RADAR) and a beam (successive images of the moving target in the case of RADAR) reaching a moving mirror that is exploited for a fine measurement – sub-wavelength – of displacement.[2]

In previous studies, we only analyzed the magnitude of the signal emitted at frequency $f$ and reflected by targets at a range $r$. However, the baseband correlation – after down conversion by I/Q mixer to eliminate $f$ – is a complex quantity characterized by a phase $\varphi$. This phase is related to the propagation time of the wave $\tau = 2r/c$ (2 for the round trip) by $\varphi = 2\pi f\tau$, and any change in the target's position $dr$ resulting in a change in time of flight $d\tau = 2dr/c$ between two successive measurements is observed as a phase change $d\varphi = 4\pi f \cdot dr/c$, assuming that the images are always taken from the same location. Two consequences can be noted from this expression:

- The displacement resolution is not dependent on $B$; only the ability to separate two targets at different ranges from the RADAR is determined by $B$ (range resolution). Therefore, we can detect fine movements even with a RADAR not scanning a wide range of frequencies.
- Any displacement of the target greater than $\lambda/2$ (phase shift of $\pi$ on the way there and $\pi$ on the way back, resulting in a total of $2\pi$) induces a phase rotation of $2\pi$ that cannot be differentiated from no displacement. Therefore, measurements must be frequent enough to ensure that the displacement between two measurements remains less than half a wavelength. In the case of illumination around 2.45 GHz, this distance is 6.1 cm. For a target at 60 m, this corresponds to a relative displacement of 0.1% or 1000 ppm.

This measurement technique has been widely used in geophysics for remote monitoring of landslides [Matsumoto et al., 2013; Takahashi et al., 2013], swelling of volcanoes under the pressure of the magma chamber [Malassingne et al., 2001], observation of the displacement of bridge decks constrained by the passage of heavy vehicles, or deformation of dams under the pressure of the water they retain [Monserrat et al., 2014]. This is GB-SAR, also used to observe the sliding of glaciers lubricated by the flow of meltwater [BBC, 2020; Dematteis et al., 2017]. While a few commercial

---

2 Care must be taken not to confuse InSAR (interferometric SAR) with iSAR (inverse SAR), which exploits micro-variations in Doppler shift to improve the resolution of a moving target or in general proposes to move the target instead of moving the RADAR antennas. An example of iSAR is an aircraft rotating in front of a RADAR: the wing furthest from the center of rotation moves faster than the wing closest to the RADAR, allowing the shape of the aircraft to be recovered once its overall motion has been subtracted from the measurement.

versions exist, such as the IBIS from IDS in Italy, MetaSensing in the Netherlands, or GroundProbe in Australia, we will extend the functionalities of our RADAR model to add this feature. Compared to airborne SAR, GB-SAR offers the advantage of a high refresh rate (a measurement takes less than a minute on a professional system) and simple processing if the antenna positions can be reproduced without correction needed to account for the variation in the RADAR's trajectory during successive acquisitions. We will address this point first.

## 4.15 Reproducible Positioning of the Receiving Antenna: Motorized Rail

The advantage of GB-SAR compared to its airborne or spaceborne version, which requires sub-wavelength resolution knowledge of the flight vector position to correct the trajectory of the ground wave (as a reminder, at 1 GHz, the wavelength is 30 cm), is that it is possible to place the rail that moves the antenna on a permanent support guaranteeing reproducibility of positioning. This is what we have done, somewhat clumsily, by equipping a rail (Igus, double drylin series WS 10 40 rail available in a 2 m length) with a threaded rod that is rotated by a stepper motor. Under the assumption of not losing any displacement steps of the rotor, i.e. by performing rotations slow enough, each forward or backward movement (except for the backlash in the threaded rod) repositions the antenna to better than a millimeter, well below the wavelength at 2.45 GHz (Figure 4.22).

**Figure 4.22** Experimental setup with the fixed transmission antenna on the balcony floor (bottom of the photograph), the cart-mounted receiving antenna on the motorized rail fixed to the balcony railing, with the emission by the PlutoSDR configured as a pseudo-random phase generator and the Ettus Research B210 receiver acquiring both the emitted reference signal and the signal reflected by the targets. The motor is controlled by an X-NUCLEO-IHM14A1 interface driven by a dedicated microcontroller receiving its commands through a virtual serial port over USB, which is visible in the inset on the right.

Our initial goal to bring GNU Radio to the Raspberry Pi4 was to provide a stand-alone system capable of USB 3 data acquisition from the B210, generating signals with the PlutoSDR transmitter or USB-WiFi converter, and having the necessary GPIOs for motor control, a functionality absent from personal computers since the disappearance of the Centronics-compatible parallel port. Ultimately, we opted for a dedicated microcontroller for stepper motor control with an ST Microelectronics X-NUCLEO-IHM14A1 interface board.

> **Image positioning in QGIS**
>
> As mentioned before, there are no degrees of freedom to position RADAR maps on an aerial image in a geographical information system like QGIS, since the source-target distance is determined by the bandwidth and the azimuth by the rail displacement step and orientation of the antenna (in our case, the balcony railing). The following figure illustrates the use of the "Freehand raster georeferencer" plugin (icons surrounded), which allows importing a non-georeferenced bitmap image and interactively manipulating its position, scale, and rotation (more convenient than manipulating the associated World file of a georeferenced bitmap image until an acceptable solution is achieved). In the following screenshot,
>
>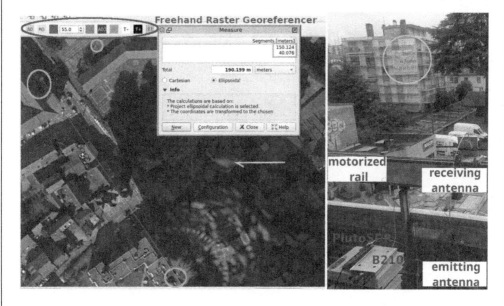
>
> we can see that the ruler is used along each axis to verify that the image scale is correct (150 m scale in the abscissa and 40 m scale in the ordinate ), while the image has already been aligned with the balcony railing. The arrow indicates the reflection of a building surrounded by scaffolding during the repainting of its walls, improving its RADAR cross section.

## 4.16 The Radio Frequency Corner Reflector

In order to demonstrate the ability to finely measure the displacement of a target, we need a controllably mobile target in a static environment (capturing an image requires one hour of acquisition

**Figure 4.23** (a) Photograph of the corner reflector (square trihedron) with a 30 cm edge after assembly (inset) and installation about 40 m from the RADAR. Notice the presence of vehicles that require a sufficiently large corner reflector RADAR cross section for the latter to be visible. (b) Principle of the corner reflector that reflects the incident wave back to its source, and the impact of a 5° error compared to the ideal 90° angle between the two faces. (c) The case of illumination at a different angle from the axis of symmetry, again verifying the proper functioning of the corner reflector and the impact of the 5° error during its assembly.

with our setup). Moreover, this target must have a RADAR cross section comparable to rooftops or vehicles that reflect electromagnetic waves and must be easily movable in steps of 1 cm (a displacement step chosen arbitrarily to be under the half wavelength of 6 cm). Various RADAR targets have reflection capabilities far superior to their physical surface (e.g. Lüneburg sphere, which refocuses the incident signal toward the illumination source), but the simplest one to produce is the corner reflector.

Everyone knows about catadioptric reflectors (Figure 4.23) – mandatory equipment for bicycle visibility – which are installed on bikes and reflect the light back toward the source. The principle is exactly the same in radio frequency [Robertson, 1947], except that the wavelength has changed from a few hundred nanometers to a few centimeters, removing the constraints on the surface condition of the mirror faces that form a half cube. A catadioptric reflector formed by three square surfaces of edge length $d$ assembled at right angles has a RADAR cross section of $\sigma = 12\pi \frac{d^4}{\lambda^2}$, which decreases to $\sigma = \frac{4}{3}\pi \frac{d^4}{\lambda^2}$ if half of each square is removed to form a triangle. For a working frequency of 2450 MHz or a wavelength $\lambda = 12.25$ cm, the RADAR cross section of a 30-cm-edge-length catadioptric reflector is 20 m$^2$, which would require a sphere with a radius of 2.5 m (projected surface of 20 m$^2$) to achieve the same result. Our catadioptric reflector therefore has a RADAR surface that is approximately equivalent to that of a car but much simpler to manipulate. The catadioptric reflector is not recommended for multi-static RADARs since by principle it reflects the incident signal in the direction of its source (unlike a sphere which reflects in all directions), but our RADAR is close enough to the monostatic condition to effectively observe the catadioptric reflector located a few tens of meters from the balcony.

The constraint on the angle between the faces of the cube is, however, drastic [Garthwaite, 2017], and having failed to form an effective catadioptric reflector by assembling FR4 printed circuit boards (too flexible and deforming during soldering), it was necessary to call upon the skill of a welder to assemble three rigid steel plates with an angle close enough to a right angle to function properly. In fact, Garthwaite et al. [2015, p. 27] indicate that even a 5° deviation from the ideal angle of 90° between the faces of the catadioptric reflector results in a loss of its RADAR cross section of 2.2 dB or a reduction of a factor of 2. The effective surface area of the corner reflector is $A_{eff} = d^2 \times \sqrt{3}$ for square faces or $A_{eff} = d^2/\sqrt{3}$ for triangular faces. With an error of 1° in the angle between the faces of the cube, the loss of the catadioptric reflector's RADAR cross section becomes negligible at less than 0.1 dB.

## 4.17 Fine Displacement Measurement

A RADAR image is mostly composed of uninteresting areas with no reflectors. A first filtering step to detect sub-wavelength displacements of targets between two measurements is therefore to eliminate areas whose reflected power is below a certain threshold.

The identification of pixels containing the information associated with the reflection of the catadioptric reflector was obtained by subtracting an image acquired in the presence of the catadioptric reflector from an image acquired in the absence of the reflector. The subtraction of the two amplitude maps provides the location of the catadioptric reflector (Figure 4.24). The analysis of the phase of the signal at this location is presented in Figure 4.25, which shows the nominal displacement of the catadioptric reflector on the $x$-axis – centimeter by centimeter – and the phase converted to displacement on the $y$-axis as $dr = d\varphi \cdot c/(4\pi f) = \lambda \times d\varphi/(4\pi)$. We intentionally included the case of a phase rotation of $2\pi$ which is unwrapped to obtain a set of curves corresponding to the various points in the {distance, azimuth} map containing the corner reflector information. The same

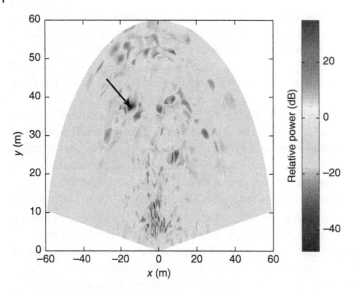

**Figure 4.24** Difference in the modules of the {distance, azimuth} maps acquired in the presence and absence of a reflector. The pixels identified as containing the signal reflected by the corner reflector are indicated by the arrow: the phase of these pixels is analyzed in Figure 4.25.

analysis is performed on the echo from the roof at 47.5 m (light gray arrow in Figure 4.25 bottom right) – assumed to be static relative to the balcony – of the house facing the RADAR. This reference measurement indicates the noise on the measurement, which we estimate to be 4 mm over the duration of the acquisitions.

We now need to evaluate the source of this phase measurement uncertainty. A first aspect concerns the impact of electronics and, in particular, fluctuations in the phase of the local oscillator (Figure 4.26) between the time it takes for the wave to travel back and forth between the transmitter-target-receiver and the integration time during correlations. Considering that we only measure targets up to a 100 m and that we acquire at a rate of 2.7 Msamples/s (maximum data transfer rate without loss on two channels), a number of samples, arbitrarily selected at around 135 000, are used to calculate each correlation. The time range during which the phase of the local oscillator fluctuates is $1/2.7 = 370$ ns to $135\,000/2.7 \times 10^6 = 50$ ms or, in other terms, a frequency offset from the carrier of 2.7 MHz to 20 Hz. Assuming that the local oscillator impacts the receiver as the transmitter, the digital integration of the curve acquired at 2450 MHz in Figure 4.26, which represents a measurement of phase noise $S_\varphi = \sigma_\varphi^2/B$, where $\sigma_\varphi$ is the phase fluctuation in each frequency band $B$ away from the carrier, indicates that the phase fluctuates by $\sigma_\varphi \simeq 1.3°$, contributing to approximately $\delta r = \frac{c}{f} \times \frac{\sigma_\varphi}{720} \simeq 0.1$ mm. This value does not match the fluctuation observed over the course of a day, and we must also consider the impact of weather conditions that cause variations in the speed of the electromagnetic wave in the air.

## 4.18 Impact of the Atmosphere

We mentioned that at 2.45 GHz, a $2\pi$ phase rotation is observed every 6 cm, which corresponds to a relative displacement of 1000 ppm at 60 m. However, the electromagnetic wave propagates through the air, which is not a vacuum but a gas whose relative permittivity depends mainly on

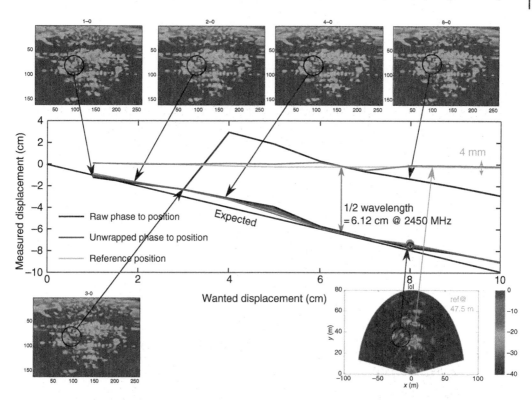

**Figure 4.25** Evolution of the phase of the observed signal at the location identified as representative of the corner reflector echo, as this target is moved in 1 cm steps (abscissa). The phase is converted to displacement (ordinate) knowing the central wavelength of the transmitted signal. The raw images {distance, azimuthal angle} with phase encoded in color – saturation eliminating terms if the reflected power is less than −40 dB – eliminate background noise and highlight relevant targets, including the static roof. The position of this reference signal is also displayed to estimate the measurement uncertainty over a day, assuming the house is static during this period.

the temperature $T$ (absolute temperature in Kelvin), air pressure $p$ (in millibars), and water vapor partial pressure $p_e$ (also in millibars). It is known [Luzi et al., 2004] that weather conditions impact the estimation of distances to targets by varying the propagation velocity of the electromagnetic wave. We rely on Smith and Weintraub [1953] for our analysis, which, although not the most up-to-date documentation on the subject, is clear and applicable. Thus, we learn that at the nominal atmospheric pressure of 1013 hPa = 1013 mbar, the impact of temperature on the refractive index of air is $dK_1 = 77.6 \times p/T = 77.6 \times 1013/273 = 287$ ppm, and the variation of this coefficient with temperature is $dK_1/dT = 287/T \simeq 1$ ppm/K around the ambient temperature.

The impact of meteorological conditions is widely discussed in the literature, with our estimations of a few hundred ppm in absolute terms and a few ppm daily variations compared to the speed of light in a vacuum in agreement with published data [Fabry, 2004]. Thus, placing a corner reflector on a known static support – for example, a rocky base near a landslide – allows to evaluate these variations in the velocity of the electromagnetic wave and compensate for them in post-processing [Luzi et al., 2004]. The 4 mm variation in position of the roof of the house at a distance of 47.5 m corresponds to 84 ppm. Even though not all of these fluctuations can be exclusively attributed to meteorological variations (e.g. movement of the rail during successive antenna sweeps), their contribution cannot be neglected.

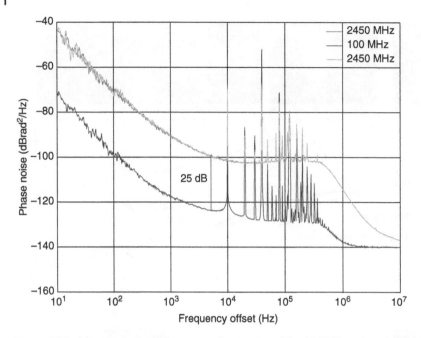

**Figure 4.26** Measurements of the output phase noise of the AD9361 equipped with an Ettus Research B210 software-defined radio platform at 100 (bottom) and 2450 MHz (top). The two curves are well separated by $20\log_{10}(2450/100)$ as expected for phase noise that increases with the square of the frequency.

## 4.19 Time of Flight Measurement with Sub-sampling Period Resolution and the Use of a Surface Acoustic Wave Cooperative Target for Reproducible Range Simulation

Spectrum spreading using a pseudo-random sequence, as used in the noise RADAR, provides fascinating signal processing perspectives, including measuring time of flight with sub-sampling period resolution. Throughout this discussion, we estimate the cross-correlation $xcorr(p, s)(\tau)$ with $\tau$ being a delay defined as an integer delay between $p$ and $s$. We have then seen how sub-half wavelength displacements can be measured by using the phase, with the drawback that the measurement is performed modulo half the wavelength since the $2\pi$ phase rotation induced by the two-way trip of half a wavelength displacement wraps to 0.

However, the correlation peak shape is not used in this analysis; only its maximum is searched for and identified with sampling period resolution. The correlation peak can be fitted, e.g. by a parabolic shape between the maximum and its two neighbors, and the position of the fitted parabola provides a fine estimate of the time delay, assuming the correlation peak is unique, as is the case of a well-defined target illuminated by a pseudo-random sequence. This approach was used in timing two-way satellite time and frequency transfer signals emitted from ground toward a geostationary satellite, which in turns broadcasts back toward the ground a frequency-transposed copy of the incoming signal, as described at https://github.com/oscimp/amaranth_twstft. The parabolic peak-fitting, custom-processing block, since not found as a native GNU Radio block, is described to demonstrate how to write a Python block in Chapter 9.

Since the correlation is computed as a product in the Fourier domain, a natural interpolation method is zero-padding (adding vectors of zeros in the high-frequency domain range of the

Fourier transform) in the Fourier domain, after the multiplication but prior to calculating the inverse Fourier transform. By creating more samples in the frequency domain, the same number of samples is generated in the time domain after inverse Fourier transform, acting as interpolation. The zero-padding function is also not available as a native block in GNU Radio, so we introduce here the custom Python block achieving this result:

```
"""
Embedded Python Blocks: zero padding
see https://wiki.gnuradio.org/index.php?title=Python_Block_with_Vectors
"""

import numpy as np
from gnuradio import gr

class blk(gr.sync_block):  # other base classes are basic_block, decim_block
    , interp_block

    def __init__(self, vectorSize=16383, M=2):  # only default arguments
    here
        gr.sync_block.__init__(
            self,
            name='Zero Padding',
            in_sig=[(np.complex64,vectorSize)],
            out_sig=[(np.complex64,vectorSize*(2*M+1))]
        )
        self.vectorSize = vectorSize
        self.M = M

    def work(self, input_items, output_items):
        for vectorIndex in range(len(input_items[0])):
            z=np.zeros(self.vectorSize*self.M, dtype=np.complex64)
            output_items[0][vectorIndex][:] = np.concatenate( (z,
    input_items[0][vectorIndex], z) , axis=0)
        return len(output_items[0])
```

where the __init__ is the class constructor with two arguments, the incoming vector size consistent with the output of the FFT block, and the padding factor **on both ends** so that the output size is $(2 \cdot M + 1) \times vectorSize$. The work function, which is executed with each new batch of samples, concatenates a vector of 0s at the beginning and end of the incoming dataset. This operations **assumes** that the FFT was computed with the zero frequency at the center, the first sample representing minus the Nyquist frequency (minus half the sampling frequency bin) and the last sample representing the Nyquist frequency (half the sampling frequency bin). Notice that this convention **requires activating the shift flag** of the FFT block, since otherwise (as would be the case in Octave or Numpy) the first sample is the DC component (0 Hz) bin and the last sample is the sampling frequency bin. These two representations hold the same information through the discrete Fourier transform periodicity, but the right convention must be used when zero padding. Under Octave, shifting from one convention to the other is achieved with the fftshift() function.

In order to demonstrate the sub-sampling period resolution, we wish to use a reproducible, tunable source of delay. Introducing a sub-sampling delay in the digital domain is not trivial, as it requires a fast sampler, fine delay introduction, and a fast synthesizer of the delayed signal. In the analog domain, a delay is introduced with a length of cable. However, introducing a 1 μs delay would require a 200-m-long RG-58 coaxial cable, and tuning the delay would require adding or removing cable length. An alternative solution is to convert the electromagnetic wave into an acoustic wave, which is propagating $10^5$ times slower, so that the 200-m-long cable becomes a 2-mm-long acoustic delay line. The conversion of an electromagnetic wave to an acoustic wave is best achieved using piezoelectric substrates, and the field of surface acoustic wave (SAW) [Hashimoto, 2000; Morgan, 2010] generation is based on patterning interdigitated electrodes with the appropriate period on a piezoelectric wafer: the inverse piezoelectric effect converts the electromagnetic wave to an acoustic wave, which is confined to the surface of the substate when using the right cut and polarization (so-called Rayleigh wave, from the similarly polarized earthquake wave), and the direct piezoelectric effect converts this acoustic wave back to an electromagnetic wave when reaching a similar interdigitated electrode pattern. From an electronics engineer's perspective, the SAW delay line is an electrical component, but for the physicist, it is a fascinating tool for generating long delays in a compact format.

SAW delay lines can be commercially obtained in the form of transmission filters, and many frequencies and bandwidths can be found from manufacturers such as Taisaw (sold in Europe by Golledge), EPCOS/TDK (e.g. the B3607 filter centered on 140 MHz exhibiting a 1.35 μs nominal group delay), or Murata. However, these filters are all designed to exhibit minimal sensitivity to the environment, including temperature and stress, whereas we wish to introduce a tunable delay. For this demonstration, a dedicated transmission delay line was designed at the time and frequency department of the FEMTO-ST Institute in Besançon (France) by W. Daniau and manufactured in its cleanroom facility. The YXl/128° lithium niobate substrate exhibits a strong, 70 ppm/K [Soluch and Lysakowska, 2005], temperature sensitivity, making the device ideally suited for tuning the delay with nanosecond resolution using temperature. Since the Rayleigh wave propagates at 3980 m/s on this substrate, the 2.13-cm-long delay line exhibits a 5.33 μs delay, varying by 373 ps/K or 37.3 ns for a 100 K temperature variation [Friedt et al., 2024]. The same demonstration can be reproduced by the reader using one of the commercial references cited earlier [Friedt, 2016].

The flowgraph for implementing these principles is shown in Figure 4.27. In this flowchart, a pseudo-random sequence of length $2^N - 1$ is generated by a Galois Linear Feedback Shift Register (GLFSR) of length $N$, interpolated and emitted by one of the channels of the B210 transceivers. Because both transmitters must be active, the second one emits no signal on the same carrier frequency (both emitter and receiver channels are controlled by the same local oscillator) and is loaded on a matched 50Ω resistor to avoid radiating unwanted signals. Since the Analog Devices AD936x does not use the same local oscillator for emitting and receiving, a coupler at the output of the emitter sends a fraction of the emitted signal into the receiver, and the signal propagating through the delay line is recorded by the second B210 input. Since both inputs are subject to the same local oscillator offset of the receiver with respect to the emitter, their cross-correlation will cancel the frequency offset and exhibit a fine correlation peak. This correlation peak is zero-padded using the custom block. However, the Python block is inefficient, using an interpreted language, and the output data rate would be far too fast when monitoring temperature so that a `Keep 1 in N` block throws away 29 out of 30 measurements, allowing the interpolating block not to run out of computational power. Once zero-padded, the resulting vector is inverse-Fourier-transformed and the correlation peak fitted. Since this demonstration runs on the headless Raspberry Pi 5, all graphical interfaces were removed and a `No GUI` flowchart is generated. The resulting file is read and the measurements plotted using Octave with

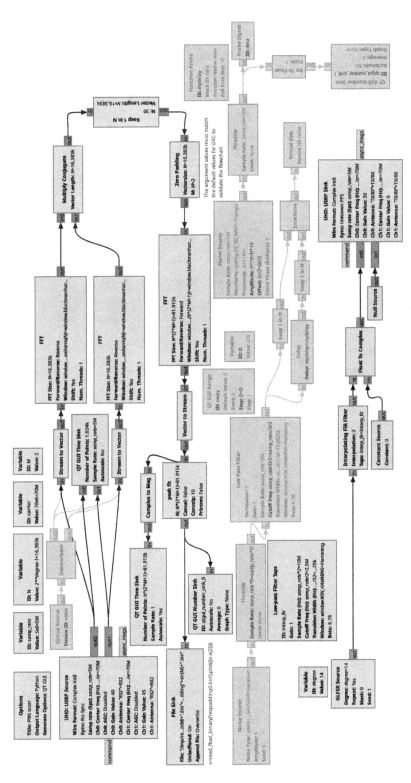

**Figure 4.27** Flowchart for spectrum spreading a carrier wave using a Galois linear feedback shift register (GLFSR) pseudo-random generator (bottom left) to drive the output of an Ettus Research B210 emitter. Since both channels of the emitter must be active, the second is configured not to output any signal (`Null Source`). On the receiver side (top left), the signal is sampled during this time frame), and the resulting vector is decimated to reduce computational load and output data rate, before oversampling by zero-padding in another custom `Python Block` and computing the inverse Fourier transform. The resulting cross-correlation is fitted by a parabolic function in another custom `Python Block` before being saved in a file for post-processing. When running on the headless Raspberry Pi 5 single-board computer, the graphical outputs are commented out and the flowgraph is configured as `No GUI`. The commented blocks (gray) are for simulating using synthetic signal (`Virtual Sink` to the right feeding a `Virtual Source` to the top left) using an oversampled noise generator, introducing a fine delay driven by the `dela` probe signal varying as a triangle wave, and decimating. The low-pass filter spreads energy among adjacent samples for the correlation to accumulate energy even though the decimation only keeps one in 200 initial samples for fine delay tuning.

```
fs=5e6/16383/30; % samp_rate/code_length/decimation_factor
x=read_float_binary('respadding2.bin');plot([0:length(x)-1]/fs,x);
```

where `read_float_binary()` is provided in the `gr-util/octave` directory of the GNU Radio source code. The time axis is given by the length of the correlation (here the length of the pseudo-random sequence, selected as a trade-off between improving the signal-to-noise ratio, thanks to the coherent accumulation of energy in the integral, and computational power needs) multiplied by the decimation factor between the FFT product and the iFFT, divided by the sampling rate. Figure 4.28 illustrates the experimental setup and a measurement example in which the oversampling and parabolic fit allow for identifying the delay with a resolution improvement over the sampling period equal to the signal-to-noise ratio. It is quite fascinating to realize that even sampling as slow as 5 Msamples/s (200 ns period) allows for measuring delays with picosecond resolution [Friedt and Goavec-Merou, 2023].

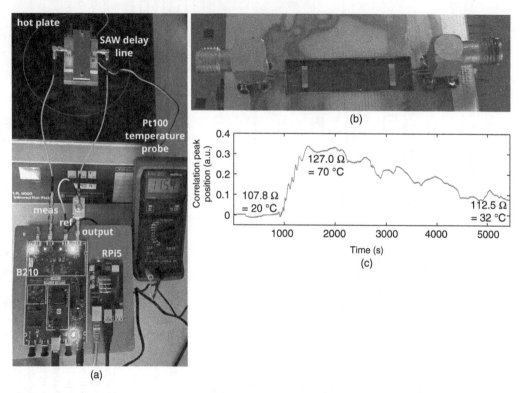

**Figure 4.28** The temperature variation during this experiment was 50 K, so that a temperature sensitivity of 70 ppm/K on a 5.33-µs-long delay line induces a delay variation of 18.7 ns or about half the sampling period of the 5 Msamples/s (200 ns period) oversampled by a factor of 5 (40 ns/period), consistent with the observed delay variation considering the distance between the actual SAW delay line and the Pt100 temperature probe. (a) Experimental setup, with a SAW delay line located on a hot plate for tuning its temperature and hence the delay, since the acoustic velocity of the lithium niobate substrate is highly temperature-sensitive for the selected crystalline cut. The B210 (bottom) emits the pseudo-random sequence generated as a 14-bit-long Galois Linear Feedback Shift Register (GLFSR). A fraction of the emitted signal is coupled to a reference input channel ("ref."), while the second input channels ("meas.") measure the same signal delayed by the SAW delay line. (b) Picture of the 2.13-cm-long transmission SAW delay line before packaging (manufactured by W. Daniau, FEMTO-ST, Besançon, France). (c) Cross-correlation maximum position after oversampling and parabolic fit as the SAW delay line is heated by about 50 K and returns to room temperature.

Notice that this use of environmental sensitive SAW transducers as passive wireless sensors has been the subject of intensive research by one of the authors [Friedt et al., 2010; Friedt, 2017] but never caught up when facing the silicon-based radio frequency identifier (RFID) chips despite their poorer technical performance.

## 4.20 Conclusion

This chapter was devoted to the use of SDR for measuring time of flight to static targets, demonstrating various implementations of covert RADAR systems using WiFi transmitters as noncooperative radio frequency sources. Starting from a range measurement using cross-correlation between the recorded broadcast waveform and the time-delayed copies reflected by distant static targets, we have added azimuth (DoA) resolution by moving the receiver in a bistatic configuration and finally have considered the use of corner reflectors as cooperative targets for fine distance measurement using interferometric techniques.

The final setup, detailed at https://github.com/jmfriedt/SDR-GB-SAR, achieved with dedicated hardware, is shown in Figure 4.29 [Friedt, 2024]. Since azimuth compression requires reproducible location with sub-wavelength accuracy, a dedicated 2-m-long rail system was acquired, allowing reproducible positioning of the transmitter and receiver with ±0.08 mm (or ±0.5° phase variation at 5.5 GHz where the wavelength is 54 mm) repeatability and a long enough path for high-resolution DoA measurement. The 5.5 GHz WiFi emitter allows for covert and dual use (RADAR and communication) of the broadcast system, while keeping the high-gain directional horn antennas small and light to be moved by the rail system. The Raspberry Pi 4 single-board computer and B210 dual-channel coherent receivers are both fitted behind the antennas on the payload moving along the rail to avoid phase fluctuations that would result from bending the radio frequency cables. In this setup, only the Ethernet cable carrying digital data is moved as the RADAR system scans the rail length. Finally, with respect to the main text where the emitter was static and the receiver moving, leading to a displacement step of half wavelength to meet the spatial equivalent of the sampling

(a)

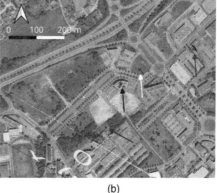
(b)

**Figure 4.29** (a) Final experiment setup assembled using dedicated hardware. The rail is controlled by the Raspberry Pi 4 located on the moving payload on the rail behind the antennas, as is the B210 dual-coherent channel receiver. Inset: The 24 V rail control circuit adapter for the Raspberry Pi 4 using an open collector Darlington transistor ULN2803, and on the background, the B210 in its shielding aluminum case.
(b) Example of targets mapped with this system. The light arrow points to a building not yet visible on the aerial map.

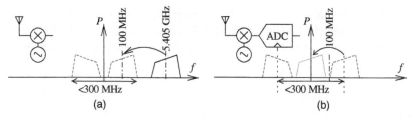

**Figure 4.30** Signal processing scheme used in Sentinel-1 spaceborne RADAR to compensate for IQ imbalance calibration issues: first, the signal (a, black) is transposed from the microwave C-band to an intermediate frequency high enough to include all relevant spectral components of the frequency-shifted real signal using a single mixer (a, dashed). This signal is sampled at high enough a sampling rate so that both positive and negative frequency components are included in the record, and the digital complex frequency transposition brings one of the intermediate frequency images to baseband while the other is filtered out with a low-pass filter (b, dotted). Since the digital frequency transposition is not subject to IQ imbalance (different amplitude of the sine and cosine components or frequency offset slightly different from 90° of the cosine and sine), no additional correction is needed when processing the published IQ records.

theorem, here moving both the emitter and receiver requires moving with quarter-wavelength step to avoid angle of arrival aliasing.

It is worth mentioning in this conclusion that all these techniques are applicable to any IQ stream resulting from SDR measurements for RADAR measurements, including the datasets collected by the spaceborne Sentinel-1 satellites and provided as raw datasets by the European Space Agency. Indeed, Sentinel-1 was designed as an SDR receiver for getting rid of the classical IQ imbalance issue where the I and Q channels do not exhibit the same gain and orthogonality conditions (90° phase shift between cosine and sine). As stated in Piantanida [2019], "For Sentinel-1 however, the instrument's receive module performs the demodulation in the digital domain, therefore the I/Q gain imbalance and I/Q non-orthogonality corrections are no longer necessary." Sentinel-1 is broadcasting a 100-MHz wide chirped waveform centered on the 5.405 GHz C-band (within the carrier range of the Ettus Research B210 SDR receiver): this signal is frequency-shifted using a real mixer at an intermediate frequency sampled at 300 MS/s, and the resulting record is frequency transposed by the intermediate frequency (IF) to reach baseband using a software multiplication insensitive to IQ imbalance (Figure 4.30). This knowledge of the Sentinel-1 low-level signal structure allows for decoding the raw IQ streams as described at https://github.com/jmfriedt/sentinel1_level0 or for passive RADAR applications using ground-based data collection of the direct reference signal and the backscattered measurement signal [Feng et al., 2022].

## References

BBC. Mont Blanc: Glacier collapse risk forces Italy Alps evacuation, 2020.

R. Bourret. A proposed technique for the improvement of range determination with noise radar. *Proceedings of the Institute of Radio Engineers*, 45(12):1744–1744, 1957.

M. Braun. *OFDM radar algorithms in mobile communication networks*, 2015. publikationen.bibliothek.kit.edu/1000038892/2987095.

N. Dematteis, G. Luzi, D. Giordan, F. Zucca, and P. Allasia. Monitoring alpine glacier surface deformations with GB-SAR. *Remote Sensing Letters*, 8(10):947–956, 2017.

F. Fabry. Meteorological value of ground target measurements by radar. *Journal of Atmospheric and Oceanic Technology*, 21(4):560–573, 2004.

W. Feng, J.-M. Friedt, G. Goavec-Merou, and F. Meyer. Software-defined radio implemented GPS spoofing and its computationally efficient detection and suppression. *IEEE Aerospace and Electronic Systems Magazine*, 36(3):36–52, 2021.

W. Feng, J.-M. Friedt, and P. Wan. SDR-implemented passive bistatic SAR system using Sentinel-1 signal and its experiment results. *Remote Sensing*, 14(1):221, 2022.

H. Forstén. Synthetic-aperture radar imaging, 2019. https://hforsten.com/synthetic-aperture-radar-imaging.html.

J.-M. Friedt. Wideband measurement strategies: From RADAR to passive wireless sensors… and how passive wireless sensors were/are used by intelligence agencies. In *FOSDEM*, Brussels, Belgium, 2016. https://archive.fosdem.org/2016/schedule/event/radar/.

J.-M. Friedt. Passive cooperative targets for subsurface physical and chemical measurements: A systems perspective. *IEEE Geoscience and Remote Sensing Letters*, 14(6):821–825, 2017.

J.-M. Friedt. Covert ground based synthetic aperture RADAR using a WiFi emitter and SDR receiver. In *FOSDEM*, Brussels, Belgium, 2024. https://fosdem.org/2024/schedule/event/fosdem-2024-2050-covert-ground-based-synthetic-aperture-radar-using-a-wifi-emitter-and-sdr-receiver/.

J.-M. Friedt and G. Goavec-Merou. Time of flight measurement with sub-sampling period resolution using software defined radio. In *13th GNU Radio Conference (GRCon)*, volume 8, Tempe, AZ, USA, 2023. https://pubs.gnuradio.org/index.php/grcon/article/view/142/120.

J.-M. Friedt, C. Droit, G. Martin, and S. Ballandras. A wireless interrogation system exploiting narrowband acoustic resonator for remote physical quantity measurement. *Review of Scientific Instruments*, 81(1):014701 2010.

J.-M. Friedt, B. Chupin, M. Lours, É. Meyer, O. Chiu, F. Meyer, W. Daniau, and J. Achkar. Results of a software defined radio (SDR) implementation of Two Way Satellite Time and Frequency Transfer (TWSTFT) emitter and receiver system. In *European Frequency and Time Forum (EFTF)*, Neuchatel, Switzerland, 2024.

M. C. Garthwaite. On the design of radar corner reflectors for deformation monitoring in multi-frequency InSAR. *Remote Sensing*, 9(7):648, 2017.

M. C. Garthwaite, S. Nancarrow, A. Hislop, A. Thankappan, J. H. Dawson, and S. Lawrie. The design of radar corner reflectors for the Australian geophysical observing system, 2015.

R. Gold. Optimal binary sequences for spread spectrum multiplexing. *IEEE Transactions on Information Theory*, 13(4):619–621, 1967.

K.-Y. Hashimoto. *Surface Acoustic Wave Devices in Telecommunications*, volume 116. Springer, 2000.

B. M. Horton. Noise-modulated distance measuring systems. *Proceedings of the IRE*, 47(5):821–828, 1959.

K. A. Lukin and R. M. Narayanan. Historical overview and current research on noise radar. In *2011 3rd International Asia-Pacific Conference on Synthetic Aperture Radar (APSAR)*, pages 1–2. IEEE, 2011.

G. Luzi, M. Pieraccini, D. Mecatti, L. Noferini, G. Guidi, F. Moia, and C. Atzeni. Ground-based radar interferometry for landslides monitoring: Atmospheric and instrumental decorrelation sources on experimental data. *IEEE Transactions on Geoscience and Remote Sensing*, 42(11):2454–2466, 2004.

C. Malassingne, F. Lemaître, P. Briole, and O. Pascal. Potential of ground based radar for the monitoring of deformation of volcanoes. *Geophysical Research Letters*, 28(5):851–854, 2001.

M. Matsumoto, K. Takahashi, and M. Sato. Long-term landslide monitoring by GB-SAR interferometry in Kurihara, Japan. In *Conference Proceedings of 2013 Asia-Pacific Conference on Synthetic Aperture Radar (APSAR)*, pages 529–532. IEEE, 2013.

O. Monserrat, M. Crosetto, and G. Luzi. A review of ground-based SAR interferometry for deformation measurement. *ISPRS Journal of Photogrammetry and Remote Sensing*, 93:40–48, 2014.

D. Morgan. *Surface Acoustic Wave Filters: With Applications to Electronic Communications and Signal Processing*. Academic Press, 2010.

M. Müller. 2020. https://lists.gnu.org/archive/html/discuss-gnuradio/2020-06/msg00125.html.

R. Piantanida. Sentinel-1 level 1 detailed algorithm definition. Technical report, European Space Agency, 2019.

S. Prager, T. Thrivikraman, M. S. Haynes, J. Stang, D. Hawkins, and M. Moghaddam. Ultrawideband synthesis for high-range-resolution software-defined radar. *IEEE Transactions on Instrumentation and Measurement*, 69(6):3789–3803, 2019.

S. D. Robertson. Targets for microwave radar navigation. *Bell System Technical Journal*, 26(4):852–869, 1947.

G. N. Saddik, R. S. Singh, and E. R. Brown. Ultra-wideband multifunctional communications/radar system. *IEEE Transactions on Microwave Theory and Techniques*, 55(7):1431–1437, 2007.

S. Scholl. DC9ST, Introduction and experiments on transmitter localization with TDOA. In *Software Defined Radio Academy*, 2017. https://www.youtube.com/watch?v=Km4TU17b05s and http://www.panoradio-sdr.de/tdoa-transmitter-localization-with-rtl-sdrs/.

E. K. Smith and S. Weintraub. The constants in the equation for atmospheric refractive index at radio frequencies. *Proceedings of the IRE*, 41(8):1035–1037, 1953. https://nvlpubs.nist.gov/nistpubs/jres/50/jresv50n1p39_A1b.pdf.

V. Soluch and M. Lysakowska. Surface acoustic waves on X-cut LiNbO/sub_3. *IEEE Transactions on Ultrasonics, Ferroelectrics, and Frequency Control*, 52(1):145–147, 2005.

H. Sun, Y. Lu, and G. Liu. Ultra-wideband technology and random signal radar: An ideal combination. *IEEE Aerospace and Electronic Systems Magazine*, 18(11):3–7, 2003.

K. Takahashi, M. Matsumoto, and M. Sato. Continuous observation of natural-disaster-affected areas using ground-based SAR interferometry. *IEEE Journal of Selected Topics in Applied Earth Observations and Remote Sensing*, 6(3):1286–1294, 2013.

# 5

## Digital Communications in Action: Design and Realization of a QPSK Modem

### 5.1 Digital Communication Concepts

#### 5.1.1 What Is Digital Information?

Defining information for a digital transmission system is crucial for several reasons, but let us focus on some key ones (Figure 5.1). When designing a communication system, you are typically allocated a power and spectrum occupancy budget, so it is essential not to squander it by transmitting unnecessary data – data that carry no useful information. What constitutes unnecessary data depends on the nature of the information source, denoted as $X$. For instance, when transmitting voice, it has been demonstrated that there is a significant amount of redundancy that can be eliminated. This redundancy is a characteristic of human language, facilitating comprehension by the majority of listeners. Similarly, in the case of video, it is obvious that when there is no motion, the images exhibit minimal changes. These examples underscore the possibility of extracting essential information from a data source, known as **entropy**, which can be quantified for any digital source. However, let us revisit the concept of information itself. According to Shannon [1948], the information produced by a digital source $X$ when the $j$th message is transmitted is given by

$$I_j = -\log_2\left(p_j\right) \tag{5.1}$$

where $p_j$ is the probability of occurrence of the $j$th message. $\log_2$ accounts for the fact that we use binary data. The other key concept is the probability of occurrence. The lower the probability, the higher the information. Therefore, at the reception side, the more surprise, the more information! Indeed, when you cannot predict what you will receive, you get a lot of information. If you take into account all the $M$ messages from the source $X$, you get its entropy expressed in **bits** (binary units) and defined by

$$H(X) = -\sum_{j=1}^{M} p_j \cdot \log_2\left(p_j\right) \tag{5.2}$$

The $M$ messages represent the different brightness levels of a pixel in the case of a black-and-white digital image. An example provided in the following box can help grasp this concept more effectively.

Moving forward, to transmit our data efficiently, we will represent it with digital codes that closely approximate the source entropy. In other words, we aim to answer the question, "How many bits are needed to represent the digital elements of my source?" The interested reader can explore Huffman or Lempel-Ziv coding techniques for more insights. This process, defined by Shannon as "source coding," allows us to optimize the representation of our data.

*Communication Systems Engineering with GNU Radio: A Hands-on Approach,*
First Edition. Jean-Michel Friedt and Hervé Boeglen.
© 2025 John Wiley & Sons, Inc. Published 2025 by John Wiley & Sons, Inc.
Companion website: www.wiley.com/go/friedtcommunication

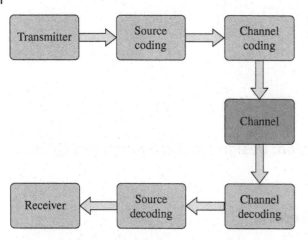

**Figure 5.1** The digital communication system.

Now equipped with the ability to evaluate the quantity of information conveyed by a digital source, we are prepared to transmit it through a channel. It is important to highlight a fundamental aspect of digital communication: all the techniques employed in today's communication systems are justified by the specific characteristics of the transmission channel. Sophisticated techniques are necessitated by the presence of "challenging channels," such as a wireless, noisy, and time- and frequency-selective channels like the mobile radio communication channel. In such scenarios, we are typically constrained by transmission bandwidth, which is related to the capacity $C$ of the channel, expressed in bits per second. This aspect was also addressed in Shannon's landmark paper. Hence, the third Shannon theorem, pertaining to the capacity of the average white Gaussian noise channel (AWGN), is given by the following equation:

$$C_{bits/s} = B \cdot \log_2\left(1 + \frac{S}{N}\right) \tag{5.3}$$

with $B$ the channel bandwidth and

$$\frac{S}{N} = \frac{E_b}{N_0} \cdot \frac{R}{B} \tag{5.4}$$

with $E_b$ being the bit energy (J), $N_0$ the noise variance per two dimensions (W/Hz), $R$ the data rate (bits/s), and $\rho = R/B$ the **spectral efficiency** (bits/s/Hz). Let us now have a look at the relation between the spectral efficiency $\rho$ and the channel capacity $C$. This is shown in Figure 5.2. You can notice in this figure that the x-axis variable is $E_b/N_0$, which is called the **normalized SNR** (because it does not take into account the bandwidth). As we will see later on, this parameter is really important as it determines the performance of digital modulations. Figure 5.2 holds significant importance, as it provides insight into the achievable capacity for any modulation scheme relative to the Signal-to-Noise Ratio (SNR).

---

**Example of entropy and capacity calculation**

A black-and-white television picture consists in 300 000 pixels. These pixels have a brightness level ranging from 0 to 9 with equal probability. We consider a transmission rate of 30 images/s and a transmission $S/N = 30$ dB. Calculate the required bandwidth for the transmission of this signal.

---

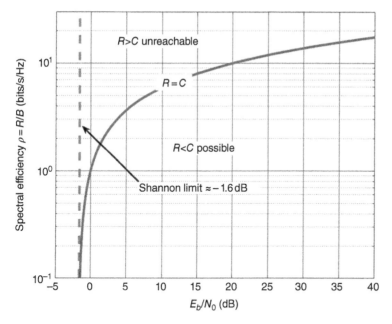

**Figure 5.2** Capacity of an AWGN channel.

---

We first calculate $H(X)$ using equation 5.2:

$$H(X) = \sum_{j=1}^{10} \frac{1}{10} \cdot \log_2\left(\frac{1}{10}\right) = 3.32 \text{ bits} \tag{5.5}$$

Next the data rate $D$:

$$R = H(X) \cdot 30 \cdot 3 \cdot 10^5 = 29.9 \text{ Mbits/s} \tag{5.6}$$

Finally, using equation 5.3, we determine the required bandwidth $B$:

$$B = \frac{R}{\log_2(1001)} \approx 3 \text{ MHz} \tag{5.7}$$

---

### 5.1.2 From Digital Data to Electrical Pulses

Modulation is the operation allowing to use a high-frequency signal to carry information from the transmitter to the receiver. It is well known that this is practically realized by modifying a parameter of the carrier frequency $f_c$ at the rate of the modulating signal. This can be amplitude, frequency, or phase. In the case of a digital modulation, the modulating signal is digital, i.e. directly related to bits. There exists a large number of digital modulation schemes having specific features that will have advantages for different applications. In what follows, we will concentrate on M-PSK and M-QAM modulations, which are the most used in practice. $M$ corresponds to the number of possible states of the modulation with $M = 2^{nbits}$. For example, a Quadrature Phase Shift Keying (QPSK) modulation has $M = 4$ states, meaning that each state can carry $nbits = \log_2(M) = 2$. Let us build a

QPSK modulator with GNU Radio and study the key points we need to understand for that purpose. The signal space representation of the QPSK signal over a symbol duration $T_s$ is given by

$$s_j(t) = \begin{cases} \sqrt{\frac{2E_s}{T_s}} \cdot \cos\left[2\pi f_c t + (2j-1)\frac{\pi}{4}\right], & 0 \leq t \leq T_s \\ 0, & \text{elsewhere} \end{cases} \quad (5.8)$$

with $j = 1, 2, 3, 4$; $E_s$ the signal energy per symbol, and $T_s$ the symbol duration. Using a simple trigonometric identity, equation 5.8 can be rewritten as

$$s_j(t) = \sqrt{E_s} \cdot \cos\left[(2j-1)\frac{\pi}{4}\right] \cdot \sqrt{\frac{2}{T_s}} \cdot \cos(2\pi f_c t)$$
$$-\sqrt{E_s} \cdot \sin\left[(2j-1)\frac{\pi}{4}\right] \cdot \sqrt{\frac{2}{T_s}} \cdot \sin(2\pi f_c t) \quad (5.9)$$

with

$$\phi_I(t) = \sqrt{\frac{2}{T_s}} \cdot \cos(2\pi f_c t) \quad (5.10)$$

$$\phi_Q(t) = \sqrt{\frac{2}{T_s}} \cdot \sin(2\pi f_c t) \quad (5.11)$$

$\phi_I(t)$ and $\phi_Q(t)$ are quadrature carriers having unit energy (because $f_c = N/T_s$ with $N$ integer). From equation 5.9, the following remarks can be made. First, we notice that this equation naturally leads to our now well-known In-phase/Quadrature (IQ) modulator structure. Moreover, we have four message points that can be placed on a complex plane diagram represented in Figure 5.3. This diagram is very useful to analyze digital modulation schemes and is called a **constellation** (see Section 2.5.3). As we have four points, it means that each point actually carries two bits. Table 5.1 summarizes QPSK modulation parameters. At this point of the discussion, we can ask ourselves how to make the connection between the dibits carried by the QPSK modulation and their duration $T_s$. This is done by using a pulse of duration $T_s$ having a defined shape. One of the most straightforward shapes for encoding bits into electrical pulses is the rectangular shape:

$$h(t) = A \cdot \text{rect}\left(\frac{t}{T_s}\right) \quad (5.12)$$

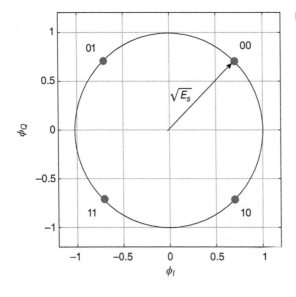

Figure 5.3 QPSK modulation constellation.

**Table 5.1** QPSK parameters.

| Gray-encoded dibit | Phase (radians) | Value of message point |
|---|---|---|
| 00 | $\pi/4$ | $\sqrt{E_s/2}\,(1+j)$ |
| 01 | $3\pi/4$ | $\sqrt{E_s/2}\,(-1+j)$ |
| 11 | $5\pi/4$ | $\sqrt{E_s/2}\,(-1-j)$ |
| 10 | $7\pi/4$ | $\sqrt{E_s/2}\,(1-j)$ |

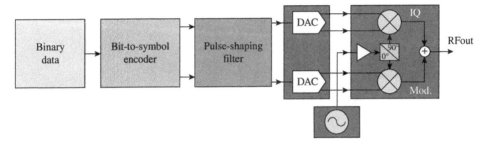

**Figure 5.4** QPSK modulator.

where

$$rect(t) = \begin{cases} 1, & \text{when } -\frac{1}{2} < t < \frac{1}{2} \\ 0, & \text{otherwise} \end{cases} \quad (5.13)$$

Unfortunately, this is not a good idea since a rectangular pulse in the time domain translates into an infinite spectrum in the frequency domain. Moreover, sending such shaped waveforms on a bandlimited channel will yield a dramatic consequence called inter-symbol interference (ISI). Practically, at the receiver side, it means that several pulses will interfere with each other, resulting in decoding errors. In effect, what is called **pulse shaping** is not a trivial operation, as we will see in Section 5.1.3. Currently, we possess all the necessary components to construct our QPSK modulator, beginning with a sequence of bits and progressing through spectrum translation to the carrier frequency. This is summarized in Figure 5.3, where Gray encoding is defined as adjacent symbols only changing by a single bit (Figure 5.4).

### 5.1.3 Occupied Bandwidth and Spectral Efficiency

We have just learned the importance of carefully selecting the shape of a pulse when associating digital modulation symbols. The first aspect we have to take into account is ISI that will occur on bandlimited channels. It is often mistakenly believed that this issue is irrelevant when transmission takes place over a radio channel. Even if the radio channel seems to be ultra wideband, you are not alone. Moreover, the electronic transceiver is made of several elements which are bandlimited. Finally, at the receiver side, we have to minimize the amount of noise to maximize the SNR: a bandpass filter will thus limit the reception spectrum to the useful signal. These were just a few arguments to convince the reader that ISI is indeed a concern that communication system designers must address. Coming back to pulse shaping, it is well known that one has to select a pulse shape that respects the Nyquist ISI criterion [Proakis and Salehi, 2008]. Choosing the right pulse in the time domain has to translate into individual spectra that do not overlap in the frequency domain

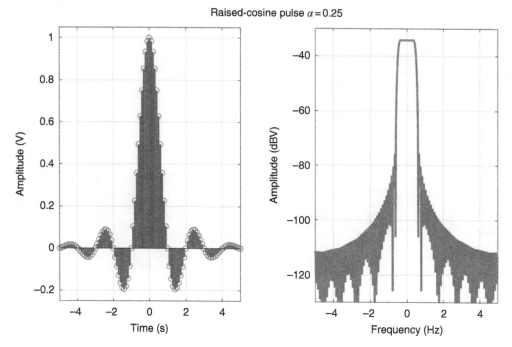

**Figure 5.5** Raised-cosine pulse.

and combine to form a flat continuous spectrum. There are many pulse shapes that will agree. One of the most used in practice is the **raised cosine** (RC) pulse.

Looking at Figure 5.5, one can think about the cardinal sine pulse (sinc). This is not the case as the latter is not physically realizable (not causal). The RC pulse is realizable but has a drawback: it occupies a bandwidth that can be as wide as twice that of the sinc pulse. This depends on the $\alpha$ parameter called the **roll-off factor** which can vary between 0 and 1. In practice, values between 0.25 and 0.5 are found (going too low, i.e. $\alpha < 0.25$, will pose synchronization issues at the receiver). There is an important thing we have to keep in mind regarding pulse shaping, and this is **matched filtering**. Indeed, the optimal receiver theory states that to maximize the SNR at reception, we have to convolve the received pulses with a matched filter, i.e. a filter having the same shape as the pulse used for transmission. If $x(t)$ is our pulse shape and $s(t)$ is the matched filter waveform, the received signal $y(t)$ is then given by

$$y(t) = \int_{-\infty}^{+\infty} x(\tau) s(\tau - t) d\tau = x(t) * s(-t) \tag{5.14}$$

How does this affect our pulse-shaping scheme? Well, quite simply, for achieving zero ISI, we require an RC pulse throughout the transmission chain. Additionally, to attain the highest SNR at reception, it is imperative to maintain the same pulse shape as used during transmission. Therefore, we need a squared-root RC (RRC) at transmission and a squared-root RC at the reception side! Indeed, the characteristics of these pulses make them quite an interesting aspect of signal processing. However, we have saved the best for last. Pulse shaping also has to do with spectrum occupancy and the concept of **spectral efficiency** of digital modulations. With this concept, we answer the question: "How many Hertz does my transmission system require?" Looking again at Figure 5.5, we can notice that the null of the main lobe occurs at time $t = T_s = 1$ s. This translate to a bandwidth $B = (1 + \alpha)/T_s = 1.25$ Hz. To evaluate spectrum efficiency of different modulation schemes, one has to use the same bit rate $R_b$ as reference. Let us compare Binary Phase Shift Keying (BPSK) and QPSK with a bit rate $R_b = 500$ kbit/s with a GNU Radio flowgraph. The result is shown in Figure 5.6. On the upper left diagram you get the QPSK and BPSK RRC pulses. On the

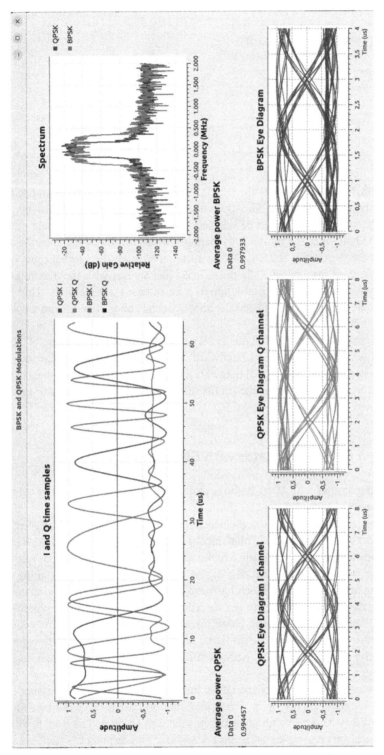

**Figure 5.6** BPSK–QPSK spectrum efficiency comparison.

**Table 5.2** Spectral efficiency $\eta$ (bits/s/Hz) for four digital modulations.

| Modulation | Symbol time | Spectral efficiency $\eta$ |
|---|---|---|
| BPSK | $T_s = T_b$ | 1 |
| QPSK | $T_s = 2T_b$ | 2 |
| 8PSK | $T_s = 3T_b$ | 3 |
| 16QAM | $T_s = 4T_b$ | 4 |

upper right diagram, the spectrum of both modulation is represented. One can notice that BPSK spectrum occupancy is twice that of QPSK (700 versus 350 kHz). As a result, we can conclude that QPSK is twice as bandwidth-efficient as BPSK. Notice also a common type of diagram called the **eye diagram** at the bottom of the figure. This diagram, which allows to visualize the amount of ISI at reception, is constructed as follows: Over a duration of twice the symbol time, i.e. $2T_s$, we superimpose the received time pulses. The result looks like an eye, with the maximum opening occurring at the optimum sampling instant, which is an integer multiple of $T_s$. This diagram is very useful to evaluate the amount of ISI and the SNR at the reception side: the more open the eye, the less ISI.

In our example, it can be observed that the BPSK symbol time (2 µs) is half that of QPSK (4 µs). This is the reason why QPSK is twice as bandwidth efficient as BPSK while carrying the same amount of data. In conclusion, with $T_b$ being the bit time, $M$ the number of symbols of the digital modulation, and $T_s = \log_2(M)T_b$, we get the results shown in Table 5.2 for four well-known digital modulation schemes.

## 5.2 Building a QPSK Modulator with GNU Radio

Building a QPSK transmitter with GNU Radio is really easy. It can be achieved by using just five blocks, as can be seen in Figure 5.7.

The place where most operations are realized is the `Constellation Modulator` block. This is a hierarchical block (abbreviated "hier" block) using several important GNU Radio blocks. Hier blocks are nice when it comes to simplify the display of a dense flowgraph, but as we aim to comprehend the primary components of a digital modulator, it is more intriguing to explore its constituent elements. When a hier block has been created by GNU Radio Companion (GRC), it is possible to go down the hierarchy to see the constituent blocks. It is not possible to do so with the `Constellation Modulator` block since it was created with Python code. Reading `generic_mod_demod.py` file from the gr-digital module allows to build the flowgraph shown in Figure 5.8. As the reader can notice, without surprise, we find the same building blocks as depicted in Figure 5.4.

Let us set up the main parameters of the QPSK modulator. First, we set the symbol rate $R_s$ at 250 ksymb/s, i.e. 500 kbits/s. This yields a sampling rate of 2 Msps since we take an oversampling factor of $sps = 8$. A suitable value ranges from 4 to 12 in practice. The RRC filter is realized using a polyphase arbitrary resampler using $nfilts = 8$ filters. This last block is used in conjunction with the `firdes.root_raised_cosine()` function. The function-calling parameters are the following:

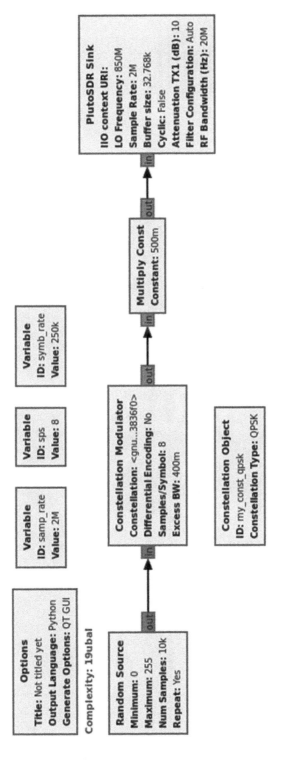

**Figure 5.7** GRC QPSK modulator with hierarchical ("hier") block.

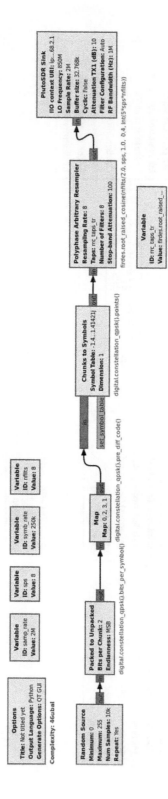

**Figure 5.8** GRC QPSK modulator.

- gain = *nfilts*/2
- sampling_freq = sps
- symbol_rate = 1.0
- alpha = 0.4
- ntaps = *D* ∗ *nfilts*

The filter gain value is set so as to limit the amplitude of the pulses to ±1 V at the entry of the `Pluto Sink` block (to avoid saturation). This way, the transmitted power only depends on the tunable Radio Frequency (RF) gain of the Adalm-Pluto. In general, it is always a good idea to normalize the baseband signals before they enter or come out of an RF spectrum translation hardware. Indeed, most signal processing blocks involved in a digital transceiver expect input values normalized to ±1 V (more on this in the section dedicated to the receiver).

What about the *D* parameter? *D* defines the number of symbols spanned by the impulse response of the filter. A larger *D* implies a better approximation of the ideal response but at the expense of a higher complexity. In practice, a value of *D* ranging from 5 to 20 will do the job. As shown in Appendix A of [Rice, 2009] the smallest acceptable value of *D* is primarily driven by the minimum stop-band attenuation requirement. This is especially true for small values of $\alpha$ ($\alpha < 0.15$).

The last parameter of interest is *nfilts*, which represents the number of filters used in the polyphase decomposition to realize the filter. The polyphase interpolator implementation used in the QPSK modulator (the `Polyphase Arbitrary Resampler` block) is very efficient in terms of computation time (see the box on the next page). Last but not least, please note that GNU Radio Companion provides the user with a tool called Filter Design Tool (Tools menu), which allows to design digital filters using a template and, in particular, RRC filters. Figure 5.9 shows an example of an RRC filter design using this tool.

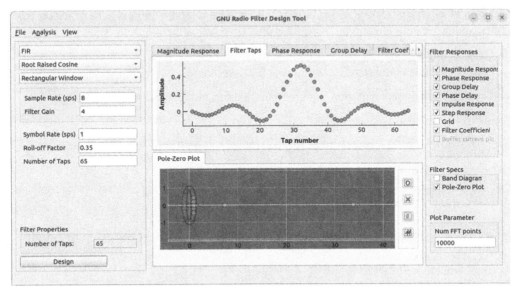

**Figure 5.9** GRC filter design tool. Source: GNU Radio project.

## 164 | 5 Digital Communications in Action: Design and Realization of a QPSK Modem

---

**Is GNU Radio signal processing efficient?**

For someone not used to multirate signal processing, some GNU Radio blocks can be a bit confusing. Indeed, many blocks dealing with filters, and in particular interpolating and decimating ones, utilize the polyphase implementation. The polyphase decomposition for the implementation of digital filters is very efficient in terms of computation time as it allows to parallelize the calculation of the convolution operation. Let us take the example of the RRC transmit filter included in the QPSK modulator. If the constellation modulator hier block is used, we cannot change the number of filters used in the polyphase decomposition, which is by default set to 32. This is the same for the symbol span $D$ parameter, which is set to 11. The filter coefficients are generated, thanks to the `firdes.root_raised_cosine()` function, yielding to 353 coefficients ($32 \times 11 + 1$), and are input to the `filter.pfb_arb_resampler_ccf()` function to upsample by a factor of sps (8 in our example). If we were using a traditional implementation, this would result in a convolution calculation implying $32 \times 11 \times sps + 1 = 2817$ coefficients. Convolution is computationally intensive as it implies "multiply and accumulate" (MAC) operations. Moreover, if we look carefully at this calculation, we can notice that 7/8 of the 2817 multiplications needed at each time instant are useless since they are multiplications by 0 (upsampling by a factor of 8 implies inserting 7 zeros between samples). To eliminate these redundant operations, one can transform the natural structure into a more efficient one, as depicted of Figure 5.10(b). This streamlined structure is achieved by applying one of the "Noble identities" [Zielinski, 2021]. Furthermore, because the filter coefficients are now rearranged to the low-sampling side of the network, this structure necessitates a computational workload reduced by a factor of $M = sps$ compared to the structure depicted on the left of Figure 5.10!

The following Octave listing will finish to convince the reader about the polyphase decomposition efficiency in terms of computation time. Here we compare the calculation time for the classical implementation and the polyphase one. On a typical desktop computer, this yields

- **Classical implementation.** Elapsed time is 0.288 seconds.
- **Polyphase implementation.** Elapsed time is 0.012 seconds.

The polyphase implementation is more than 24 times faster than the classical one!

```
pkg load communications
M = 32; sps = 8;
% SRRC filter coefficients
h = rcosfir(0.4, [-5*M-M/2 5*M+M/2], sps, 1, 'sqrt');
% Input signal: 1000 dirac pulses
x = repmat([1 zeros(1,sps-1)], 1, 1000);
Nx = length(x);
% Classical upsampling and filtering
tic()
xup = upsample(x,M);
y = filter(h,1,xup);
toc()
% plot the signal
stem(y), hold
```

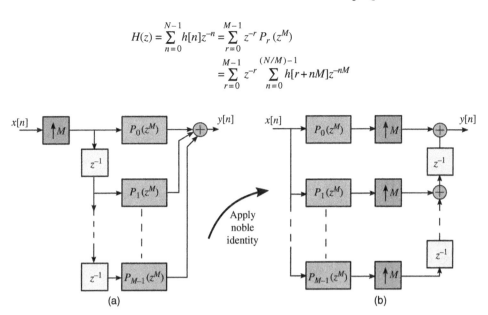

**Figure 5.10** Polyphase interpolator structures. The natural implementation is displayed on (a). The efficient one, obtained after applying one of the "Noble identities," is shown on (b).

```
tic()
% Polyphase decomposition
xipp = zeros(1, length(x));
% output initialization
for k = 1:M
% Polyphase filtering
xipp(k:M:M*Nx) = filter( h(k:M:end), 1, x ); % k-th polyphase
    component filtering
end
toc()
% plot the signal
stem(xipp, 'r')
```

## 5.3 Building a QPSK Demodulator with GNU Radio

We now have a fully functional QPSK modulator built with GNU Radio. What about the receiving side? To illustrate what lies ahead in this section, let us have a look at Figure 5.11 where we have the spectrum at reception (a) and the associated constellation (b). In this case, we have a transmission between an Adalm-Pluto and an RTL2832 dongle. The transmit power is −30 dBm at 850 MHz and the antennas are 2 m apart. The first obvious observation is that the constellation is not recognizable and has nothing to do with a QPSK one. The second important thing to notice is that the spectrum is not centered: it is left-shifted by an amount of roughly 50 kHz. Could this be the source of our messy constellation?

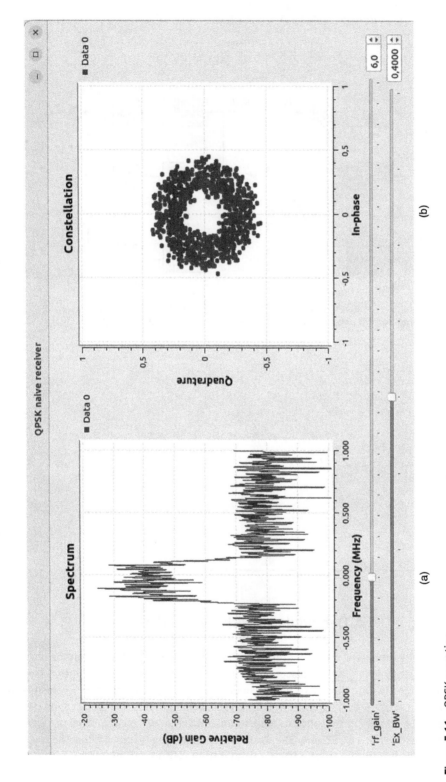

**Figure 5.11** QPSK reception.

### 5.3.1 Synchronization

Indeed, in the case of a real system, the transmitter and the receiver are not synchronized. Synchronization is a word derived from ancient Greek: "syn" ($\sigma \upsilon \nu$) meaning "together," and "chronos" ($\chi \rho o \nu o \varsigma$) means "time" (Chronos was the god of time in the mythology). Therefore, synchronization means to coordinate several operations as a function of time. In the case of our receiver, we have to consider the following operations:

- Carrier phase and frequency synchronization.
- Timing synchronization, i.e. sampling at the optimal instants of the symbol boundaries.

The first item is directly observed in Figure 5.11 with a left 50 kHz shift of the received spectrum. This was indeed expected. Although you may think you set exactly the same frequency on the two TX and RX devices, this does not guarantee that the hardware will be able to achieve it. Indeed, all oscillators on RF devices have a precision expressed in ppm (parts per million). For the Adalm-Pluto, the standard 40 MHz oscillator (which is the source for the AD9361 radio chip) is given with a precision of $\pm25$ ppm. For a standard (V1) RTL2832 dongle, this can go up to $\pm100$ ppm. Therefore, for our system, the worst case can be $\pm106.25$ kHz! Anyway, you can correct this simply by adding the frequency difference of 50 kHz to the carrier frequency of 850 MHz at the transmitter. Although this correction was performed here, centering the spectrum, the constellation is still messy, and no decoding seems to be possible. Other operations have to be performed before we can observe a proper constellation.

In order to understand the structures GNU Radio uses to perform synchronization, it is important to recall the following key concepts. First, a digital receiver uses two main phases to solve synchronization issues, i.e. **"acquisition"** and **"tracking."** At first, a quick and coarse correction is performed (acquisition phase), followed by a fine correction which tracks the remaining variations. Taking frequency correction as an example using Figure 5.11, we would first perform a coarse frequency correction of 50 kHz and then correct the remaining tens of Hz using a tracking loop (more on this in the following). These processes use algorithms that can make use of the data to perform the correction. This is referred to as **"data aided"** and **"decision directed"** or **"non-data-aided"** (or blind) algorithms. Moreover, according to the way the data is transmitted, i.e. in **"burst"** or in **"continuous"** modes, we will find either **"feedforward"** or **"feedback"** structures to implement the algorithms. Using our previous example, coarse frequency correction will be implemented in a feedforward manner, whereas the fine correction phase uses a feedback structure called a phase-locked loop (PLL).

We start with the estimation and correction of the coarse frequency offset. The feedforward algorithm chosen is derived from the maximum likelihood (ML) estimation of a single complex tone. It is important to specify to the reader that most synchronization algorithms, although designed using ad hoc methods, can be analyzed theoretically within the framework of ML estimation theory. This is precisely the point of view of the excellent book by Mengali and d'Andrea, "Synchronization Techniques for Digital Receivers." We refer the reader to this reference [Mengali and D'Andrea, 1997] for a comprehensive treatment of the matter. It can be shown [Wasenmuller et al., 2008] that the following estimator is a simplified version of the ML estimator of a single complex tone:

$$X(k) = \sum_{n=0}^{N-1} x[n] \cdot \exp\left(-j\frac{2\pi \cdot k \cdot n}{N}\right) \tag{5.15}$$

with $x[n] = \hat{r}[n] = r[n]^M$ is the received signal raised to the $M$th power in order to remove the $M$-ary phase-shift keying (MPSK) data influence on the signal. Then the estimated residual

**168** | *5 Digital Communications in Action: Design and Realization of a QPSK Modem*

frequency offset $\hat{f}_r$ is given by spectral analysis of the Fast Fourier Transform (FFT) output. The frequency bin $k_{f_r}$ corresponding to the frequency offset is therefore

$$k_{f_r}, s.t. : \left| X \left( k_{f_r} \right) \right| = max \, |X(k)| \, k = 0, 1, \ldots, N-1 \tag{5.16}$$

Finally, the frequency offset $f_r$ is given by

$$\hat{f}_r = \frac{k_{f_r}}{M \cdot N} \quad \text{for} \quad k_{f_r} < \frac{N}{2}$$
$$\hat{f}_r = \frac{k_{f_r} - N}{M \cdot N} \quad \text{for} \quad k_{f_r} \geq \frac{N}{2} \tag{5.17}$$

$M$ represents the modulation order and $N$ the number of points of the FFT. This is rather simple to implement for the QPSK modem ($M = 4$) in GNU Radio as can be seen in Figure 5.12. This operation is generally performed only once during the acquisition phase. However, even if you evaluate the right value of $k_{f_r}$, this is not going to be satisfying as this value will vary over time. Therefore, a tracking and correction step has to be performed using a PLL.

#### 5.3.1.1 The Digital PLL

PLLs, and in particular digital ones (DPLL), often appear as somewhat mysterious phenomena to many people. For example, on the GNU Radio mailing list, one can find several posts with questions about the parameter values required for the `Costas Loop` or `Symbol Sync` blocks. So let us try to demystify a little bit this matter in the following by designing a carrier phase tracking loop. This example will give us the basic elements we need to consider for the design of the frequency and timing tracking loops of the QPSK modem. The DPLL structure for this case study is presented in Figure 5.13. The DPLL is composed of three main blocks:

- **The phase error detector (PED).** This element delivers an error signal $e[n]$ proportional to the phase difference between the input signal $s[n]$ and the reference signal $r[n]$ issued from the numerically controlled oscillator (NCO). For this example, it consists of multiplying the input signal $s[n]$ by the complex conjugate of the NCO output $r[n]$ and taking the argument of the result:

$$e[n] = arg\{x[n] \cdot (r[n])^*\} = arg\{e^{j\Theta[n]} \cdot e^{-j\hat{\Theta}[n]}\} = \Theta[n] - \hat{\Theta}[n] \tag{5.18}$$

An important parameter of the PED is its gain $K_P$, which can be obtained from the so-called S-curve (see Figure 5.14). Since we have $\bar{g}(e[n]) = e[n]$, the PED is linear over the interval $-\pi \leq e[n] \leq \pi$ and has a slope of 1. Therefore, its gain $K_D = 1$.

- **The Numerically Controlled Oscillator (NCO).** It is the digital equivalent of the analog voltage controlled oscillator (VCO). Starting from the equation of the VCO and applying Laplace transform, we get

$$\phi(s) = \mathscr{L} \left\{ k_0 \cdot \int_0^t x(\tau) \, d\tau \right\} = \frac{k_0}{s} \cdot x(s) \tag{5.19}$$

Going to the $z$-domain and using the backward difference to approximate the integral, we obtain

$$\phi(z) = k_0 \cdot Ts \cdot \frac{z^{-1}}{1 - z^{-1}} \cdot X(z) \tag{5.20}$$

This yields the following difference equation:

$$\phi[n] = K_0 \cdot x[n-1] + \phi[n-1] \tag{5.21}$$

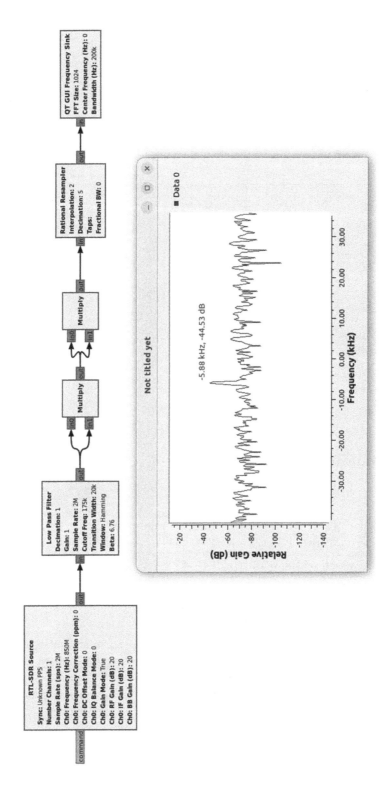

**Figure 5.12** Carrier frequency offset estimation.

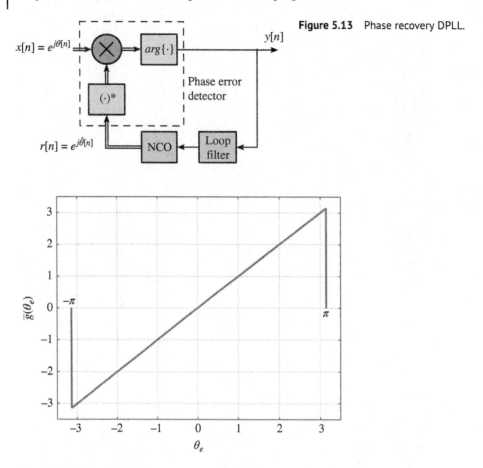

Figure 5.13 Phase recovery DPLL.

Figure 5.14 Linear PED S-curve.

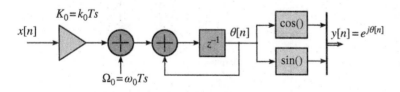

Figure 5.15 Numerically controlled oscillator.

Finally, the NCO structure can easily be deduced from this equation and is depicted in Figure 5.15. Please note that the structure allows to oscillate around a constant frequency $f_0$ whose phase $\omega_0$ is added to the constant $K_0$.

- **The loop filter.** This element is highly significant as it conditions the dynamic performance of the loop on the one hand and filters out the noise and the signal's high-frequency components on the other hand. The most used structure in practice is the proportional plus integrator (PI) filter. The adoption of this structure leads to a second-order loop which has been well studied in the literature. For the detailed calculation of these elements, we refer the reader to Rice [2009] and Chaudhari [2018]. As can be seen in Figure 5.16, we are left with two more constants to be

**Figure 5.16** Loop filter.

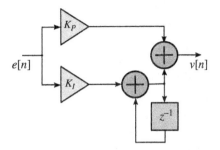

determined, namely $K_P$ and $K_I$. The associated difference equation which we will use to program the filter is given by

$$v[n] = (K_P + K_I) \cdot e[n] + K_I \cdot e[n-1] \qquad (5.22)$$

**Computing the Loop Constants**  Designing a PLL for radio applications starts with defining:

- **The damping factor** $\zeta$. Large values of $\zeta$ imply no overshoots but long convergence time. Typical values of $\zeta$ range from 0.5 to 2 in practice.
- **The loop bandwidth** $B_n$. A small $B_n$ allows to filter most of the noise, whereas a large $B_n$ allows to track fast-phase variations. Therefore, the choice of $B_n$ will be a trade-off. Moreover, in the digital domain, the usage is to normalize $B_n$ to the sample rate $Ts$. A practical value of $B_n Ts$ commonly found in digital receivers is 1%, hence the value of $2\pi/100$ found in some GNU Radio blocks ($2\pi$ being a normalization factor). At this point, we are left with $K_0$, $K_D$, $K_P$, and $K_I$ to be determined:
- $K_0$: The gain of the NCO, which can be set to 1.
- $K_D$: The PED gain, which depends on the type of detector.
- $K_P$ and $K_I$: When $K_0$ and $K_D$ are determined, we are left with these two unknown constants. To compute them, we need the normalized natural frequency of the PLL, which is given by

$$\theta_n = \frac{B_n Ts}{\zeta + \frac{1}{4\zeta}} \qquad (5.23)$$

It can be shown [Rice, 2009] that

$$K_P = \frac{1}{K_D K_0} \cdot \frac{4\zeta \theta_n}{1 + 2\zeta \theta_n + \theta_n^2}$$

$$K_I = \frac{1}{K_D K_0} \cdot \frac{4\theta_n^2}{1 + 2\zeta \theta_n + \theta_n^2} \qquad (5.24)$$

In GNU Radio, for most of the synchronization blocks (**not for the new Symbol Sync block**), only the loop bandwidth $B_n Ts$ is required. This is explained by Tom Rondeau in Rondeau [2011]. **Additionally, if the inputs of the blocks are normalized values, $K_0$ and $K_D$ can be assumed to be 1.** In the old synchronization blocks, the parameters "alpha" and "beta" correspond to $K_P$ and $K_I$, respectively. A further simplification can be made on equation 5.24 since $B_n Ts \ll 1$ and considering that $\zeta = K_0 = K_D = 1$:

$$\begin{aligned} K_P &= 4\theta_n \\ K_I &= 4\theta_n^2 \\ \Rightarrow K_I &= \frac{K_P^2}{4} \end{aligned} \qquad (5.25)$$

**172** | 5 Digital Communications in Action: Design and Realization of a QPSK Modem

---

**DPLL summary exercise**

We aim to design a second-order DPLL for phase recovery with a closed-loop equivalent normalized noise bandwidth $BnTs = 0.5\%$ and a damping constant $\zeta = 0.7071$. The structure is the one depicted in Figure 5.13. The input signal is a real sinusoid with discrete frequency $k/N = 1/20$ cycles/sample with a phase shift of $0.75\pi$:

$$s[n] = \cos\left(2\pi\frac{1}{20}n + 0.75\pi\right) \tag{5.26}$$

- Determine the values of $K_0$, $K_D$, $K_P$, and $K_I$.
- Write an octave script that simulates the behavior of this DPLL.

```
clear;
close all;

%Number of samples to simulate
Nsp = 4000;

%Input signal and NCO parameters
w0 = 2*pi*(0.005 + 0.0005);
w0dds = 2*pi*.005;

%Loop parameters
theta = -0.75*pi;
zeta = 0.707;
BnTs = .005;

% Calculate the loop constants. See Rice's book Appendix C (C.58)
K0 = 1;
KD = 1;
den = 1 + (2*zeta)*(BnTs/(zeta + 1/(4*zeta))) + ...
          (BnTs/((zeta + 1/(4*zeta))))^2;

num = (4*zeta)*(BnTs/(zeta + 1/(4*zeta)));
Kp = num/den;
Kp = Kp/(K0*KD);

num = 4*(BnTs/(zeta + 1/(4*zeta)))^2;
Ki = num/den;
Ki = Ki/(K0*KD);

%Several approximations of the analog PI filter
%b = [ (Kp+Ki) -Kp ];
b = [Kp (Ki-Kp)];
%b = [(Kp + Ki) Ki];
a = [ 1 -1 ];
```

```
%a = 1;

in = exp(1j*((1:Nsp)*w0 + theta));
out = zeros(size(in));
error = zeros(size(in));

% Initial conditions
enm1 = 0;
ddsom1 = 0;
ncoout = 1;

for n = 1:Nsp
    %Phase comparator
    pedo = angle(in(n)*conj(ncoout));
    error(n) = pedo;

    %Loop filter
    en = pedo*Ki + enm1;
    lpo = pedo*Kp + en;

    %DDS
    ddso = lpo*K0 + w0dds + ddsom1;

    %Update previous values
    enm1 = en;
    ddsom1 = ddso;

    %NCO output
    ncoout = exp(1j*ddso);
    out(n) = ncoout;
end

%% Plot data
figure(1);
plot(1:Nsp, real(in), 'r', 'linewidth', 2), xlabel('Sample
    number'), ylabel('Amplitude (V)');
hold, plot(1:Nsp, real(out), 'b--', 'linewidth', 2);grid;legend
    ('Input', 'Output');
title('Input and output of the DPLL');
hold off;
figure(2), plot(1:Nsp, error, 'linewidth', 2), grid; title
    ('Evolution of phase error');
xlabel([ '$ \hat{\theta} $ = ' num2str(ddso) ], 'Interpreter', '
    latex');
```

*(Continued)*

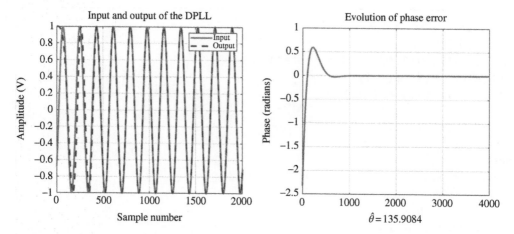

**Figure 5.17** DPLL exercise simulation results.

---

**(Continued)**

We get the following results:

- $K_0 = K_D = 1$
- $K_P = 0.0132$
- $K_I = 8.83 \cdot 10^{-5}$
- The input and output of the DPLL and the evolution of the phase error are represented in Figure 5.17.

---

Having addressed the coarse frequency offset correction and armed with the knowledge about DPLLs developed previously, we are now prepared to address the tracking and correction of residual carrier and timing offsets.

### 5.3.1.2 Maximum Likelihood Estimation and the Costas Loop

This step is called fine phase and frequency offset correction and is performed by a closed-loop system in the form of a special type of PLL. Phase and frequency offset compensation are generally two separate operations. However, when frequency offset is small compared to the symbol rate (less than 5%), it can be treated as a time-dependent linear phase error so that the phase and frequency error can be compensated for using the same loop. The majority of synchronization algorithms were initially developed using a heuristic approach. It was only later that researchers demonstrated they could be analyzed within the framework of estimation theory, particularly by employing ML estimation techniques. Going back to QPSK phase recovery, it can be shown that the decision-directed phase error is given by [Rice, 2009]:

$$e[n] = x_Q[nTs] \cdot \hat{a}_I[n] - x_I[nTs] \cdot \hat{a}_Q[n] \tag{5.27}$$

with $x_I[nTs]$ and $x_Q[nTs]$ being the de-rotated $I$ and $Q$ matched filter outputs (see Figure 5.18). The decisions for QPSK are given by

$$\begin{aligned}\hat{a}_I[n] &= A \cdot \mathrm{sgn}\left\{x_I[nTs]\right\} \\ \hat{a}_Q[n] &= A \cdot \mathrm{sgn}\left\{x_Q[nTs]\right\}\end{aligned} \tag{5.28}$$

since $(a_I[n], a_Q[n]) \in \{\pm A, \pm A\}$.

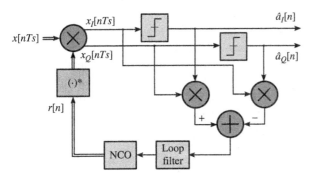

**Figure 5.18** ML QPSK phase recovery loop.

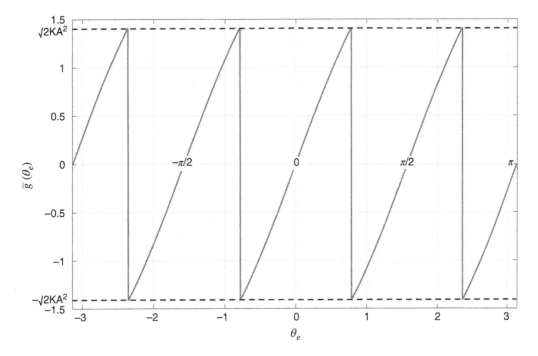

**Figure 5.19** QPSK phase detector S-curve.

It can be demonstrated that the average S-curve of the PED, as depicted in Figure 5.19, can be derived from Rice [2009]:

$$\bar{g}(\theta_e) = \begin{cases} -2\chi A^2 \sin(\theta_e) & -\pi < \theta_e < -\frac{3\pi}{4} \\ 2\chi A^2 \cos(\theta_e) & -\frac{3\pi}{4} < \theta_e < -\frac{\pi}{4} \\ 2\chi A^2 \sin(\theta_e) & -\frac{\pi}{4} < \theta_e < \frac{\pi}{4} \\ -2\chi A^2 \cos(\theta_e) & \frac{\pi}{4} < \theta_e < \frac{3\pi}{4} \\ -2\chi A^2 \sin(\theta_e) & \frac{3\pi}{4} < \theta_e < \pi \end{cases} \quad (5.29)$$

From this S-curve, two important elements can be observed. First, the gain of the PED depends on A and $\chi$, the amplitude of the signal at the input of the loop. As already stated, this conditions the gain $K_p$ of the PED, which equals $2\chi A^2$. In GNU Radio, setting it to 1 implies a method to keep

**176** | *5 Digital Communications in Action: Design and Realization of a QPSK Modem*

the input signal constant by utilizing an automatic gain control (AGC) system. Second, there are four stable lock points for this DPLL, namely $-\pi/2, 0, \pi/2, \pi$. This means that the system can lock on one of these four points, which results in a $\pi/2$ **phase ambiguity** that needs to be resolved.

The skilled reader must have noticed that this structure can be viewed as a generalization of the **Costas loop**. This is precisely the name used by the block found in GNU Radio, which implements exactly this structure for QPSK. However, there is a mistake in the documentation of the block which says: "The Costas loop locks to the center frequency of a signal and downconverts it to baseband." This is not true: the block does not perform any downconversion to baseband since it works directly at baseband. Finally, and this gives us a transition to Section 5.3.1.3, the ML estimator used to set up the phase recovery loop assumes that only the signal phase is unknown. The other parameters are supposed to be known; i.e. after matched filtering, we have to perform perfect timing estimation! This is the beginning of a story that has been much discussed: should phase recovery be performed before or after timing recovery?

### 5.3.1.3 QPSK Timing Recovery

Although the receiver knows the symbol rate of the data to recover, it has to find the optimum sampling instants for each RRC pulse coming in. To help us understand the process, the eye diagram is very useful, as can be observed in Figure 5.20. The optimum sampling instant $\tau_{opt}$ is located at the eye maximum opening. If this is the case, we get the black points shown on the QPSK constellation diagrams. If the sampling instants are $\tau_1$ or $\tau_2$ as depicted in the figure, this produces a square group of points (in light gray) around the optimum constellation points. The size of the groups is proportional to the difference between the chosen sampling instant and the optimum sampling instant $\tau_{opt}$.

What we want to set up now is a timing recovery loop using as much as possible the elements we have used for the phase recovery loop. In particular, we would like to design a phase detector equivalent that produces an error signal proportional to the timing error (i.e. the difference between the chosen value of $\hat{\tau}$ and $\tau_{opt}$). This structure is known as a **timing error detector (TED)** and corresponds to the first parameter of the GNU Radio Symbol Sync block. It is possible to choose between nine different TED algorithms, which have been the subject of research papers. As this is a key element that will determine the operation of the timing recovery loop, we are going to explain in the following the key points the reader must have in mind for the optimum design of the loop. First, it is important to understand the behavior of the TED. By referring again to the eye diagram, let us design the ML TED for BPSK. In Figure 5.21, we can see four different cases considered for the design:

(a) Transition from a negative to a positive symbol. $\hat{\tau}$ is too early. The slope is positive.
(b) Transition from a positive to a negative symbol. $\hat{\tau}$ is too early. The slope is negative.
(c) Transition from a negative to a positive symbol. $\hat{\tau}$ is too late. The slope is positive.
(d) Transition from a positive to a negative symbol. $\hat{\tau}$ is too late. The slope is negative.

The output of the TED should be negative when $\hat{\tau}$ is too early and positive when it is too late. If we consider only the sign of the slope, cases (a) and (d) work as expected, but cases (b) and (c) do not. However, if we incorporate the sign of the considered symbol in the calculation, it works for the four cases. This leads to the following TED equation:

$$e[n] = a[n] \cdot \dot{x}[nTs + \hat{\tau}[n]] \tag{5.30}$$

for the data-aided case, and

$$e[n] = \hat{a}[n] \cdot \dot{x}[nTs + \hat{\tau}[n]] \tag{5.31}$$

for the decision-directed case, $\dot{x}$ being the time derivative of the TED input signal. The generalization for QPSK is straightforward, considering the same equation for the quadrature signal. In the

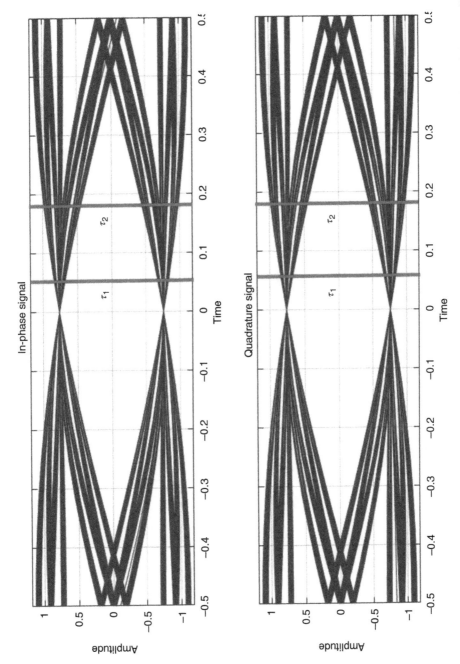

**Figure 5.20** Timing synchronization using the eye diagram. The two selected time instants $\tau_1$ and $\tau_2$, respectively, produce the left and right constellations.

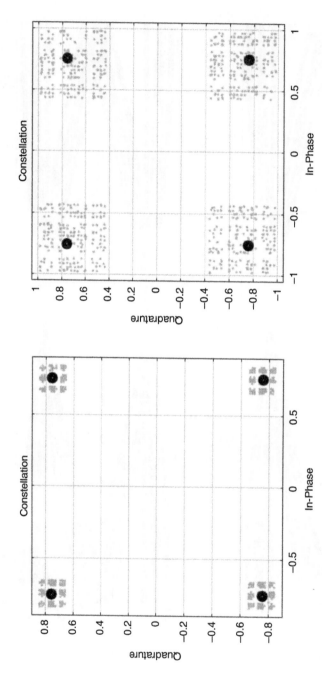

**Figure 5.20** (*Continued*)

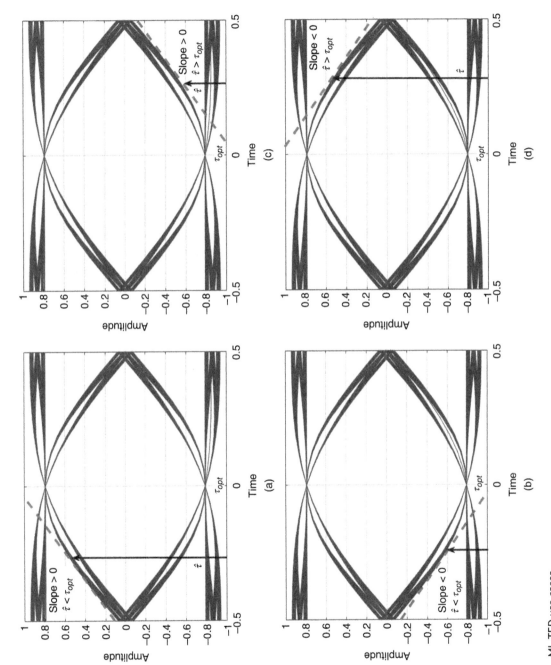

**Figure 5.21** ML TED use cases.

digital domain, it is common to approximate the derivative by a difference of $\Delta Ts$. Interestingly, applying this operation to the previous equations yields the equations of the so-called **early-late TED**:

$$e[n] = a[n] \cdot (x[nTs + \hat{\tau}[n] + \Delta Ts] - x[nTs + \hat{\tau}[n] - \Delta Ts]) \qquad (5.32)$$

for the data-aided case, and

$$e[n] = \hat{a}[n] \cdot (x[nTs + \hat{\tau}[n] + \Delta Ts] - x[nTs + \hat{\tau}[n] - \Delta Ts]) \qquad (5.33)$$

for the decision-directed case.

The Symbol Sync block-implemented TED algorithms can be decision-directed (e.g. Mueller ç Müller) or blind (e.g. Gardner). In the case of the decision-directed ones, the user must provide the reference constellation as a parameter (digital.constellation_qpsk().base() in our case). Moreover, they can work either on a "two samples per symbol" or a "one sample per symbol" basis. This has to be taken into account during the different resampling processes implemented in the flowgraph. The Symbol Sync block also requires the $K_p$ parameter value. This implies the calculation of $\bar{g}(\tau_e)$ in the interval $-Ts/2 < \tau_e < +Ts/2$. It can be shown [Rice, 2009] that $\bar{g}(\tau_e)$ not only depends on the TED algorithm chosen but also on the symbol pulse shape. Igor Freire has made significant contributions in this area by providing the community with a collection of Matlab scripts. These scripts enable the determination of $K_p$ for the majority of algorithms implemented by the Symbol Sync block [Freire, 2016]. We have completed his code to cover all the algorithms of the Symbol Sync block and any pulse shape (only the RRC was implemented). Using this code, we have obtained $K_p = 2.6$ for the zero crossing TED and an RRC pulse filter of $\alpha = 0.4$, which is used in the QPSK receiver GNU Radio flowgraph. The corresponding S-curve is shown in Figure 5.22.

When designing a fully digital symbol timing recovery loop, the most efficient way (implemented in GNU Radio) consists in decoupling the Analog to Digital Converter (ADC)/Digital to Analog Converter (DAC) clock from the loop voltage-controlled clock (VCC; which replaces the NCO). At the reception side, the ADC clocked asynchronously from the DPLL clock delivers the samples to

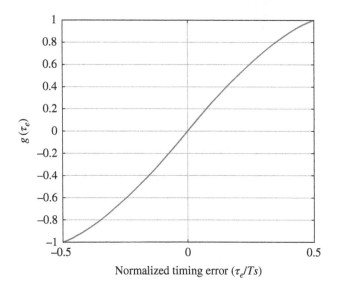

**Figure 5.22** Zero crossing TED S-curve ($\alpha = 0.4$).

the matched filter block preceding the DPLL. Most of the time, the DPLL works on fewer samples per symbol than the preceding stages (remember: the fewer samples the SDR handles, the more computationally efficient). In this perspective, it means that to align the samples to $\tau_{opt}$ requires a way to "create" samples that are not present in the sampled signal. The way to do this is to use an interpolator. Thus, the digital interpolator, controlled by the filtered TED error, produces a sample aligned with $\tau_{opt}$. The adopted structure is shown in Figure 5.23. The reader interested in the design of the interpolator (using Farrow filters) and its controller may refer to Chapter 8 of Rice's book [Rice, 2009]. Moreover, an associated Octave/Matlab script is also available on this book companion website [Friedt and Boeglen, 2024].

Although the digital interpolator implementation works quite well in practice, there exists another very efficient way to replace this structure by acting directly on the matched filter-generated samples. To be able to do so, the matched filter is implemented using a polyphase decomposition. This structure has been comprehensively described in a paper by Harris and Rice [2001] and is available in the GNU Radio Symbol Sync block. Thus, instead of calculating an interpolant between two samples, the desired time delay is selected by connecting the output to the appropriate filter in the filterbank. If you do select this implementation in the Symbol Sync block, you have to disable the existing matched filter located at the input of the block and provide the filter coefficients generated by the firdes.root_raised_cosine() function.

### 5.3.2 Automatic Gain Control (AGC)

Because of the RF channel propagation conditions (distance, fading effects due to multipath propagation, and Doppler effect), the amplitude at the input of the receiver can vary over a wide dynamic range. We have just seen that for the synchronization functions to work properly, a signal amplitude of $\pm 1$ V is necessary. Here is where the AGC comes into play. Its role is to provide a relatively constant output amplitude so that the blocks following the AGC require less dynamic range. AGC structures have been well described in the literature, starting with the analog radio receivers. The digital structure implemented in GNU Radio builds up on the structure represented in Figure 5.24.

It is again a feedback loop. There are two parameters to consider which are $R$, the value of the reference signal, and $\alpha$, which conditions the convergence rate of the loop. We have

$$|y[n]| = \sqrt{y_I^2[n] + y_Q^2[n]} = |b[n-1] \cdot x[n]| \tag{5.34}$$

The error signal $e[n]$ is given by

$$e[n] = R - |b[n-1] \cdot x[n]| \tag{5.35}$$

so that the amplitude update equation can be written as

$$b[n] = b[n-1] + \beta \cdot e[n] \tag{5.36}$$

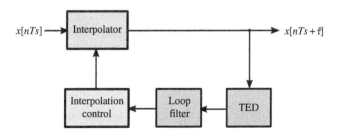

**Figure 5.23** Interpolator-based timing recovery DPLL.

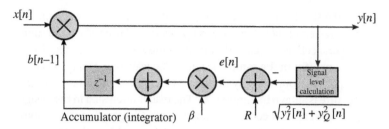

**Figure 5.24** Automatic gain control (AGC) loop.

It can be shown that the convergence time of the loop depends on $\beta$ and the input signal level. Therefore, the convergence time of the AGC is long when the signal level is low and small when the signal level is high. To avoid nonlinearity concerns, it is recommended to limit the loop gain to a finite value. This is what is done for the GNU Radio AGC block with the value of 65 536. It is also possible to differentiate the growth (attack) and decay rates of the AGC as in the AGC2 GNU Radio block. The curious reader is encouraged to consult Chapter 9 of [Rice, 2009] for an in-depth exploration of this topic.

### 5.3.3 Assembling All the Components: The Ultimate QPSK Receiver Flowgraph

Now, it is time to integrate all the elements presented earlier to construct a QPSK receiver flowgraph and experiment with it. We will not delve into the parts that mirror those found in the transmitter, as they are straightforward. The QPSK receiver flowgraph consists of the following three main parts:

- Automatic coarse frequency offset calculation and correction. This operation is represented in Figure 5.25.
  This part of the flowgraph works as follows: The signal at the output of the AGC block is raised to the fourth power (since $M = 4$), and the sampling rate is lowered to a factor of 5, resulting in a rate of 480 kHz. This effectively selects a search range of 120 kHz (see equation 5.17). Subsequently, a 512-point FFT is executed, providing a correction resolution of approximately 234 Hz. Afterwards, the square of the FFT module is calculated and converted in dB. From these values, we search for the index corresponding to the frequency where the maximum magnitude occurs and determine the associated frequency value. This process is handled by the `Carrier Offset Calculation` embedded Python block. The output of this block is probed and injected into the `RTL-SDR Source` block, thereby modifying the `Ch0: Frequency (Hz)` parameter. This step applies coarse frequency correction at the beginning of the flowgraph.
- AGC, timing, and carrier phase recovery loops. This is represented in Figure 5.26.
  After a filtering stage, AGC is performed on the input QPSK signal, which then enters the timing recovery loop (`Symbol Sync` block). After a downsampling to one sample per symbol, the signal enters the carrier phase recovery loop ("Costas Loop" block). At the output of this block, you should observe a perfectly aligned QPSK constellation. The user can interact with this part of the flowgraph since the main parameters of the loops can be varied, thanks to QT Graphical User Interface (GUI) Range cursors (this is highly recommended!).
- QPSK demodulation. This is depicted in Figure 5.27. These steps simply mirror the operations performed at transmission.
  It is worth noticing the differential decoding stage, which is the easiest way to handle the $\pi/2$ phase ambiguity of the timing recovery loop in GNU Radio. This scheme's robustness can be readily demonstrated by examining the flowgraph depicted in Figure 5.28. It facilitates Radio

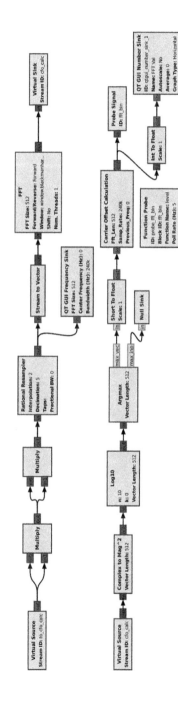

**Figure 5.25** Coarse frequency estimation and correction.

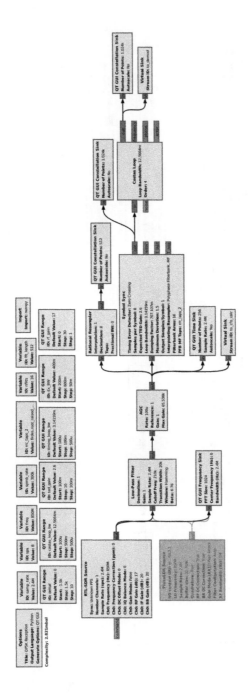

**Figure 5.26** AGC, timing, and carrier phase recovery loops.

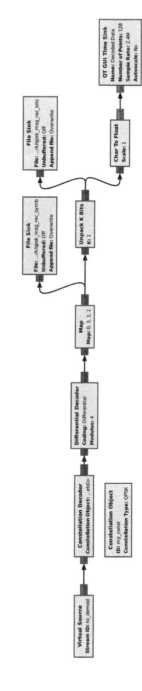

**Figure 5.27** QPSK demodulation and data saving.

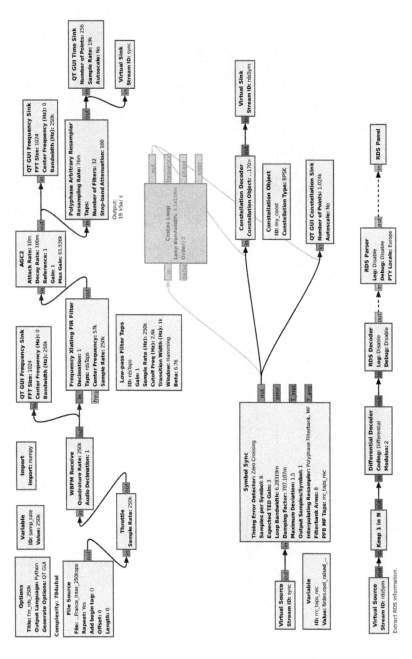

**Figure 5.28** RDS decoding showing the robustness of differential encoding with the "Costas Loop" block deactivated. This flowgraph requires the gr-rds Out Of Tree (OOT) module available at https://github.com/bastibl/gr-rds.

Data System (RDS) decoding of an Frequency Modulation (FM) broadcast station recorded to a file. RDS, as presented in Chapter 2, employs differential BPSK modulation. Upon executing the flowgraph shown in Figure 5.28, readers will observe that despite the deactivation of the "Costas Loop" block, RDS decoding performs effectively even when the received constellation undergoes slow rotation. This resilience is a result of the differential encoding, which mitigates phase rotation induced by carrier frequency offset, provided the rate of phase change remains within the range of at least two QPSK symbols. Despite its robustness, it is important to note that differential encoding is susceptible to the "double error phenomenon." This occurs because of the differential encoding process, where encoding the current symbol relies on the preceding one. Hence, if an error occurs during the decoding stage, it leads to a two-bit error, thereby imposing a 3 dB penalty on the BER curve compared to non-differentially encoded QPSK. Another commonly used method in practice, which is not susceptible to the double error phenomenon, is to employ a "unique word" (UW) [Nguyen, 1989]. This method involves integrating a suitable bit pattern at the beginning of the QPSK data, robust enough to prevent false detection. From this pattern, when there is no phase rotation, or from the phase-shifted copies of the original pattern ($\pi/2$, $\pi$ or $3\pi/2$), the receiver can deduce and correct the phase rotation if necessary.

After the demodulation stages mentioned previously, the QPSK symbols and the associated bits are saved to files for decoding outside GNU Radio.

Experiencing this flowgraph and its associated QPSK transmitter can be a bit frustrating. Although the received signal constellation looks nice, the transmitted data is generated outside of GNU Radio using an Octave script. This also applies to the generated received data files. Performing these two steps inside GNU Radio will necessitate to modify the QPSK modem so as to handle packet data and to find a means to synchronize with the beginning of the packets. We will show how to handle this in Chapter 7 dedicated to tags and messages.

## 5.4 Conclusion

In this chapter, we have provided an overview of digital communications techniques, which have enabled us to build a fully functional QPSK modem with GNU Radio. We have noticed that the framework is indeed very complete since the majority of the necessary functions have been implemented and, most of the time, in a very efficient way (e.g. polyphase decomposition). Although it is possible to perform simulations with GNU Radio, you should reserve it for real-world implementation as it has been designed for that purpose. It is the reason why it provides all the synchronization blocks needed in a real digital communication system. Building a complete QPSK modem with GNU Radio has drawn our attention on the following key points:

- Digital synchronization blocks work on normalized signals ($\pm 1$ V).
- Because of channel-induced impairments, signal levels at the receiver input have a high dynamic that has to be maintained as constant as possible around $\pm 1$V, thanks to an AGC block.
- The phase recovery algorithm (Costas derived) has a $\pi/2$ phase ambiguity. This must be corrected using either differential encoding (which implies a $-3$ dB penalty on the BER) or by using the Unique Word (UW) technique (see [Rice, 2009]).
- Synchronization blocks have an optimal operating range. Do not expect a Costas loop to work efficiently when the frequency offset exceeds 20% of the symbol rate. If it is above this level, then perform a coarse frequency offset correction first!

**188** | *5 Digital Communications in Action: Design and Realization of a QPSK Modem*

- The order of the synchronization blocks is not chosen randomly. The right order is coarse frequency offset correction followed by timing synchronization (**ML derived Costas loop assumes perfect timing synchronization**) and finally fine frequency offset correction with Costas Loop.

We have not covered all the aspects we wanted to in this chapter and have left some of them as homework for the reader. In particular, we have not discussed the error correcting codes (ECC) which exist in GNU Radio and are a very useful part of an efficient digital communication system. In any case, this latter topic will be addressed in Chapter 8, where convolutional and Reed–Solomon codes will be discussed in detail. Finally, although you will find some code on the companion website of the book that implements this, we have not addressed explicitly the type of the data transmitted by the QPSK modem and how we synchronize on the beginning of this data. As can be observed, there are still numerous digital communication aspects left to explore and experiment with in GNU Radio.

## References

Q. Chaudhari. *Wireless Communications from the Ground Up: An SDR Perspective*. CreateSpace Independent Publishing Platform, 2018.

I. Freire. https://igorfreire.com.br/tag/timing-recovery/, 2016.

J. M. Friedt and H. Boeglen. https://gitlab.xlim.fr/gnuradio_book, 2024.

F. Harris and M. Rice. Multirate digital filters for symbol timing synchronization in software defined radios. *IEEE Journal on Selected Areas in Communications*, 19(12):2346–2357, 2001.

U. Mengali and A. N. D'Andrea. *Synchronization Techniques for Digital Receivers*. Springer, New York, NY, 1997.

T. M. Nguyen. *Phase-Ambiguity Resolution for QPSK Modulation Systems. Part 1: A Review*, volume 89-4. NASA JPL Publication, 1989.

J. G. Proakis and M. Salehi. *Digital Communications*. McGraw-Hill, 2008.

M. Rice. *Digital Communications: A Discrete-Time Approach*. Prentice-Hall, 2009.

T. Rondeau. http://www.trondeau.com/blog/2011/8/13/control-loop-gain-values.html, 2011.

C. E. Shannon. A mathematical theory of communication. *The Bell System Technical Journal*, 27(3):379–423, 1948.

U. Wasenmuller, T. Brack, and N. Wehn. Analysis of communications and implementation performance of FFT based carrier synchronization of BPSK/QPSK bursts. *U.R.S.I. Advances in Radio Science*, 6:95–100, 2008.

T. P. Zielinski. *Starting Digital Signal Processing in Telecommunication Engineering –A Laboratory-Based Course*. Springer, 2021.

# 6

## Messages, Tags, and Packet Communications

### 6.1 Introduction

It is well known that the GNU Radio scheduler works on continuous streams of data [Rondeau, 2013]. The flowgraph we build to set up a communication chain moves data continuously from source blocks to sink blocks. The scheduler ensures uninterrupted streams by allocating appropriate buffer sizes between blocks. This way of working is appropriate for continuous communication schemes (e.g. Digital Audio Broadcasting (plus) [DAB+]) but not for packet communications (e.g. WiFi). Although it is not currently possible to change the way the scheduler works, several mechanisms have been introduced to handle bursty communication cases. The first mechanism concerns "tags." Tags are designed to hold metadata and control information. They are associated with a particular sample in the data stream (e.g. the start of a 96 ms DAB+ frame). If not blocked, they flow downstream from block to block. In this manner, subsequent blocks following the one that has added the tag can be notified that a specific event has occurred, such as reaching a definite voltage threshold. The major limitations with tags is that they are only accessible inside a block work function and that they can only propagate in one direction, i.e. from source to sink. On the other hand, the interesting feature of the tag interface is that it is isosynchronous with the data stream. To cope with tag limitations, a more general message passing interface has been introduced. Built on an asynchronous basis, it first allows blocks downstream to communicate back to blocks upstream and, second, it permits to communicate easily back and forth between external applications and GNU Radio. Both interfaces, tag and message, use a particular type of data called "polymorphic types" or PMTs. Although this can look as a computer scientist technobabble, it is a way to extend the strict typing of C++ with something more flexible. The best way to understand the way these two interfaces work is undoubtedly to study several use cases where they are implemented.

The remainder of this chapter is structured as follows: Section 6.2 will provide an introduction to PMTs and their common applications, particularly in tags and messages. Following that, Section 6.3 will present the message interface, while Section 6.4 will focus on the utilization of tags. In Section 6.5, we will delve into case studies to illustrate practical applications of tags, focusing on packet communications. Finally, Section 6.6 will conclude the chapter. Since this chapter heavily relies on writing custom Python and C++ blocks, the reader might be interested in probing Chapter 9 dedicated to these topics.

*Communication Systems Engineering with GNU Radio: A Hands-on Approach,*
First Edition. Jean-Michel Friedt and Hervé Boeglen.
© 2025 John Wiley & Sons, Inc. Published 2025 by John Wiley & Sons, Inc.
Companion website: www.wiley.com/go/friedtcommunication

## 6.2 Polymorphic Types

In GNU Radio, PMTs are used to carry data from one block/thread to another. They are extensively used in stream tags and message passing interfaces. PMTs can represent a variety of data, ranging from Boolean values to dictionaries and can be considered as an extension of the strict typing policy of C++. For a complete list of PMTs features, please refer to the source code and in particular the header file pmt.h at https://gnuradio.org/doc/doxygen/pmt_8h.html. To use them in Python, you have first to import the pmt module:

```
import pmt
```

In C++, the header pmt.h has to be added in your source code:

```
#include <pmt/pmt.h>
```

For the two languages Python and C++, it is possible to directly print the content of PMTs since they have built-in capabilities to cast their value to a string. Moreover, PMTs know their type, so it is possible to query for it using appropriate methods like, for example, is_integer(). This is illustrated by the following example in Python:

```python
import pmt

P_int1 = pmt.to_pmt(12)
P_int1
P_int2 = pmt.to_pmt(22)
P_int2
i1 = pmt.to_python(P_int1)
i2 = pmt.to_python(P_int2)

P_double = pmt.to_pmt(0.75)
P_double
d = pmt.to_double(P_double)

P_cplx = pmt.make_rectangular(1, -1)
P_cplx
z = pmt.to_complex(P_cplx)

P = P_int1

if pmt.is_integer(P):
    d = pmt.to_long(P)
elif pmt.is_real(P):
    d = pmt.to_double(P);
else:
    print("Expected an integer or a double!")

if pmt.eq(P_int1, P_int2):
    print("Equal!")
```

All PMTs are of the type `pmt::pmt_t`. To manipulate PMTs, special PMT functions have to be used. Except for PMT vectors, PMTs are **immutable**. In computer science, mutable and immutable refer to the state of an object and how variables can be assigned and changed. Immutable objects are unchangeable so that you never change the data in a PMT: you have to create a new PMT with new data. As explained in the GNU Radio wiki,[1] the main reason is thread safety. In the GNU Radio multithreaded environment, since it is not possible to write to a shared PMT object, there is no possibility of guaranteeing thread safety of one block reading the PMT data while another is writing the data.

Let us summarize the main features of the PMT library with the types it can represent:

- Boolean values of true/false
- Strings (as symbols)
- Integers (long and uint64)
- Floats (as doubles)
- Complex (as two doubles)
- Pairs
- Tuples
- Vectors (of PMTs)
- Uniform vectors (of any standard data type)
- Dictionaries (list of key:value pairs)
- Any (contains a boost::any pointer to hold anything)

To manipulate PMTs, the library also defines the following set of functions:

- Equal/equivalence between PMTs
- Length (of a tuple or vector)
- Map (apply a function to all elements in the PMT)
- Reverse
- Get a PMT at a position in a list
- Serialize and deserialize
- Printing

The library also defines three important constants:

- pmt::PMT_T – a PMT True
- pmt::PMT_F – a PMT False
- pmt::PMT_NIL – an empty PMT (think Python's 'None')

Let us now describe in more details some unusual but important PMT types. The first one concerns dictionaries. In computer science, a dictionary is an abstract data type that stores a collection of key:value pairs, such that each possible key appears at most once in the collection. Remember that this type is immutable. Here is an example of using PMT dictionary methods in Python.

```python
import pmt
key0 = pmt.intern("int")
val0 = pmt.from_long(123)
val1 = pmt.from_long(234)
```

---

1 https://wiki.gnuradio.org/index.php/Polymorphic_Types_(PMTs).

# 6 Messages, Tags, and Packet Communications

```python
key1 = pmt.intern("double")
val2 = pmt.from_double(5.4321)

# Make an empty dictionary
a = pmt.make_dict()

# Add a key:value pair to the dictionary
a = pmt.dict_add(a, key0, val0)
print a

# Add a new value to the same key;
# new dict will still have one item with new value
a = pmt.dict_add(a, key0, val1)
print a

# Add a new key:value pair
a = pmt.dict_add(a, key1, val2)
print a

# Test if we have a key, then delete it
print pmt.dict_has_key(a, key1)
a = pmt.dict_delete(a, key1)
print pmt.dict_has_key(a, key1)

ref = pmt.dict_ref(a, key0, pmt.PMT_NIL)
print ref

# The following should never print
if (pmt.dict_has_key(a, key0) and pmt.eq(ref, pmt.PMT_NIL)):
    print "Trouble! We have key0, but it returned PMT_NIL"
```

Another very useful type is the BLOB ("binary large object"). It is basically a wrapper around a u8vector. There are two PMT BLOBs of interest. The first one concerns "pairs" and "cons" concepts originated from Lisp language. Simply, if you combine two PMTs, they form a new PMT, which is a pair (or cons) of those two PMTs. Concerning the weird names, the best thing to keep in mind is that "cons" actually should be thought of as a "construct." Pairs are just a useful way of packing several data objects into a single PMT. For example:

```python
import numpy, pmt
P_f32vector = pmt.to_pmt(numpy.array([2.0, 5.0, 5.0, 5.0, 5.0],
    dtype=numpy.float32))
print(pmt.is_f32vector(P_f32vector)) # Prints 'True'
P_pair = pmt.cons(pmt.string_to_symbol("taps"), P_f32vector)
P_pair
print(pmt.is_pair(P_pair)) # Prints 'true'

# car and cdr allow to deconstruct a pair by accessing the key (car)
    and values (cdr) of the pair
```

```
P_key = pmt.car(P_pair)
P_key # Prints "taps"

P_f32vector2 = pmt.cdr(P_pair)
P_f32vector2
```

Finally, the Protocol Data Unit (PDU) BLOB stands out as perhaps the most tangible object, as it readily relates to packet communications. A "PDU" is a PMT pair consisting of a dictionary and a uniform vector. The following snippet builds a packet carrying the message "Hello World!":

```
import numpy, pmt
dict = pmt.dict_add(dict, pmt.intern("packet"), pmt.from_long(1))
pdu = pmt.cons(dict, pmt.pmt_to_python.numpy_to_uvector(numpy.array
    ([ord(c) for c in "Hello world!"], numpy.uint8)))
```

While PMTs might initially seem like a computer science gadget, they actually offer an efficient means of representing data that can be shared between GNU Radio blocks in a thread-safe manner. We will delve into their usage further in Sections 6.3 and 6.4 when we discuss the message and tag interfaces.

## 6.3 Messages

Message passing is a mechanism to asynchronously communicate events and data between blocks. Unlike streams and tags, messages can be passed to upstream blocks. Moreover, it adds a way to communicate back and forth between external applications and GNU Radio. The message passing interface makes an extensive use of PMTs, a concept discussed in Section 6.2. The message passing interface is defined at https://gnuradio.org/doc/doxygen/classgr_1_1basic__block.html in the gr::basic_block class. This class is the parent of all GNU Radio blocks. It defines several methods to deal with messages in the blocks, i.e. to handle incoming messages and to post messages to other blocks. Hence, to use the message interface, a block has to declare its input and output message ports in the constructor method. The message ports are described by PMTs which represent the name of the port. The Application Programming Interface (API)text calls to define a port are implemented in C++ as follows:

```
void message_port_register_in(pmt::pmt_t port_id)
void message_port_register_out(pmt::pmt_t port_id)
```

and in Python:

```
self.message_port_register_in(pmt.intern("port name"))
self.message_port_register_out(pmt.intern("port name"))
```

Thanks to the port name, other blocks can post or receive messages on this port by subscribing to it. To send a message, the following method is invoked in C++:

```
void message_port_pub(pmt::pmt_t port_id, pmt::pmt_t msg);
```

and in Python:

```
self.message_port_pub(pmt.intern("port name"), <pmt message>)
```

## 6 Messages, Tags, and Packet Communications

Subscription is actually achieved by connecting message ports as part of the flowgraph, which internally translates to a call to the `gr::basic_block::message_port_sub` method. To process the messages posted to it, a subscriber block must declare a message handler function. After the declaration of a subscriber port, it has to be bound to the message handler. Starting with GNU Radio 3.8 using C++11, this is realized by a lambda function. A "lambda function" is a function that can be written inline in source code, typically used to pass into another function, akin to the concept of "function pointer."

```
set_msg_handler(pmt::pmt_t port_id,
    [this](const pmt::pmt_t& msg) { message_handler_function(msg); });
```

and in Python:

```
self.set_msg_handler(pmt.intern("port name"), <msg handler function>)
```

When a new message is pushed onto a port's message queue, this function is utilized to process the message. The prototype for all message handling functions is

```
void block_class::message_handler_function(const pmt::pmt_t& msg);
```

and in Python:

```
def handle_msg(self, msg):
```

The connection of messages through a flowgraph is performed with the `gr::hier_block2::msg_connect` method. In this case, PMTs for the port names are not required; simple strings can be used instead. For example, in Python, to connect the block **blk1** output port called "cmd" to the block **dbg** port called "print", we write

```
self.tb.msg_connect(src, "msg", dbg, "print").
```

Before delving into the examples, let us highlight two important features of the message passing interface. First, it enables the intake of messages from an external source. We can directly call a block's `gr::basic_block::_post` method and pass it a message. Consequently, any block with an input message port can receive messages from the outside using this method. A notable example of this feature is the "PDU to tagged stream" block, designed to await PDUs posted to it and convert them to a normal stream, a functionality we will utilize shortly. Second, messages can be employed to send commands to blocks. For instance, the signal amplitude of the `Signal Source` block can be altered using the following command added to a "Message Strobe" block:

```
pmt.dict_add(pmt.make_dict(), pmt.intern("ampl"), pmt.from_double(0.5))
```

As can be noticed, the aforementioned command creates a dictionary and adds a PMT pair of ("ampl", 0.5) to the dictionary. This is expected by the `Signal Source` block message interface. Now, let us summarize the presented message passing interface concepts with two practical examples.

- **First example**. Using the message passing interface to change the frequency of the `Signal Source` block in real time.

  The flowgraph is shown in Figure 6.1. The objective is to dynamically adjust the frequency of the `Signal Source` block in real time, incrementing it by 10 kHz steps from 10 to 50 kHz – a result similar to those described in Section 3.3 when using an embedded Transmission Control Protocol (TCP) server or Extensible Markup Language (XML)-Remote Procedure Call (RPC) but here using a more generic technique. This interaction with the user is facilitated through a `Qt GUI`

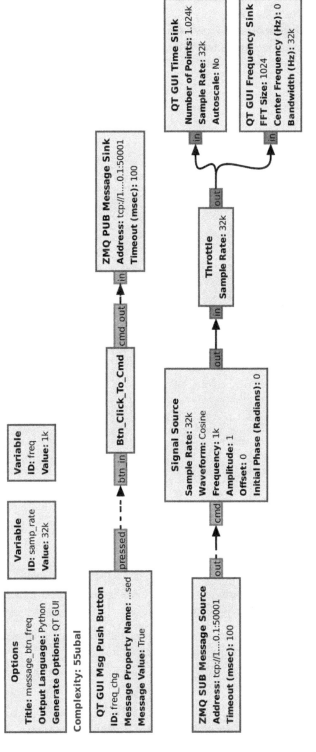

**Figure 6.1** Changing the frequency of the Signal Source block in real time.

**196** | *6 Messages, Tags, and Packet Communications*

`Msg Push Button`: with each click, the frequency increases by 10 kHz. Once it reaches 50 kHz, it loops back to 10 kHz. The button clicks are handled by the `Btn_Click_To_Cmd` Python block which generates the command for the `Signal Source` block. This block accepts messages carrying commands in the form of a PMT key:value pair. To demonstrate that this functionality can be achieved from two separate computers within the same network, we have incorporated ZMQ message blocks. The code for the `Qt GUI Msg Push Button` is straightforward:

```python
"""
Embedded Python Block
"""

import numpy as np
from gnuradio import gr
import pmt

class blk(gr.sync_block):
    """Btn_Click_To_Cmd"""

    def __init__(self):
        gr.sync_block.__init__(self,
            name = "Btn_Click_To_Cmd",
            in_sig = None,
            out_sig = None)
        self.message_port_register_in(pmt.intern('btn_in'))
        self.message_port_register_out(pmt.intern('cmd_out'))
        self.set_msg_handler(pmt.intern('btn_in'), self.handle_msg)
        self.counter = 1
        self.fvalue = 1000.0
        self.freq = 0.0

    def handle_msg(self, msg):
        key = pmt.car(msg)
        value = pmt.cdr(msg)

        if key == pmt.intern("pressed"):
            self.counter = self.counter + 1
            if self.counter == 6:
                self.counter = 1
            self.freq = self.counter * self.fvalue

        self.message_port_pub(pmt.intern('cmd_out'), pmt.cons(pmt.intern("freq"),pmt.from_float(self.freq)))
```

- **Second example**. Packet communications. The associated flowgraph is shown in Figure 6.2. It represents the basics of packet communications in GNU Radio.
  We start with a `Message Strobe` block which constructs the payload and sends it at a rate of one second. The message has to be a PMT pair:

```python
pmt.cons(pmt.make_dict(), pmt.pmt_to_python.numpy_to_uvector(numpy
    .array([ord(c) for c in "Hello GNU Radio!"], numpy.uint8)))
```

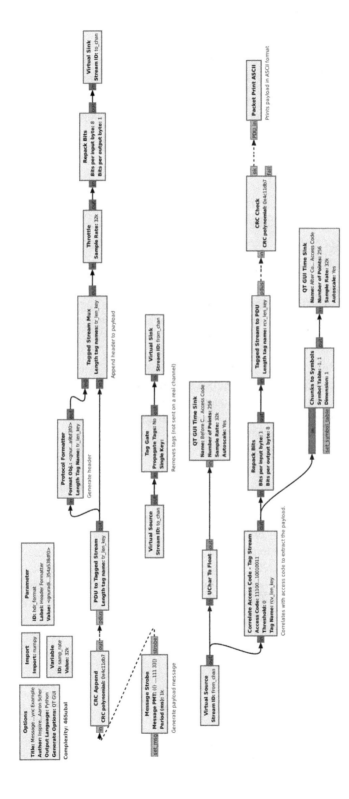

**Figure 6.2** Packet communication using message and stream blocks.

**198** | *6 Messages, Tags, and Packet Communications*

Next, we append a cyclic redundancy check (CRC) to the payload using the CRC Append block. This CRC will be checked upon reception with the associated CRC Check block. At this point, we transition from the message passing interface to the stream interface using the PDU to Tagged Stream block. Then, a header is added to the payload by the Protocol Formatter block. This block is configured by a Parameter block defining the structure of the header:

```
digital.header_format_default('11100001010110101110100010010011', 0)
```

Utilized by the Protocol Formatter block, this defines a header with the following structure:

```
| access code | hdr |
```

where the access code = '11100001010110101110100010010011' and hdr is the length of the payload encoded as a 16-bit value repeated twice:

```
|  0 — 15 | 16 — 31 |
| pkt len | pkt len  |
```

It is important to mention that hdr is used by the Correlate Access Code - Tag Stream block to set the packet length tag. Finally, the payload is appended to the header by a Tagged Stream Mux block to form the following packet structure:

```
| access code | hdr | payload |
```

We do not propagate the tr_len_key tag and block it with a Tag Gate block, which replicates the transmission over a channel. This requires synchronization at the receiving end with the start of a packet, thus leading to the introduction of the Correlate Access Code - Tag Stream block. This block retrieves the payload size, thanks to the "hdr" field after correlating the incoming data with the access code. You may have noticed that, with this example, we have switched back to the stream interface and that synchronization is performed, thanks to the use of tags. This is precisely the subject of the Section 6.4.

- **Third example**. A C++ block detects an input signal zero-crossing and sends a Boolean message accordingly. While out-of-tree (OOT) C++ modules are detailed in Chapter 9, we introduce here their basic structure with a constructor (initialization), a work function (an infinite loop fed with batches of input stream provided by the scheduler), and a destructor. The message port is initialized in the constructor, and the work function of the emitter tests the input signal for the zero-crossing condition for sending the message: in addition to the template created by gr_modtool newmod message and from the newly created gr-messsage directory, gr_modtool add boolean, we edit in the lib directory the C++ source code and add

```cpp
#include "boolean_impl.h"
#include "pmt/pmt.h"
#include <gnuradio/io_signature.h>

namespace gr {
namespace message {
using input_type = float;
boolean::sptr boolean::make()
  {return gnuradio::make_block_sptr<boolean_impl>();}
```

```
boolean_impl :: boolean_impl ()        // constructor
    : gr :: sync_block ( "boolean" ,
                          gr :: io_signature :: make(1 , 1, sizeof(
    input_type )) ,
                          gr :: io_signature :: make(0, 0, 0))
{ message_port_register_out (pmt :: mp( "boolean" )) ;}

boolean_impl :: ~ boolean_impl () {} // destructor

int boolean_impl :: work( int noutput_items , // work function (
    infinite loop)
                          gr_vector_const_void_star& input_items ,
                          gr_vector_void_star& output_items)
{ int index ;
  pmt :: pmt_t P = pmt :: from_bool( true );
  auto in = static_cast<const input_type*>(input_items [0]);
  for (index =0;index<noutput_items -1;index++)
     if (( in [ index ]<=0)&&( in [ index +1]>0))
        { printf( "send \n" );
          message_port_pub (pmt :: mp( "boolean" ) , P);
        }
  return noutput_items ;
}
}}
```

so that a message named `boolean` holding a true Boolean value is emitted when the zero-crossing condition is met. GNU Radio Companion is informed that no output (sink block) is produced, but a tag is generated by editing in the `grc` directory the Yet Another Markup Language (YAML) block definition and modifying `message_boolean.block.yml` with

```
inputs :
- label : in
  domain : stream
  dtype : float
  vlen : 1
  optional : 0

outputs :
- label : boolean
  domain : message
  optional : 0
```

making sure that the `label` of the `output` is the **same name** as the message we are broadcasting from the block. This zero-crossing detection block can be connected to the `Message Debug` block to assess the message emission, but we also create a second block for receiving the message. By `gr_modtool add boolean_recv`, a second block is created, and its source code in `lib` is updated with

```
#include "boolean_recv_impl.h"
#include "pmt/pmt.h"
#include <gnuradio/io_signature.h>

namespace gr {
namespace message {

boolean_recv::sptr boolean_recv::make()
  {return gnuradio::make_block_sptr<boolean_recv_impl>();}

void boolean_recv_impl::message_handler_function(const pmt::pmt_t& msg)
  {printf("receive\n");}

boolean_recv_impl::boolean_recv_impl()
  : gr::sync_block("boolean_recv",
                     gr::io_signature::make(0, 0, 0),
                     gr::io_signature::make(0, 0, 0))
{ message_port_register_in(pmt::mp("boolean"));
  set_msg_handler(pmt::mp("boolean"),
    [this](const pmt::pmt_t& msg) { message_handler_function(msg); });
}

boolean_recv_impl::~boolean_recv_impl() {}

int boolean_recv_impl::work(int noutput_items,
                              gr_vector_const_void_star& input_items,
                              gr_vector_void_star& output_items)
{return noutput_items;}
}}
```

with no input and no output stream, an empty work function since no sample is to be processed, the connection of an input message port to the `boolean` message in the constructor, and the definition of the callback function asynchronously handling the messages. The callback function prototype is defined as a private function in the header file. The absence of input and output stream but an input message is described to GNU Radio Companion through the YAML configuration in the `grc` directory as

```
inputs:
- label: boolean
  domain: message
  optional: 0
outputs:
```

connecting an input message port named with the same name as the received message.

The final flowchart is displayed in Figure 6.3 and connects the `boolean` block input to a slow sine wave generator (to limit the number of produced messages) while the output feeds both a `Message Print` and `boolean recv` blocks. The console output displays

```
send
****** MESSAGE DEBUG PRINT ********
#t
**********************************
receive
```

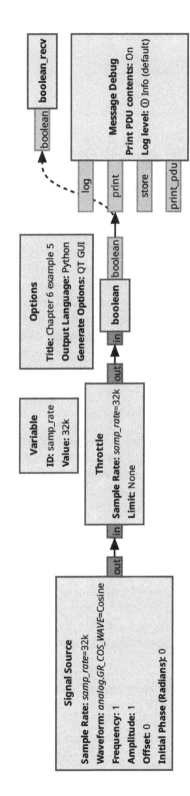

**Figure 6.3** C++ block implementation of message passing demonstration.

## 6 Messages, Tags, and Packet Communications

once every second (frequency of the signal source), with `send` printed by the `boolean` block, `#t` for true value (or `#f` if broadcasting a false value) by the `Message Print` block, and `receive` by the `boolean recv` block.

## 6.4 Tags

Stream tags are an isosynchronous data stream that operates concurrently with the main data stream. A stream tag is generated by a block's work function, for example after a signal processing operation like a correlation. It then flows downstream alongside a specific sample until it reaches a sink or is halted from further propagation by another block. Tags are formed as a key:value pair. The key identifies the meaning of the value, while the value holds the actual data contained within the tag. Both the key and the value are PMTs, with the key being a PMT symbol and the value being any type of PMT. This flexibility allows us to handle any data we wish to pass.

In order to enable tags, the original GNU Radio `gr::block` API had to be extended to include **absolute** item numbers. In the data stream scheduler model, each block's work function of a flowgraph is allocated a buffer that is referenced from 0 to $N-1$. This corresponds to a **relative** offset into the data stream. The **absolute** reference begins from the start of the flowgraph and increments with each subsequent item. Each input stream is linked to the concept of the "number of items read," while each output stream is associated with the "number of items written." These values are obtained during runtime through two API calls:

```
unsigned long int nitems_read(unsigned int which_input);
unsigned long int nitems_written(unsigned int which_output);
```

Therefore, each tag corresponds to an item in the absolute time scale, which is calculated using these functions. Similar to the rest of the data stream, the counts for the number of items read-/written are updated only once during the call to the work function. Hence, within a work function, nitems_read/written will reflect the state of the data stream at the beginning of the function. Therefore, the current relative offset in the data stream has to be added to this value. For instance, if we are iterating overall output items with index `i`, we would write the stream tag to output ports at `nitems_written(0) + i` for the 0th output port.

The stream tag API is divided in two main parts: adding tags to a stream and getting tags from a stream. Please note that the functions described here must be accessed within a call to the general_work/work function. This ensures that we obtain an accurate understanding of the item counts in the buffers.

To add a tag to an output stream one, has to use the `gr::block::add_item_tag` function. It can be called by passing a `gr::tag_t` structure or by specifying the tag values. Tag values are given by

- **Offset**. The offset of the tag in the data stream given in **absolute** item time.
- **Key**. The PMT symbol identifying the type of tag.
- **Value**. The PMT holding the data of the tag.
- **srcid**. The PMT symbol identifying the block which created the tag (optional).

The `gr::tag_t` structure comprises the same members as the in the aforementioned list. Hence, the C++ tag-adding functions are the following:

```cpp
void add_item_tag(unsigned int which_output, const tag_t &tag);
```

```cpp
void add_item_tag(unsigned int which_output,
                  uint64_t abs_offset,
                  const pmt::pmt_t &key,
                  const pmt::pmt_t &value,
                  const pmt::pmt_t &srcid=pmt::PMT_F);
```

In Python we can use one of the following:

```python
add_item_tag(which_output, abs_offset, key, value)
add_item_tag(which_output, abs_offset, key, value, srcid)
```

To get tags from a stream, there are two following possibilities:

- `gr::block::get_tags_in_range`: This function retrieves all tags from a particular input port between a certain range of items (in **absolute** item time).
- `gr::block::get_tags_in_window`: This function retrieves all tags from a particular input port between a certain range of items (in **relative** item time within the work function).

The principal difference between these two functions is that the first one considers **absolute** item time whereas the second considers **relative** item time. The returned argument is a vector of `gr::tag_t` which contains all the tags found in the considered range. Please note that these functions allow to select a particular tag if the user specifies its key. For example, in C++, to get all the tags between the given range of items, we use:

```cpp
void get_tags_in_range(std::vector<tag_t> &v,
                       unsigned int which_input,
                       uint64_t abs_start,
                       uint64_t abs_end);
```

Adding an argument to the previous function allows to retrieve only the specified `key`:

```cpp
void get_tags_in_range(std::vector<tag_t> &v,
                       unsigned int which_input,
                       uint64_t abs_start,
                       uint64_t abs_end,
                       const pmt::pmt_t &key);
```

In Python, you can only retrieve a list of tags using the following method:

```python
def work(self, input_items, output_items):
    ....
    tags = self.get_tags_in_window(which_input, rel_start, rel_end)
    ....
```

If a specific tag must be selected from the list, it is simply a matter of utilizing classical Python functions to achieve the result.

From Section 6.3 second example, we observed that tags do not propagate from a transmitter flowgraph to a receiver flowgraph when a channel exists between them. Additionally, we identified a block designed to halt tag propagation. We can also ask ourselves about what happens to unused tags. In fact, tag propagation depends on the tag policy chosen. There are four policies available in GNU Radio:

**204** | *6 Messages, Tags, and Packet Communications*

- **All-to-All**. All tags from any input port are replicated to all output ports (default).
- **One-to-One**. Tags from input port *i* are only copied to output port *i* (this depends on num inputs = num outputs).
- **Don't**. Does not propagate tags. Tags are either stopped here or the work function recreates them in some manner. The tag propagation policy is set in the block's constructor using the following method:
- **Custom**. It is similar to "Don't," but signals the block that it should implement application-specific forwarding behavior.

```
void set_tag_propagation_policy(tag_propagation_policy_t p);
```

The GNU Radio scheduler handles tags as they propagate through rate-changing blocks such as decimators (`gr::sync_decimator`) and interpolators (`gr::sync_interpolator`), adjusting tag positions by incorporating an offset implied by the rate change. Similarly, synchronous blocks (`gr::sync_block`) with predefined relative rates pose no issues. Regarding the standard `gr::block`, the default relative rate is set to 1.0, although it is the user's responsibility to adjust this if it does not meet their requirements. Nevertheless, GNU Radio adeptly manages tag propagation for a wide range of scenarios. In specific cases, users can set the policy to `Don't` or `Custom` and manually handle tag propagation.

Tags represent a potent capability within GNU Radio, offering a mechanism to synchronize data streams with critical events in communication scenarios, particularly in digital communication systems. Consider packet communication schemes such as WiFi or Zigbee, where accurately detecting the start of an incoming packet is crucial for successful decoding. For continuous communication strategies like DAB+ or Digital Video Broadcasting [DVB], pinpointing specific segments of the data within the stream is essential for decoding associated sound or image data accurately. While tags prove invaluable in such scenarios, it is important to acknowledge that they introduce overhead due to the operations required to extract or propagate them through the stream, necessitating vector manipulations. A prime example of judicious tag usage is evident in the Universal Software Radio Peripheral (USRP) source, which generates a time tag at the onset of communication. Notably, there is no necessity to timestamp each sample, as the start time allows easy calculation of time at any given instant. The USRP adds a new time tag to the stream only following the discovery of discontinuities in received packets.

Let us now develop some practical examples about tags.

- **First example**. Adding a start tag to a DAB+ multiplex frame. Detecting the start of the DAB+ multiplex frame is the initial crucial step in decoding the data. Further elaboration on this aspect will be provided in Chapter 7, which is dedicated to DAB+. In this specific scenario, utilizing a correlation algorithm enables the detection of the Orthogonal Frequency-Division Multiplexing (OFDM) null symbol within the DAB+ signal. This detection process should occur every 96 ms, corresponding to the size of a DAB+ frame. Adding a tag to mark the beginning of the initial data symbol assists subsequent blocks in aligning with the signal, facilitating the execution of further operations required in the decoding process. Here is an excerpt of the C++ code from the `ofdm_synchronization_cvf_impl.cc` file, indicating the specific location and procedure for inserting the tag "Start":

```
if(d_tracking) {
  unsigned int num_syms = std::min(d_symbols_per_frame -
    d_symbol_count, noutput_items / (d_fft_length +
    d_cyclic_prefix_length));
```

```
for (int sym_i = 0; sym_i < num_syms; ++sym_i) {
    // Measure frequency offset for each symbol individually.
    delayed_correlation(&in[sym_i * (d_fft_length +
d_cyclic_prefix_length)], true);
    d_frequency_offset_per_sample = std::arg(d_correlation) /
d_fft_length; // in rad/sample
    if (d_symbol_count == 0) {
        d_phase = gr_complex(1,0);

        //Here is the tag!
        this->add_item_tag(0, this->nitems_written(0) + nwritten,
            pmt::mp("Start"), pmt::from_float(std::arg(d_correlation
))));
    }
```

As observed, the `Start` tag contains a float variable representing the calculated correlation value. Further downstream in the decoding chain, the `DAB Coarse Frequency Correction` block makes use of this tag, as evidenced by the following C++ code excerpt from the `ofdm_coarse_frequency_correction_vcvf_impl.cc` file:

```
int
    ofdm_coarse_frequency_correction_vcvc_impl::work(int
    noutput_items,
                                    gr_vector_const_void_star
    &input_items,
                                    gr_vector_void_star
    &output_items) {
        const gr_complex *in = (const gr_complex *) input_items[0];
        gr_complex *out = (gr_complex *) output_items[0];
        std::vector <gr::tag_t> tags;
        unsigned int tag_count = 0;

        // get tags for the beginning of a frame
        get_tags_in_window(tags, 0, 0, noutput_items);
        for (int i = 0; i < noutput_items; ++i) {
        /* new calculation for each new frame, measured at the
    pilot symbol and frequency correction
         * applied for all symbols of this frame
         */
        if (tag_count < tags.size() &&
            tags[tag_count].offset - nitems_read(0) - i == 0) {
            measure_energy(&in[i * d_fft_length]);
            measure_snr(&in[i * d_fft_length]);
            tag_count++;
        }

        // copy the first half (left of central sub-carrier) of
    the sub-carriers to the output
```

# 6 Messages, Tags, and Packet Communications

```
        memcpy(out, &in[i * d_fft_length + d_freq_offset],
    d_num_carriers / 2 * sizeof(gr_complex));
        // copy the second half (right of central sub-carrier) of
    the sub-carriers to the output
        memcpy(out + d_num_carriers / 2,
            &in[i * d_fft_length + d_freq_offset + d_num_carriers /
    2 + 1],
            d_num_carriers / 2 * sizeof(gr_complex));
        out + = d_num_carriers;
    }

    // Tell runtime system how many output items we produced.
    return noutput_items;
}
```

Finally, Figure 6.4 depicts a flowgraph performing the DAB+ start of frame detection and associated tag generation.

- **Second example**. Navigate the GNU Radio wiki and follow the instructions provided on the *Python Block Tags*[2] page. We consider this tutorial to be excellent, as it effectively emphasizes the tag insertion process in a highly pedagogical manner, along with demonstrating how to retrieve tags with their corresponding sample offsets within a specific time window. Furthermore, this presents an excellent opportunity to inform readers about the existence of the GNU Radio wiki and tutorial pages, which have seen significant enrichment in recent years. This endeavor undoubtedly constitutes valuable work that greatly aids the community in understanding the key aspects of GNU Radio.

Let us transition to the case studies section, where we showcase two applications demonstrating how the utilization of tags and messages assists us in resolving practical issues.

## 6.5 Case Studies

### 6.5.1 Improving the `gr-nordic` OOT Module

The `gr-nordic` module is a GNU Radio OOT written eight years ago by Marc Newlin from Bastille Research, an American company specialized in wireless security. The module was presented [Newlin, 2016] in 2016 as a tool to show vulnerabilities in some wireless mice and keyboards. In particular, it was shown that mice positions based on "Logitech Unifying" dongle and embedding a Nordic Semiconductor NRF24L01+ IC were transmitting without encryption. Indeed, Logitech considered at the time that encryption was not necessary, as it would not give any useful information to a potential attacker. Anyway, `gr-nordic` is a useful tool to analyze Nordic Semiconductor proprietary protocol called "enhanced shockburst." The NRF24L01+ chip can be found in a large number of low-cost IoT devices or radio-controlled toys, including some low-cost drones[3] ). Moreover, since the NRF24L01+ chip implements a packet communication scheme, it constitutes a nice playground to experience with GNU Radio message and tag interfaces. Bastille Research has not upgraded the `gr-nordic` module, which is only compatible with GNU Radio 3.7. We have

---

2 https://wiki.gnuradio.org/index.php?title=Python_Block_Tags.
3 Zyma X8C Venture at https://symatoys.com/goodshow/x8c-syma-x8c-venture.html.

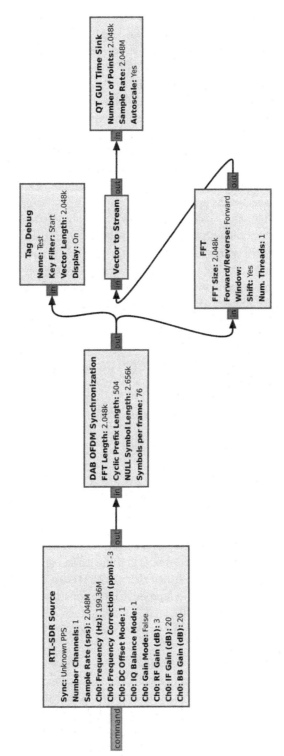

**Figure 6.4** Flowgraph illustrating the DAB+ start of frame generation.

## 6 Messages, Tags, and Packet Communications

recently restructured the module, upgrading it to version 3.8 of GNU Radio. We presented this updated work at the European GNU Radio Days 2019 conference [Boeglen, 2019]. Subsequently, we have ported the module to version 3.10 of GNU Radio, and it is now accessible at https://github .com/hboeglen. This module facilitates the construction of compatible Nordic "enhanced shockburst" transmitters and receivers.

Let us begin with the transmitter flowgraph depicted in Figure 6.5. Our objective is to build an `enhanced shockburst` packet, as illustrated by the fields in Figure 6.6. We achieve this utilizing the `nordictap_transmitter` Python block and the `nordic_tx` C++ block. The former empowers the user to provide various elements necessary for packet construction, including the channel number, the Transmission (TX) address, and, naturally, the payload. These elements are assembled in the `nordictap_transmitter` block to form a PMT message containing a byte-organized vector (`u8vector`). This vector is then sent to the `nordic_tx` block, which formats this data, calculates, and adds a CRC to produce an `enhanced shockburst` compliant packet akin to the one depicted in Figure 6.6. The output of the block is indeed a stream of bytes. To be able to synchronize the visualization with the beginning of a packet (but also to ease further treatments in the transmission chain), we have added three tags: `tx_sob` to mark the start of the data block, `tx_end` to mark the end of the data block, and `packet_len` to give the data length. They appear clearly on the time sink graph shown in Figure 6.7. The following C++ code is an excerpt of `nordic_tx_impl.cc` related to the definition of the previously mentioned tags:

```
add_item_tag(0, nitems_written(0), pmt::string_to_symbol("tx_sob"),
    pmt::PMT_T);
        add_item_tag(0, nitems_written(0) + packet->bytes_length(),
    pmt::string_to_symbol("tx_eob"), pmt::PMT_T);

        int packet_length = packet->bytes_length()*2;

        add_item_tag(0, // Port number
                nitems_written(0), // Offset
                pmt::mp("packet_len"), // Key
                pmt::from_long(packet_length) // Value
                );
```

In this example, the NRF24L01+ compatible packets are transmitted every two seconds at a 2 Mbps rate on a 2.520 GHz (channel 120) carrier. Using the same parameters as this transmitter (see the associated flowgraph), it can be verified that a NRF24L01+ can actually decode these GNU Radio built packets. The reader willing to reproduce these experiments might want to use an STM32 Nucleo board connected to a cheap NRF24L01+ module such as those found at https://az-delivery.de/fr/products/3x-nrf24l01-mit-2-4-ghz. In that case, the STM32 programs can be found at https://github.com/hboeglen.

What about performing the reception with GNU Radio? This is shown in Figure 6.8.

To exclusively capture NRF24L01+ packets, we have added a `Power Squelch` block, which transmits the input signal only if it is higher than the `Threshold (dB)` parameter we have set (here −40 dB). Moreover, the block will emit a tag with the key `pmt::intern("squelch_sob")` having the value of `pmt::PMT_NIL` on the first item it passes, and with the key `pmt::intern ("squelch_eob")` on the last item it transmits. This allows to synchronize the display to the start and end of a packet as can be noticed in Figure 6.9.

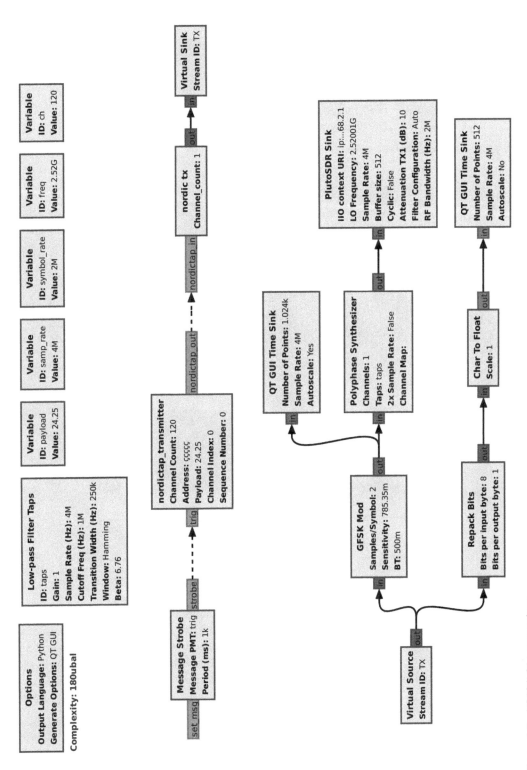

**Figure 6.5** NRF24L01+ transmitter.

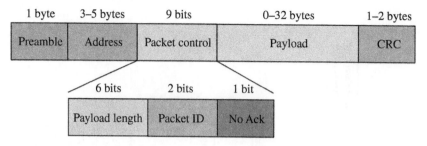

**Figure 6.6** Enhanced ShockBurst Packet format.

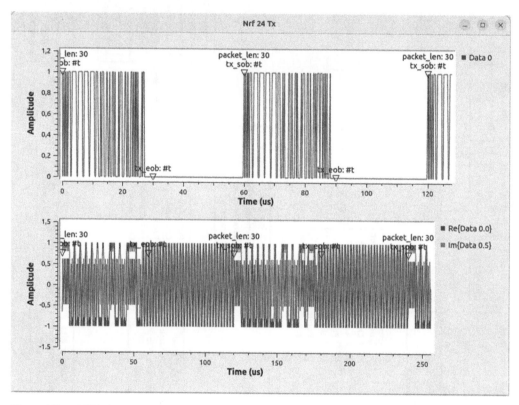

**Figure 6.7** NRF24L01+ packet transmission – time sink graphs.

After Gaussian Minimum-Shift Keying (GMSK) demodulation, the data undergoes processing in the `nordic_tx` block. This block parses the binary data to identify the commencement of a packet by searching for the predefined preamble, which could either be 0x55 or 0xAA depending on whether the first bit of the address in use is set to 0 or 1, respectively. Subsequently, the block produces an output message containing comprehensive details about the decoded packet such as

```
CH=2520
SEQ=3
ADDR=E7:E7:E7:E7:E7
PLD=32:32:2E:35:30:00:00:00
CRC=AD:EB
```

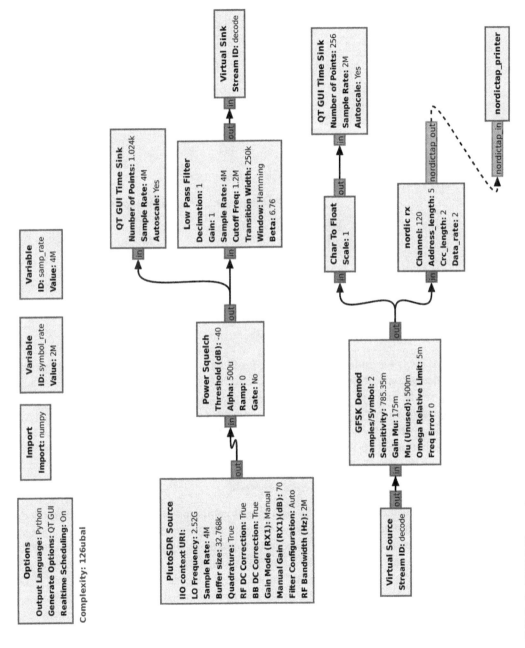

**Figure 6.8** GNU Radio NRF24L01+ receiver.

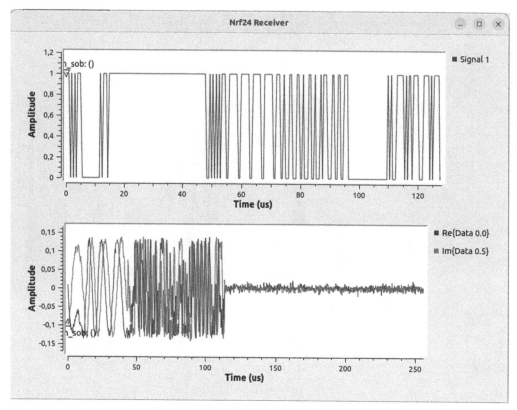

**Figure 6.9** NRF24L01+ packet reception – time sink graphs.

The `gr-nordic` OOT is an interesting module for initiating work with messages and tags. Its handling of a relatively straightforward packet scheme coupled with a quite simple physical layer utilizing GMSK renders it highly accessible. Moreover, the availability of affordable NRF24L01+ modules facilitates experimentation with low-cost embedded systems, thereby enabling validation of the GNU Radio code we are working with. Now, let us transition to the second example, which involves the Quadrature Phase-Shift Keying (QPSK) modem.

### 6.5.2 Converting the QPSK Modem to Packet Mode

In Chapter 5, we concluded with a fully operational continuous transmission QPSK modem. It transmitted a continuous stream of data, which was effectively decoded following timing and frequency recovery procedures. While this approach worked by employing a final offline decoding step outside of GNU Radio, it was not satisfactory. Therefore, if we can devise a method to synchronize with the data stream, coupled with the utilization of message and tag interfaces, we could potentially perform real-time decoding. The main concept is to transition from continuous stream communication to packet-based communication. In Chapter 5, we did not pay much attention to crucial practical transmission issues like timing and frequency recovery loop convergence time. This was because we were working with a large reception file containing the same information repeated numerous times. Therefore, we only needed to examine the initial portion of the data to identify the beginning pattern. However, in the scenario of real-time packet communication, these aspects cannot be overlooked, as neglecting them will lead to decoding failures. Moreover, as synchronizing aspects necessarily imply overhead, the packet rate has also to be taken into account. Let us elaborate on the critical design aspects in what follows.

- Modifications to the QPSK transmitter flowgraph. We need to structure the data to be transmitted in the form of packets. These packets need to incorporate elements to account for the reception time and the convergence time of the frequency recovery loop. To fulfill this requirement, a header of suitable size is designed, utilizing repeating alternating patterns. We have determined experimentally that a packet footer was also necessary. Finally, in order to synchronize with the beginning of a packet, we add an access code to the data to allow for correlation upon reception. More precisely, as we intend to use the `Correlate Access Code - Stream Tag` block, we also have to add the packet length to the packet data. To build this packet, two ways are possible. The first one is by creating a file having the requested structure using an external tool like Octave or Python. The second one, uses the message interface blocks to build the packets. The two ways are presented in the following.

  Here is the Octave code which creates the packet containing a text message. In this example, the text is "This is the GNU Radio QPSK modem working with tags!" With these settings, we get a packet rate of about 10 packets/s.

```
%QPSK modem packet generation

hdr1=repmat([85 170 85 170 85 170 85 170],1,16);
ftr1=repmat([85 170 85 170 85 170 85 170],1,128);

% 1024*7*4*(1/(symb_rate) duration
ftr2=zeros(1,1024*7);

% Enter the text message to transmit
rep=input('Enter the message to send:\n','s');
%This is the GNU Radio QPSK modem working with tags!

%Conversion from ASCII to uint8
chtoasc = uint8(rep - 0);

%Adding access code and payload length
header = uint8([hdr1 225 90 232 147 0 length(chtoasc) 0 length(chtoasc)]);
msg = [header chtoasc ftr1 ftr2];

%Create the file
write_char_binary(msg,'qpsk_msg');
xx = read_char_binary('qpsk_msg')'
```

where the functions `write_char_binary()` and `read_char_binary()` are provided by the Octave compatibility scripts by GNU Radio at https://github.com/gnuradio/gnuradio/tree/main/gr-utils/octave.

As can be noticed, the packet structure will be the one depicted in Figure 6.10.

The same result can be obtained directly in GNU Radio Companion by using the blocks shown in Figure 6.11 instead of the `File Source` block.

## 214 | 6 Messages, Tags, and Packet Communications

| Preamble | Payload | Access code + length | Footer | Zeros |

**Figure 6.10** QPSK modem packet structure.

**Figure 6.11** Packet generation using message blocks.

The command inserted in the Message Strobe block is the following:

```
pmt.cons(pmt.PMT_NIL, pmt.pmt_to_python.numpy_to_uvector(numpy.
    array([ord(c) for c in "This is the GNU Radio QPSK modem
    working with tags!"], numpy.uint8)))
```

And here is the code of the Packet Format custom Python block:

```
"""
Embedded Python Block
"""

import numpy as np
from gnuradio import gr
import pmt

class blk(gr.sync_block):
    """Packet Format"""

    def __init__(self):
        gr.sync_block.__init__(self,
            name = "Packet Format",
            in_sig = None,
            out_sig = None)
        self.message_port_register_in(pmt.intern('PDU_in'))
        self.message_port_register_out(pmt.intern('PDU_out'))
        self.set_msg_handler(pmt.intern('PDU_in'), self.handle_msg)

    def handle_msg(self, msg):
        #Generate header
        hdr1 = np.tile([85, 170, 85, 170, 85, 170, 85, 170],16)
        hdr1 = hdr1.tolist()
        #Generater footer
        ftr1 = np.tile([85, 170, 85, 170, 85, 170, 85, 170],128)
        ftr1 = ftr1.tolist()
        ftr2 = np.tile([0, 0, 0, 0, 0, 0, 0, 0], 896)
        ftr2 = ftr2.tolist()
        #Incoming message
        inMsg = pmt.to_python(msg)
        pld = inMsg[1]
        mLen = len(pld)
```

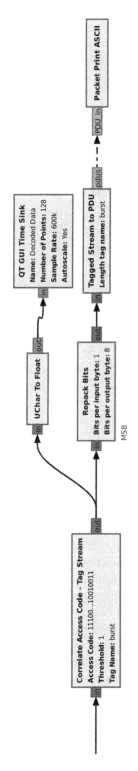

**Figure 6.12** QPSK packet reception added blocks.

```
        if (mLen > 0):
            char_list = hdr1
            #Add access code
            char_list.append(225)
            char_list.append(90)
            char_list.append(232)
            char_list.append(147)
            #Add payload length twice (as required by the 'Correlate
Access Code - Stream Tag' block
            char_list.append (mLen >> 8)
            char_list.append (mLen & 255)
            char_list.append (mLen >> 8)
            char_list.append (mLen & 255)
            char_list.extend (pld)
            char_list.extend (ftr1)
            char_list.extend (ftr2)
            out_len = len(char_list)

            self.message_port_pub(pmt.intern('PDU_out'), pmt.cons(pmt.
PMT_NIL,pmt.init_u8vector(out_len ,( char_list))))
```

Let us switch to the reception part now. The added blocks are shown in Figure 6.12. The main modification compared to Chapter 5 flowgraph is the introduction of the Correlate Access Code - Stream Tag block. This block is responsible for correlating with the 32-bit access code '11100001010110101110100010010011' incorporated into the data. Additionally, it retrieves the payload length and generates the burst tag. As evident in the QT GUI time sink graph, our data length is 51 eight-bit ASCII characters, resulting in a value of 408 bits. Finally, the message is printed in text format in the GNU Radio Companion console.

## 6.6 Conclusion

Messages and tags are essential mechanisms that have been integrated into GNU Radio to streamline the handling of continuous data streams, which form the core of the GNU Radio scheduler. The message passing interface operates asynchronously to the continuous data stream, enabling interaction with blocks at any point in the flowgraph. This feature proves particularly advantageous when dealing with packet communication schemes.

In contrast, tags function isochronously with the data stream but are only accessible from the general_work/work block function. They serve as a mechanism to synchronize with specific events within the data stream. However, excessive use of tags can introduce overhead to the processing chain and should therefore be employed judiciously.

Finally, it is worth mentioning that this chapter draws significant inspiration from the GNU Radio Wiki [2024]. Indeed, this topic is thoroughly covered across various wiki pages. Therefore, we encourage the reader to refer to them for further insights. It is also worth noting the significant contributions of Sandia Labs in the development of specialized tools for managing PDUs within GNU Radio. A portion of this work has been integrated into the GNU Radio tree under the name gr-pdu. Additionally, there are several other related modules accessible on the Sandia Labs GitHub.[4]

---

4 https://github.com/sandialabs

# References

H. Boeglen. Tutorial: Hacking a WSN protocol with GNU radio. In *European GNU Radio Days*, 2019. https://gnuradio-fr-19.sciencesconf.org/resource/page/id/2.html.

GNU Radio Wiki. 2024. https://wiki.gnuradio.org/index.php/Tutorials, accessed March 2024.

M. Newlin. MouseJack: Injecting keystrokes into wireless mice. In *Defcon*, volume 24, 2016. https://media.defcon.org/DEF%20CON%2024/DEF%20CON%2024%20presentations/DEF%20CON%2024%20-%20Marc-Newlin-MouseJack-Injecting-Keystrokes-Into-Wireless-Mice-UPDATED.pd.pdf.

T. Rondeau. http://www.trondeau.com/blog/2013/9/15/explaining-the-gnu-radio-scheduler.html, 2013.

# 7

# A Digital Communication Standard: The DAB+ Radio Broadcasting System

## 7.1 Introduction

Radio broadcasting is certainly one of the most ubiquitous forms of electronic mass media, utilized by billions of people around the globe. Since the early days of radio broadcasting in the 1920s, analog Amplitude Modulation (AM) and Frequency Modulation (FM) have been employed. In the 1970s, significant advances in the digitalization of the audio signal were made, and, in the beginning of the 1980s, the public discovered the first digital audio technology with the advent of the audio compact disk (CD). In order to port this important change to the radio broadcasting area, the European project EUREKA 147 was started in 1987 with a consortium of public research institutes (IRT in Germany and CCETT in France), broadcasters, and telecommunication industries. As will be seen in the following, the design of the new digital broadcasting system was made possible, thanks to the advent of two advanced digital techniques, namely audio bit rate reduction, ported by IRT, and coded orthogonal frequency division multiplexing (COFDM), pioneered by CCETT. This initial work led to the Digital Audio Broadcasting (DAB) system, which was standardized by ETSI in 1997 in EN 300 400 document [ETSIEN300401, 2017]. Since 1997, the DAB standard is maintained by the WorldDAB forum, an international nongovernmental organization whose objective is to promote, harmonize, and coordinate the implementation of DAB services based on the Eureka 147 DAB system. In 2005, it was decided to improve the DAB system and, in particular focusing on the advanced audio coding (AAC) CODEC and its robustness to transmission impairments. This resulted in an evolution of the standard called DAB+ which was published in 2006. Hence, the newly designed CODEC called HE-AACv2 allows increasing the compression rate while maintaining the subjective listening quality, thus permitting the transmission of about twice as more radio channels as DAB. DAB+ is indeed a very interesting case study for SDR. First, because it is an international standard that is worth studying due to its complexity! Second, as it is fully digital, it means that we will have quite many dedicated open-source codes to experiment with, and even an outdated GNU Radio Out Of Tree (OOT) module that we will have to upgrade. As stated before, there are two main aspects to study, namely the source coding part related to audio and the digital communication part related to COFDM. The latter is an interesting continuation of the digital communication aspects we have already covered, in particular, in Chapter 5. Moreover, as DAB+ is a quite old standard, there is a high probability that we can catch a real-life radio station in our living area, as is the case in the area where we live in France.

The rest of this chapter is organized as follows: Section 7.2 will be dedicated to the key technical solutions adopted in the DAB+ standard and why they have been selected. The next sections will

---

*Communication Systems Engineering with GNU Radio: A Hands-on Approach,*
First Edition. Jean-Michel Friedt and Hervé Boeglen.
© 2025 John Wiley & Sons, Inc. Published 2025 by John Wiley & Sons, Inc.
Companion website: www.wiley.com/go/friedtcommunication

be more practical. In particular, in Section 7.3, we will set up a real-time DAB+ transmitter using open-source tools. Then, in Section 7.4, we will focus on the reception side and build a complete receiver using GNU Radio. Finally, we will close this chapter by summarizing the important aspects we have learned, thanks to this digital communication standard.

## 7.2  The DAB+ Standard

### 7.2.1  Foundations of Digital Audio Coding

Audio coding consists of using algorithms to get a compact representation of high-definition audio signals. It is important to have in mind the data rate requested for a 44.1/48 kHz sampled 16-bit audio signal, which is $16 \times 44\,100/48\,000 = 705.6/768$ kbits/s for a mono signal. Considering that the gross bit rate of DAB+ is about 2.3 Mbits/s and that the net capacity falls to about 1.2 Mbits/s to account for the necessary error protection required to transmit data over time and frequency-selective wireless channels, not even one stereo channel can be transmitted without compression. It is also important to recall that DAB+ is expected to provide a subjective listening quality at least the same as the analog FM system if not better. In practice, in France, we observe that a DAB+ multiplex (a group of radio channels occupying about 1.5 MHz of bandwidth) carries 13 radio stations of 88 kbits/s, each representing a compression rate of more than 17! How is it possible to achieve a "transparent" signal reproduction with such a large compression rate? We have to clarify that "transparent" in the jargon of audio coding means that the compressed audio signal cannot be distinguished from the original signal even by a sensitive listener. However, as we will see shortly, it does not mean that the operation is lossless; it has all to do with redundancy and irrelevancy reduction. Redundancy reduction is something quite well known in the compression domain, but irrelevancy reduction is probably not, although it is particularly efficient. To understand how all this works, we have to dive a little bit into the field of psychoacoustics. As stated on Wikipedia, "Psychoacoustics is the branch of psychophysics involving the scientific study of sound perception and audiology i.e. how human auditory system perceives various sounds." So let us present the key facts that are going to help us understand how DAB+ audio coders work.

#### 7.2.1.1  The Absolute Threshold of Hearing

It is well known that the audible audio spectrum ranges from 20 Hz to 20 kHz. However, this does not mean that the human ear has the same sensitivity for all the frequencies of this spectrum, as can be observed in Figure 7.1, which represents the minimum sound pressure level (Sound Pressure Level [SPL] in dB) for an average listener to detect a pure tone stimulus as a function of frequency. Remarkably, one can notice that the best sensitivity is found for frequencies between 2 and 4 kHz, which represent the voice spectrum. Indeed, the human auditory system is optimized for communication between humans! Please also note that the human ear is less sensitive to high frequencies (higher than 10 kHz), and this behavior gets worse with age. As far as audio compression is concerned, all levels below this curve cannot be heard and can therefore be discarded.

#### 7.2.1.2  Critical Bands

The detection threshold for spectrally complex quantization noise is a modified version of the absolute threshold, with its shape determined by the stimuli present at any given time. Since stimuli

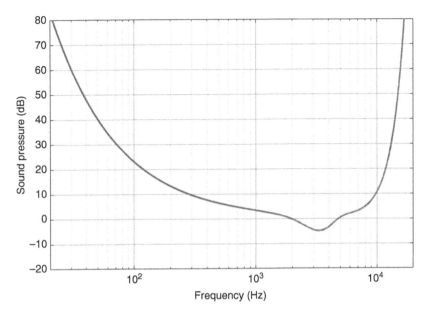

**Figure 7.1** Perceived human hearing.

are, in general, time-varying, the detection threshold is also a time-varying function of the input signal. In order to estimate this threshold, we need to understand how the ear performs spectral analysis. In fact, a frequency-to-place transformation takes place in the cochlea along the basilar membrane [Painter and Spanias, 2000]. The location of the resonant frequency associated with a sound stimulus corresponds to a specific location on the basilar membrane. There is a maximum excitation at this location with reduced sensitivity around this point that looks like a bandpass filter response. The resonant peak locations are spaced along the basilar membrane, started with high-frequency points at the input of the cochlea up to the low frequency ones at the end of the cochlea. Studies have shown that the bandwidth and the spacing of what can be seen as a **filter bank** are not equal. Moreover, there is overlapping between the adjacent band filters. The so-called critical bandwidth is a function of frequency that quantifies the cochlear filter passbands. For an average listener, critical bandwidths can be approximated by

$$bw_c(f) = 25 + 75 \cdot [1 + 1.4(f/1000)^2]^{0.69} \text{ (Hz)} \tag{7.1}$$

As can be noticed in Figure 7.2, critical bandwidths are approximately constant (about 100 Hz) up to 500 Hz and increase to around 20% of the center frequency. The distance between two adjacent critical bands is referred to as one "Bark." The **Bark scale** is a psychoacoustic scale proposed by Zwicker in 1961 on which equal distances correspond with perceptually equal distances.

The critical bandwidths defined by equation 7.1 are widely used in perceptual models for audio coding, but there are also alternative expressions. For an audio coder, the frequency resolution of the auditory filter bank largely determines which portions of a signal are perceptually irrelevant. Thus, a time-frequency analysis using an audio critical band filter bank highlights masking phenomena, which are then used to shape the coding distortion spectrum. According to the detection thresholds determined by the measurement of the energy level within a critical band, the coder can allocate the appropriate number of bits for signal components.

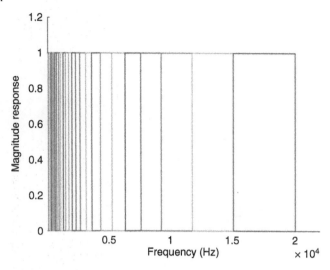

**Figure 7.2** Hearing critical bands.

### 7.2.1.3 Masking

Simultaneous masking is a frequency domain phenomenon when a lower-level signal, called the **maskee** can be made inaudible by a simultaneously stronger signal called the **masker**. It is supposed that the masker and the maskee have close enough frequencies. The influence of this type of masking is largest when it occurs in a critical band. A masking threshold can be measured, and the low-level signals below this threshold cannot be heard. This is illustrated in Figure 7.3. Please note that the slope of the masking threshold is steeper toward lower frequencies.

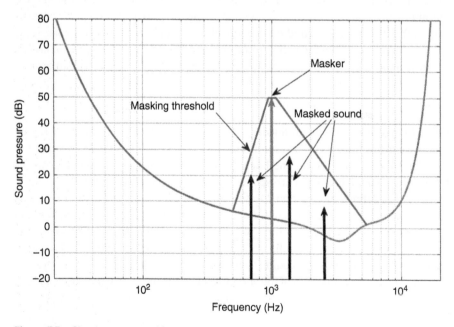

**Figure 7.3** Simultaneous masking.

**Figure 7.4** Masking thresholds and indicators. Source: Adapted from Painter and Spanias [2000].

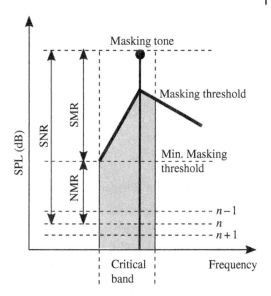

In the field of audio coding, two types of signals are considered: tonal and nontonal sounds. A **tone** is a sound of a single frequency component which emerges in intensity from other frequency components considered as nontonal and often qualified as noise. In the case of masking, it has been shown that a noise masking tone (NMT) has more influence in terms of masking than tone masking noise (TMN). Let us look at Figure 7.4 to help understand these elements. The distance between the masker level and the masking threshold is called the signal-to-mask ratio (SMR). It is minimum around the masker frequency and varies around the masking threshold curve. The minimum in the case of TMN is about 24 dB, whereas for NMT, it is around 5 dB. If we assume a quantization of $n$ bits of the audio signal, quantization noise will not be audible as long as its signal-to-noise ratio (SNR) is higher than its SMR. If we define $SNR(n)$, the SNR resulting from an $n$-bit quantization, the perceivable distortion can be measured by the noise-to-mask ratio (NMR):

$$NMR(n) = SMR - SNR(n) \text{ (dB)} \tag{7.2}$$

We observe that, within a critical band, coding noise will not be audible as long as $NMR(n)$ is negative. In practice, the source signal consists of many simultaneous maskers, each having its own masking threshold. Therefore, a global masking threshold will be computed, describing the threshold of just-noticeable distortions as a function of frequency.

In addition to simultaneous masking, there is also non-simultaneous masking which takes place in the time domain. In this case, considering a masker of finite duration, nonsimultaneous masking occurs both prior to the beginning of the masker and after the masker is turned off. Of course, humans cannot hear into the future. In fact, the built-up time it takes for our ears to perceive the sound results in sounds occurring after the test sound to mask the test sound. Whereas significant premasking tends to last only about 1–2 ms, postmasking will extend anywhere from 50 to 300 ms, depending on the strength and duration of the masker [Painter and Spanias, 2000], [Noll, 1997].

Let us now finish this short overview of psychoacoustics' main principles with Figure 7.5. It presents a generic audio encoder structure that identifies and reduces both statistical redundancies (using a filter bank) and perceptual irrelevancies (psychoacoustic model using masking thresholds). The extracted parameters then allow a bit allocation process, which shapes a quantization and coding step. Finally, a lossless entropy coding step achieves the overall bit reduction process.

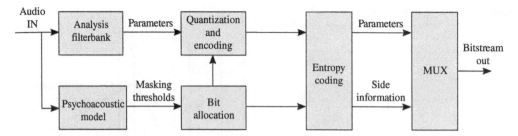

**Figure 7.5** Generic perceptual coder. Source: Adapted from Painter and Spanias [2000].

### 7.2.2 Audio Coding Standards and Their Usage in DAB and DAB+

#### 7.2.2.1 The MPEG/ISO/IEC International Audio Standards

DAB and DAB+ audio makes use of the techniques we described in Section 7.2.1. These techniques and new ones can be found in the well-known Moving Pictures Experts Group (MPEG) audio international standards. The MPEG was set up in 1988 with the aim of developing standards for coded representation of moving pictures, associated audio, and their combination. Later on, it became the working group 11 of ISO/IEC JTC 1 subcommittee 29 so that the official documents issued by this group start with the ISO/IEC JTC lISC 29/WG 11 header. Among the apparent jungle of documents for the novice, we can identify three main works corresponding to an evolution of the MPEG/ISO/IEC audio coding standards [Bosi and Goldberg, 2003]:

- **The MPEG-1 standard (ISO/IEC 11172)** finalized in 1992. It addresses the coding of joint video and audio at a total data rate of 1.5 Mbits/s. The sampling rates supported by MPEG-1 are 32, 44.1, and 48 kHz. It provides a perceptually lossless quality at 128 kbits/s, which corresponds to a compression ratio of 12:1 (for layer III). It comprises three levels called layers: I, II, and III corresponding to increasing complexity, delay, and quality. MPEG-1 layer II is sometimes called MP2, but it is not the same as MPEG-2! MPEG-1 owes much to IRT MUSICAM (layers I and II) and to Fraunhofer IIS Adaptive Spectral Perceptual Entropy Coding (ASPEC) (layer III MP3) pioneering works.
- **The MPEG-2 standard (ISO/IEC 13818)** finalized in 1994. It addresses the coding of joint video and audio at a total data rate of 10 Mbits/s. The sampling rates supported are extended downward from MPEG-1 to include 16, 22.05, and 24 kHz. MPEG-2 defines two coding standards: the backward compatible (BC) one, which preserves the compatibility with MPEG-1, and the AAC one, which does not. AAC, which we will find in DAB+, provides the highest quality at a rate that is half that of BC. The coding performance is obtained, thanks to the introduction of new tools like **gain control**, **temporal noise shaping (TNS)** controlling the temporal shape of the quantization noise, **intensity coding, and coupling** to reduce perceptually irrelevant information consisting in combining multiple channels extracted from the audio signal high-frequency zones to form a single channel. There is also a **prediction** tool whose role is to further remove the redundancies found between adjacent frames. Finally, **mid/sum (M/S)** coding eliminates stereo redundancy by encoding the sum and difference signal instead of the left and right channels. MPEG-2 AAC defines three quality–complexity trade-off profiles, namely the main profile, the low-complexity profile (LC), and the sampling rate scalable (SRS) profile. Each profile uses a particular set of different tools such as the ones listed before.
- **The MPEG-4 standard (ISO/IEC 14496)** finalized in 1998. It encompasses all the domains related to audio coding, ranging from high-fidelity speech coding and audio coding to synthetic speech and audio. The new MPEG-4 AAC decoder adopts further improvements on scalability,

error resilience, and some additional spectral processing features including perceptual noise substitution (PNS), Long-term predictor (LTP), etc. [ISO/IEC14496-3, 2019]. In 2003, a new version of AAC called high-efficiency AAC (HE-AAC) was standardized. It is targeted on low bit-rate applications. HE-AAC adopts a new tool called spectral band replication (SBR), invented by a Swedish firm called Coding Technology (now part of Dolby Digital), which is able to reconstruct high-frequency band output based on low-frequency band data and appropriate side information. Finally, in 2004, HE-AAC Version 2 (HE-AAC V2) was standardized by the MPEG community [ISO/IEC14496-3, 2019]. It uses the **parametric stereo tool (PS)** on the basis of SBR (HE-AAC), which can reconstruct stereo audio signals based on monaural downmixed signals and a limited number of additional stereo parameters.

#### 7.2.2.2 The DAB and DAB+ Audio Coders

The audio of DAB is described in ETSI TS 103 466 DAB audio coding (MPEG layer II) document [ETSITS103466, 2019]. It describes in great details the audio coding and decoding, which complies with ISO/IEC 11172-3 (1993) Part 3: Audio and ISO/IEC 13818-3 Part 3: Audio. However, in DAB, only two rates are permitted: 48 and 24 kHz. Each audio frame has either a 24 or 48 ms duration, and each frame has a constant size. The audio frames are carried in one or two DAB logical frames, respectively. A range of bit rates and audio modes are available and the addition of programme associated data (PAD) allows supplementary content to be provided. Figure 7.6 depicts the simplified block diagram of the encoder. As can be seen, it is very similar to the generic perceptual coder of Figure 7.5 and corresponds to the MPEG-1 layer II coder.

At this point, it is interesting to play with MPEG-1 layer II encoding by using an existing software implementation. This is indeed an efficient way to grasp the complexity of the different signal processing tools involved. If you want to try the Octave way, then you can use [Petitcolas, 2022]. If you prefer Python, then you can have a look at [Schuller, 2024]. Last but not least, some bits from a course called "Perceptual Audio Coding Music 422" given at Stanford university by Professor Marina Bosi, an international renowned researcher in this domain. During her course, the students write a perceptual CODEC in Python and add improvements to the baseline coder. You will find the work of former students on GitHub. She has given a nice conference about "How perceptual audio coding has shaped our lives" at the 2018 Audio Developer Conference in London,

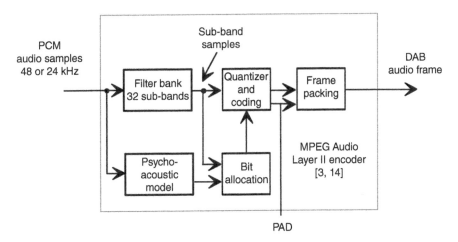

**Figure 7.6** DAB audio encoder. Source: ETSITS103466 [2019] /With permission of ETSI.

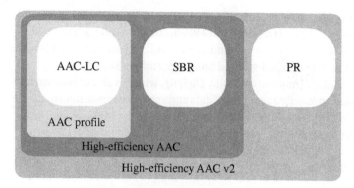

**Figure 7.7** Structure of HE-AACv2.

which is available at [Bosi, 2018]. She is also the author of an excellent book on the subject [Bosi and Goldberg, 2003].

The audio of DAB+ is described in ETSI TS 102 563 document [ETSITS102563, 2017]. DAB+ adopts the MPEG-4 high-efficiency advanced audio coding v2 (HE-AACv2) profile.

As can be seen in Figure 7.7, this profile comprises three main parts:

- **AAC-LC (low complexity)**. This is the base coder. It is a simplified version of the AAC main profile, which does not use the prediction and gain control tools. Moreover, the TNS tool used has a limited order compared to the main profile [ISO/IEC13818-7, 2006]. However, please note that, compared to MPEG-1 layer II AAC, it has the same perceptual audio quality at half the bit rate.
- **Spectral band replication (SBR)**. The principle is depicted in Figure 7.8. Used with AAC-LC, it constitutes the HE-AACv1 profile. The general idea behind SBR is that there exists a strong correlation between the high-band and the low-band frequencies of an audio signal. In the case of a 48 kHz sampled signal, the border between low-band and high-band is fixed at 8.2 kHz. The SBR coder works as follows. The input signal is first divided in two branches: the first one feeds the AAC encoder after a downsampling by two, while the second feeds a 64-channel Quadrature Mirror Filter (QMF) filter bank. The complex-valued subband signal filter bank output is then fed to an envelope estimator and various detectors. The outputs of the detectors and the envelope estimator are assembled and coded with entropy coding to form the SBR data stream, which is finally supplied to the core encoder.

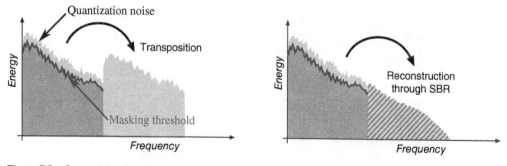

**Figure 7.8** Spectral band replication principle. Source: Adapted from Meltzer and Moser [2006].

- **Parametric stereo (PS).** This addition to the HE-AACv1 profile constitutes the HE-AACv2 profile. Parametric stereo coding aims at describing a stereo signal as a mono signal plus a set of parameters characterizing the stereo image. From the stereo input signal ($l[n], r[n]$), the time-variant stereo parameters representing the perceptually relevant spatial cues are estimated. The stereo parameters are then quantized and coded into the ancillary part of the mono bit stream, yielding a backward (mono) compatible system. This operation allows again a significant reduction of the bit stream.

Used in the DAB+ system, MPEG-4 HE-AACv2 audio will have the following restrictions [ETSITS102563, 2017]:

- The permitted output sample rates of the HE-ACCv2 decoder are 32 and 48 kHz, which means that if SBR is used, the AAC core is operated at 16 or 24 kHz per channel.
- The number of samples per channel per AU (audio unit) is 960. This allows to build an audio super frame of 120 ms, which will be compatible with the DAB framing system.
- The audio bit rate is restricted to a maximum of 192 kbits/s (including program associated data (PAD)).
- AUs are always composed into audio super frames, which always have 120 ms duration. The number of AUs per super frame can range from two to six, depending on the AAC sampling rate, which can range from 16 to 48 kHz.

Each audio super frame is carried in five consecutive logical DAB frames (Common Interleaved Frames (CIF) frames; to be described hereafter), which enables simple synchronization and management of reconfigurations. The size of the audio super frame is defined by the size of the Main Service Channel (MSC) sub-channel (see EN 300 401, clause 6.2.1), which carries the audio super frame. Sub-channels are multiples of 8 kbps in size. As can be noticed in Figure 7.9, after proper framing, the HE-AACv2 audio stream is complemented with Reed-Solomon (RS) error correction and time interleaving.

In HE-AACv2, the AAC coder encodes the low band, SBR encodes the high band, and PS encodes the stereo image in a parameterized form. As can be seen in Figure 7.10 in a typical HE-AAC implementation, the input signal is fed into a 64-band QMF bank and transformed in the QMF domain. For stereo bit rates below 36 kHz, the PS tool is used and extracts the stereo information from the QMF samples. A stereo-to-mono downmix is applied and fed into a 32-band QMF synthesis filter, which reconstructs the mono signal in the time domain at half the audio sample rate. This signal is then fed into the AAC encoder. If PS is not used, the audio signal is fed into a 2:1 resampler, which feeds the AAC encoder. As can be seen in Figure 7.10, the SBR tool also works in the QMF domain.

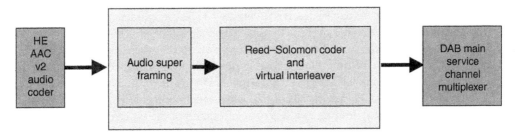

**Figure 7.9** HE-AACv2 DAB frame adaptation. Source: ETSITS102563 [2017] / With permission of ETSI.

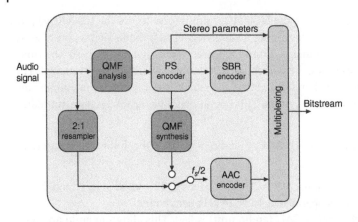

**Figure 7.10** HE-AACv2 block diagram. Source: Adapted from Meltzer and Moser [2006].

### 7.2.3 Digital Transmission over Time and Frequency-Selective Channels: The Need for COFDM

#### 7.2.3.1 Time and Frequency-Selective Wireless Channels

The DAB system was designed to ensure reliable reception on mobile systems (i.e. cars and trains). To design the appropriate digital system for these conditions, it is crucial to understand the primary physical effects encountered on a wireless channel. Let us first consider a channel with no mobility between the receiver and the transmitter. This is the case when you listen to a DAB+ station while being at home.

As depicted in Figure 7.11, the transmitted wave interacts with objects in the environment, giving rise to phenomena such as scattering, diffraction, and reflection. Consequently, the receiver detects multiple replicas of the transmitted signal, known as multipath components (MPC). The amplitudes and phases of these components vary based on their travel distance and the nature of their interactions with objects in the environment. Additionally, if the transmitter and receiver have a direct line of sight, a direct component is possible. Therefore, MPCs are typically categorized as either line of sight (LOS) or non-line of sight (NLOS). Clearly, a LOS component enhances communication. A LOS path is more likely to be present in countryside areas than in urban areas. As can be seen in Figure 7.11, MPCs arrive at the receiver with variable delays $\tau_i$. The higher the $\tau$, the longer the distance traveled by the MPC. This representation of the MPC power as a function of $\tau$ is called a **power delay profile (PDP)**. What about the impact of multipath propagation on a wireless transmitted digital signal? Although the presence of multiple copies of the original signal at the reception side could be considered as an advantage (if exploitable), it has a detrimental effect on the transmitted spectrum. As can be observed in Figure 7.11, the amplitude (and also the phase; not represented on the figure) varies over the transmitted bandwidth. Please note that the channel behaves like a filter for the transmitted signal. Hence, a convolution between the transmitted signal and the channel PDP $h(\tau)$ in the time domain consists of a multiplication between the transmitted signal spectrum and the channel transfer function $H(f)$ in the frequency domain. Therefore, $h(\tau)$ and $H(f)$ are connected by a Fourier transform. The consequence of this is a dramatic increase of intersymbol interference (ISI), leading to a severe deterioration of the bit error rate (BER). This channel impact can be compensated for by designing a filter having an inverse transfer function to that of the channel. This kind of filter is called an **equalizer**. However, it has a major drawback,

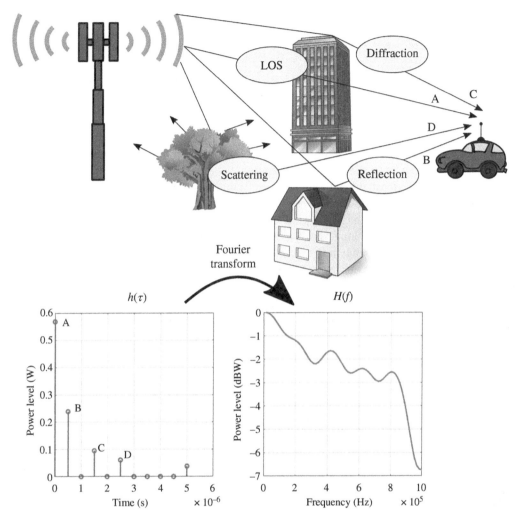

**Figure 7.11** Multipath propagation mechanisms.

which is its computational complexity when the signal bandwidth typically exceeds one megahertz. Consequently, this is not the technique used in DAB since the required bandwidth is over 1.5 MHz. To deal with this frequency selectivity aspect, we first need an indicator of the frequency stability bandwidth of the considered channel. As the wireless channel is essentially a stochastic system, we cannot expect precise values but rather statistical averages. Moreover, these will vary significantly with the environment and the type of channel considered. We have just noticed from Figure 7.11 that a PDP is made up of several MPCs having different arrival times. Rather than the maximal arrival delay time $\tau_{max}$ of the PDP, it has been shown [Lee, 1989] that the rms delay spread $\sigma_\tau$ is the parameter to consider. It has indeed a significant influence on digital communications performance (i.e. the BER). It is defined as

$$\sigma_\tau = \sqrt{\overline{\tau^2} - (\overline{\tau})^2} \tag{7.3}$$

where

$$\bar{\tau} = \frac{\sum_k P(\tau_k)\tau_k}{\sum_k P(\tau_k)} \quad \overline{\tau^2} = \frac{\sum_k P(\tau_k)\tau_k}{\sum_k P(\tau_k^2)} \tag{7.4}$$

with $\tau_k$ and $P(\tau_k)$, respectively, the MPC $k$th delay and the absolute power of the MPC $k$th delay of the PDP. This parameter is related to the 50% channel **coherence bandwidth** defined as

$$B_c \approx \frac{1}{5\sigma_\tau} \tag{7.5}$$

Practically, the coherence bandwidth is a statistical measure of the range of frequencies over which the channel can be considered "flat," meaning that it passes all the related bandwidth frequency components with roughly the same gain and a linear phase. In the case of DAB, rms delay spread values ranging from 1 to 25 μs have been measured for different channel conditions, with the two extremes being the vehicular typical urban (TU) and the countryside hilly terrain (HT) channels. This leads to a $B_c$ ranging from 8 to 200 kHz. We can therefore conclude that the transmission of the 1.5 MHz DAB bandwidth over these channels will necessarily imply high-frequency selectivity. This is the reason why the **orthogonal frequency division multiplexing (OFDM)** technique has been adopted in the standard. We will develop this aspect in the following pages.

However, this is only one part of the story. Up to now, we have considered a static channel; i.e. the receiver does not move. In the case of a moving car, the received signal spectrum is subject to the well known **Doppler effect**. The Doppler shift $f_d$ is given by

$$f_d = \frac{v}{\lambda} \cdot \cos(\vartheta) \tag{7.6}$$

with $v$ being the vehicle speed in m/s, $\lambda$ the speed of light, and $\vartheta$ the angle of arrival of the incident wave. One can notice that $f_d$ can be either positive or negative depending on whether the vehicle is traveling toward the transmitter or moving away from it. This effect, observed in the frequency domain, results in a Doppler spectrum. In the case of the transmission of a pure sine frequency $f_c$ (the carrier frequency), the Doppler spectrum will have components in the range $f_c - f_d$ to $f_c + f_d$. If we now look at the received signal in the time domain, we will observe amplitude fluctuations, which can be very important. This is depicted in Figure 7.12.

The fluctuation time rate is proportional to the maximum Doppler shift $f_{dmax}$, and it is possible to determine a channel coherence time $T_C$, which is again a statistical measure of the time duration over which the channel PDP is essentially invariant. More precisely, it is the time duration over which two received signals have a strong probability of amplitude correlation. $T_C$ is given by

$$T_C = \frac{0.423}{f_{dmax}} \tag{7.7}$$

As an example, what about the maximum Doppler shifts for DAB+ channel 8C (199.36 MHz) on Very High Frequency (VHF) band III? If a vehicle travels in France at the motorway speed limit of 130 km/h i.e. $\approx 36$ m/s, then $f_{dmax} \approx 24$ Hz and $T_C \approx 17.6$ ms. Going back to the bottom-right curve of Figure 7.11, representing the signal power at the output of the channel versus frequency, and Figure 7.12, representing the signal power at the output of the channel versus time, we can notice frequency and time zones where the signal power is very attenuated. It is said that it suffers from **fading**. In the case of deep frequency or time fades, the receiver Automatic Gain Control (AGC) will not be able to compensate for the signal attenuation, and the signal will be lost. These fading zones will have a $B_C$ frequency width and a $T_C$ time width. Finally, it is important to keep in mind that these effects occur simultaneously so that the wireless channel can definitely be seen as a two-variable system; i.e. the PDP becomes $h(\tau, t)$.

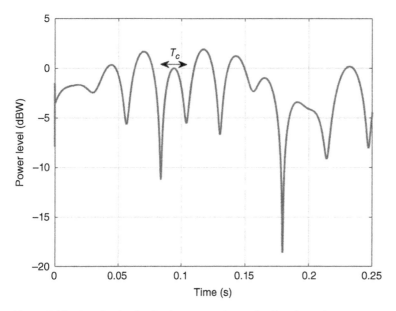

**Figure 7.12** Amplitude distribution over a time-selective channel.

It is interesting to have wireless channel simulation models to study their impact on communication systems. The most used statistical simulation models are based on the wide-sense stationary uncorrelated scattering (WSSUS) assumption, which means that the second-order moments of the simulated PDP are stationary (i.e. the MPC powers are time-independent) and that the MPCs are uncorrelated (they are simulated using separate generators) [Paetzold, 2011]. We are lucky since we can find channel model blocks in GNU Radio. Let us start with the simulation of a non-frequency-selective channel model. If not frequency selective, its PDP is made up of only one MPC whose delay $\tau = 0$. Moreover, let us suppose that this MPC is, in fact, a sum of several NLOS paths arriving at the receiver with the delay $\tau \approx 0$ and with approximately equal power. In this case, it can be shown that the power samples at the receiver follow a Rayleigh probability density function (PDF) so that it is common to say that this channel is a Rayleigh channel. If the PDP includes an LOS path, then the power samples follow a Rice PDF. **Rayleigh** and **Rice channel** models are very common in the digital communication field. Practically, when you want to set up a Rayleigh channel, you have to cascade two blocks, namely a Rayleigh fading block and an average Gaussian noise (AWGN) block. These two blocks simulate a realistic non-frequency-selective wireless channel. This is shown in Figure 7.13 with two GNU radio blocks, which you will find in the `rayleighchan.grc` flowgraph located on the companion website of the book.

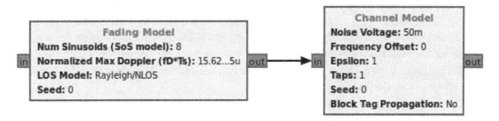

**Figure 7.13** Rayleigh channel GNU Radio blocks.

You will notice that the time variability related to the Doppler frequency is, in fact, represented in the `Fading Model` block by the **fDTs** parameter. It is quite common to normalize the Doppler frequency with respect to the sample time $Ts$. Execute the flowgraph and notice how the Quadrature Phase-Shift Keying (QPSK) constellation "dances." This is because of the amplitude and phase changes caused by the channel. Notice also that, although the amplitude of the transmitted QPSK spectrum changes with time, it is constant over all the frequencies that compose it, thus proving that this channel is not frequency-selective or frequency "flat." To obtain frequency selectivity, we must have a PDP like the one represented in the bottom left of Figure 7.11, i.e. with at least an MPC having a nonzero delay $\tau$. Go and experience with the `cost207tuchan.grc` flowgraph. This flowgraph uses a COST207 Typical Urban channel model, whose PDP is made up of 6 MPCs. COST207 channel models have been designed and standardized for the deployment of 2G telephony networks over Europe [CEC, 1989]. They define several typical environments and were obtained from intensive measurement campaigns over Europe. They have also been considered in some research papers to evaluate the DAB system in mobile conditions. Returning to the GNU Radio simulation, it is evident that the QPSK transmitted spectrum experiences amplitude (and phase, although not represented) variations that vary depending on the frequency location within the spectrum. Thus, this channel exhibits frequency selectivity. Additionally, it is important to highlight that in both studied cases, as the Doppler frequency is nonzero, the two channels also demonstrate time selectivity.

### 7.2.3.2 COFDM: Digital Communication Techniques for the Wireless Channel

A DAB/DAB+ transmission occupies roughly 1.5 MHz of bandwidth, and it has to cope with mobility. This is the reason why the wireless channel to consider is going to be time and frequency selective. Concerning the time selectivity, we came up with a $T_C \approx 18$ ms, and for the frequency selectivity, we determined a $B_c$ ranging from 8 to 200 kHz. How are these parameters used to design a digital communication scheme? In fact, for the system to be immune to time selectivity, we have to transmit modulation symbols with a symbol time $Ts \ll T_C$ i.e. 10 times smaller than $T_C$, i.e. $Ts < 1.8$ ms. To cope with frequency selectivity, applying the same method than for time selectivity, i.e. the signal bandwidth $B \ll B_C$ or $B < 800$ Hz. Here we have a problem: our bandwidth $B = 1.5$ MHz. How can we solve this? Quite simply with a technique called OFDM! Let us see how this works. As stated before, if we use a single-carrier system to transmit our 1.5 MHz bandwidth over a frequency-selective channel and if we do not implement an equalizer, we are going to lose our data. Indeed, if only some parts of a single-carrier spectrum are impacted by an amplitude fading and random phase, the signal recovery process will be compromised, resulting in the loss of all data. Let us try to send the data differently. A simple time-frequency analysis will help us get the feeling of what has to be done. Figure 7.14 shows two ways of sending the same data on a frequency-selective channel. On the left of the figure is represented the single-carrier way. We send bits with a symbol time $T_{S,SC}$, which occupies a bandwidth of $B = 1/T_{S,SC}$. As can be seen, frequency selectivity occurs. On the right side of the figure, we transmit the same data but with a much longer symbol time of $T_{S,MC}$. Each bit carried this way occupies only a small part of the total bandwidth B, and moreover, the individual occupied bandwidths $\Delta f$ are independent from each other. Seen from the individual bits, the channel is now frequency flat. Moreover, if some bits are lost because of fading in their occupied bandwidths, we can still recover the other bits of this entity. This entity represents an **OFDM symbol**.

As a start, we are left with a number of bits or modulation symbols that we want to send using the multicarrier approach. How do we "push" this data in the frequency spectrum as depicted on the right of Figure 7.14? This is shown in equation 7.8.

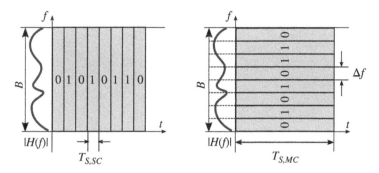

**Figure 7.14** Single-carrier versus multicarrier transmission.

$$s(t) = \frac{1}{\sqrt{N}} \sum_{-N/2}^{N/2} S_k e^{j2\pi k \Delta f t} \cdot rect\left(\frac{t}{T_{S,MC}}\right) \quad (7.8)$$

Equation 7.8 describes an OFDM signal of N subcarriers (the exponential function which "places" the data in the occupied bandwidth B) in the time domain. $T_{S,MC}$ is the OFDM symbol time duration and $S_k$ are the modulation symbols (*DQPSK* in the case of DAB). The following relations hold:

- $T_{S,MC} = N \cdot T_{S,SC}$
- $\Delta f = \frac{1}{T_{S,MC}}$
- $B = N \cdot \Delta f$

Figure 7.15 presents an example spectrum of this signal with $N = 9$ subcarriers. The left plot shows the individual subcarriers, and the right plot the OFDM signal resulting from the sum of the individual subcarriers. Please note that the carrier located at 0 Hz is not used. Also note that, although the subcarriers overlap, they do not interfere with each other. Indeed, it is seen that at each discrete subcarrier frequency location, the other subcarriers are equal to 0. This condition, in addition to being bandwidth-efficient, illustrates the orthogonality property of the subcarriers.

By discretizing equation 7.8 and setting $\Delta t = \frac{T_{S,MC}}{N}$, we end up with equation 7.9.

$$s[n] = \frac{1}{\sqrt{N}} \sum_{-N/2}^{N/2} S_k e^{j2\pi \frac{nk}{N}} \quad (7.9)$$

It can be noticed that this equation is indeed an Inverse Fast Fourier Transform (IFFT). This fact implies that the OFDM modem can be implemented with computationally efficient IFFT/Fast Fourier Transform (FFT) algorithms. We have yet to determine the key parameters of the OFDM system, i.e. how are $T_{S,MC}$, $\Delta f$ and N determined? It depends on the considered channel. We have just seen that the coherence time $T_C$ and the coherence bandwidth $B_C$ were the parameters representing the time and frequency selectivity of the channel. We can, therefore, deduce the following relations for the OFDM system of spectrum occupancy B:

$$\begin{aligned} \Delta f &\ll B_C \\ T_{S,MC} &\ll T_C \end{aligned} \quad (7.10)$$

Taking the DAB system as an example, this yields to $\Delta f \approx 800$ Hz and $T_{S,MC} = 1.8$ ms. Consequently, the number of subcarriers required to transmit the 1.5 MHz bandwidth is $N = B/\Delta f = 1875$. Using only power of two (FFT efficiency), we select 2048 subcarriers. These values are to be compared to the ones used for DAB Mode 1, which are N = 2048 (!) and $T_{S,MC} = 1$ ms.

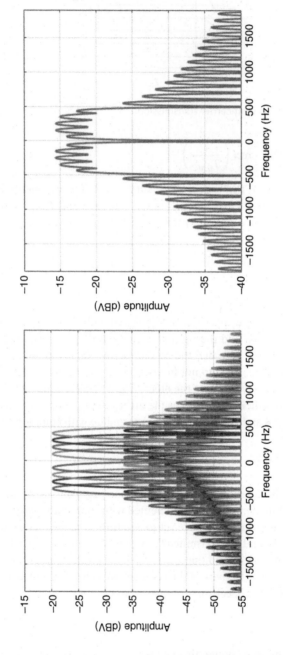

**Figure 7.15** OFDM spectrum for $N = 9$ subcarriers.

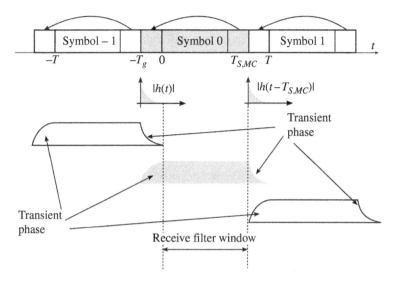

**Figure 7.16** Cyclic prefix and channel influence.

Like all digital communication systems, OFDM can be subject to inter-symbol interference, particularly on a frequency-selective channel. In order to mitigate this, a cyclic prefix (CP) is added at the front of each OFDM symbol. The CP is also called the guard interval and noted $T_g$. To avoid ISI, we must ensure that $T_g = \tau_{max}$. The operation is illustrated in Figure 7.16, where a CP is added to each OFDM symbol at the transmitter and is removed at the receiver.

One can notice that the channel PDP only influences the CP part of the OFDM symbol thus reducing significantly ISI. Finally, why is the CP formed this way, i.e. by taking samples at the end of the OFDM symbol and adding it at the beginning? Without the CP formed this way, the convolution between the OFDM symbol and the channel PDP would be linear rather than circular. This is the necessary condition for the single-tap equalization allowed by OFDM to work efficiently.

But what about the modulation carried by our OFDM system? In the case of coherent modulation using, for example QPSK modulation, the signal received from a wireless communication channel requires equalization; i.e. phase and amplitude variations induced by the channel have to be compensated for. As we have seen before, these variations would indeed prevent the recovery of coherent digital modulation schemes. In the case of coherent OFDM systems like, for example WiFi, the phase and amplitude induced by the channel are estimated, thanks to pilot tones inserted at definite positions in the OFDM spectrum. The receiver knows the position and the complex values of these pilots and can therefore use them to estimate the phase and amplitude variations induced by the channel. The main drawback of inserting pilots in the transmitted data is a decrease of the useful throughput. On the other hand, the DAB system uses a non-coherent modulation, i.e. differential QPSK or DQPSK. This scheme does not require equalization but needs a phase reference to be decoded. This is why the DAB frame includes a phase reference symbol (PRS) at the beginning of the transmitted frame. Differential modulation is applied as follows. The data bits are first QPSK modulated, after which the following operation is applied:

$$z[n] = z[n-1] \cdot y[n] \tag{7.11}$$

with $z[n]$ and $z[n-1]$ representing the current and previous DQPSK symbols, respectively, and $y[n]$ is the current QPSK symbol. This shows that it is the phase difference between the symbols that is encoded, rather than the absolute phase. This is indeed a very clever way to get rid of the phase shift induced by the channel. In fact, this phase shift is almost constant over several transmitted symbols. Therefore, this encoding scheme will definitely cancel out the constant phase shift. No technique is flawless, and DQPSK also has its drawbacks. The main one is a 3 dB penalty on the BER curve since when an error occurs, two modulation symbols are impacted.

Thanks to OFDM and Differential Quadrature Phase-Shift Keying (DQPSK), we have designed a robust transmission system for time and frequency-selective channels. However, OFDM itself is not enough to get audio-compatible bit error rates. Two useful mechanisms which represent the C of COFDM have to be added. The first one concerns classical forward error correction (FEC) capabilities. The second deals with a method to ensure that the FEC part always performs optimally.

In the case of the DAB system, FEC is achieved by a rate-compatible punctured convolutional (RCPC) code of constraint length 7 and base rate $R = 1/4$. The mother code rate $R$ can be varied between 8/9 and 1/4 by applying a puncture matrix at the output of the encoder. Although not all convolutional codes work efficiently with puncturing, this a very efficient way to obtain a variable rate $R$, and therefore a variable coding gain, with a unique structure for the encoder and for the Viterbi decoder (see Section 8.5.3). Puncturing simply means that you get rid of some of the encoded bits according to a certain pattern applied via a puncturing matrix. One might question how the Viterbi decoder can handle this. However, the solution is rather straightforward. Naturally, the receiver must be aware of the puncturing pattern applied during transmission. Then, all it needs to do is replace the eliminated bits with 0 or 1 accordingly. Hence, from the Viterbi decoder side, it is like adding errors in the received sequence; it will, in any case, succeed in decoding the most likely encoded sequence, provided that the total number of errors does not exceed the correction capability of the code. At the lowest rate $R = 1/4$ and soft decoding, this code reaches a coding gain of about 6 dB at a $BER = 10^{-5}$.

Unlike RS codes, convolutional codes perform very badly in correcting bursts of errors (i.e. on several consecutive symbols). A way to break these bursts and scatter the errors has therefore to be found. Moreover, concerning OFDM, error bursts have to be considered both in the frequency (deep fades on adjacent subcarriers) and in the time (deep fades on several OFDM consecutive time symbols) domains. The solution to this problem is interleaving the symbols in the time and frequency domains. Figure 7.17 illustrates this principle with the transmission of a piece of text subject to time and frequency burst errors.

### 7.2.3.3 The ETSI EN 300 401 DAB Standard

The DAB system provides a signal carrying a **multiplex** of several digital services simultaneously, for example stereo audio services. This multiplex of services is known as an **"ensemble."** The system bandwidth is about 1.5 MHz, providing a total transport bit-rate capacity of about 2.4 Mbit/s. The DAB system is mainly described in ETSI EN 300 401 [ETSIEN300401, 2017] document, which is freely available online. This 124-page document references several important related documents which detail particular elements of the standard. For example, the use of HE-AACv2 audio is described in ETSI TS 102 563 document [ETSITS102563, 2017]. The reading of such standard document can be a bit tedious for the non-specialist. This is the reason why we are going to present its main features in the following lines. Please note that the reader's comprehension should be facilitated by the preceding sections, which have already presented the key technical features that can be found in the standard.

DAB comprises four transmission modes which allow the use of a wide range of transmitting frequencies, up to 3 GHz for mobile reception. These modes have been designed to cope with Doppler and delay spreads, to allow for mobile reception in the presence of multipath. Table 7.1 summarizes

| After demodulation | After deinterleaving | After error correction |
|---|---|---|
| `.rirfcFFFrgeltndao`<br>`es yr FFFAic cei c`<br>`ernn FFFdueVet hi`<br>`sn dr FFF it ner T`<br>`TTTTTTFFFTTTTTTTTTT`<br>`    nearFFFdtsa onre`<br>`  eiefFFFriixrc ui`<br>`l ndhqFFFrtb lodml`<br>`  vnh FFFc sama ot`<br>`yrtoezFFFeec nimtf`<br>`geianvFFFnateeea i`<br>`  rniitFFFah.grseat`<br>`o-eyaFFFseuiott b`<br>`  emoiFFFrdni nr b`<br>`n ennFFFvbl ient`<br>`rpoixFFFudarbedee` | `This is aF FxampFe`<br>`of text cFnFainiFg`<br>`error burFtF TotF`<br>`in columnF FndTiFT`<br>`lines cauFeF by F`<br>`TTannel sFlFctiTF`<br>`fading. AFTFr deF`<br>`interTeavFnF in F`<br>`time anT FrFquenFy`<br>`theTe errFrFThavF`<br>`been randFmFzed Fs`<br>`indiTaTedF FfterF`<br>`error corFeFtionF`<br>`by a VitTFbF de-F`<br>`coder theFoFiginFl`<br>`teTt is rFcFvereF.` | `This is an example`<br>`of text containing`<br>`error bursts both`<br>`in columns and in`<br>`lines caused by`<br>`channel selective`<br>`fading. After de-`<br>`interleaving in`<br>`time and frequency`<br>`these errors have`<br>`been randomized as`<br>`indicated. After`<br>`error correction`<br>`by a Viterbi de-`<br>`coder the original`<br>`text is recovered.` |

**Figure 7.17** Interleaving explained. Source: Van De Laar et al. [1993]/with permission of EBU Technical Review.

**Table 7.1** Limiting planning parameter values for DAB transmission modes.

| System parameter | I | II | III | IV |
|---|---|---|---|---|
| Guard interval duration | $\approx 246\,\mu s$ | $\approx 62\,\mu s$ | $\approx 31\,\mu s$ | $\approx 123\,\mu s$ |
| Nominal max. transmitter separation for SFN | 96 km | 24 km | 12 km | 48 km |
| Nominal frequency range | $\leqslant 375\,MHz$ | $\leqslant 1.5\,GHz$ | $\leqslant 3\,GHz$ | $\leqslant 1.5\,GHz$ |

the limiting parameter values for each transmission mode. These parameters are valid in the context of single frequency networks (SFN), a concept which will be discussed hereafter. Moreover, we will limit our discussion to the most common mode, i.e. Mode I broadcasting in the VHF frequency band III (i.e. 174–230 MHz). Finally, although it is still possible to use MPEG-1 layer II audio, we will only consider the DAB+ version in the following.

The DAB standard introduces the important concept of **SFN**: i.e. a transmitter network in which all transmitters transmit identical information simultaneously at the same frequency. Thus, the distribution of a broadcasting program over a whole country presents the advantage of requiring only one frequency block. Moreover, thanks to the guard interval mechanism introducing resistance to MPCs, from the OFDM receiver point of view, it is impossible to distinguish a reflection from a second transmitter when both transmitters send exactly the same signal. For a SFN to work efficiently, the following two conditions have to be met:

1. Each transmission frame must start at the same absolute time for all transmitters.
2. The transmission frequencies must be identical so that the OFDM subcarriers align properly. SFNs are normally specified to keep transmitter frequencies within 1% of the carrier spacing and timing within a few percent of the guard interval [Hoeg and Lauterbach, 2009].

In order to achieve these an independent and ubiquitous time and frequency reference is required. Global positioning system (GPS) receivers are commonly used for this purpose.

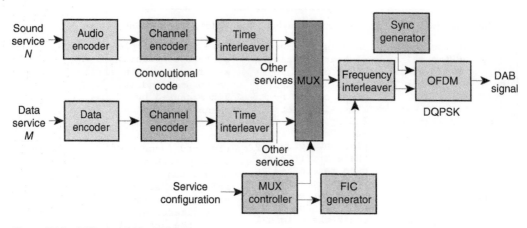

**Figure 7.18** DAB system block diagram.

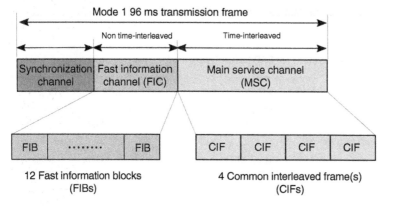

**Figure 7.19** DAB Mode I transmission frame.

Figure 7.18 presents the conceptual block diagram of the DAB system where one can find the digital communication blocks discussed in Section 7.2.3.2. As can be noticed, the multiplex is made up of several services, for example, audio radio stations. For a DAB+ audio service, the audio encoder will be the HE-AACv2 discussed previously. Audio data comprises a double Error Correction Code (ECC) protection with RS block codes completed by a convolutional code of variable rate $R$. These data are then time and frequency interleaved, DQPSK modulated (2 bits per symbol), and finally conveyed by OFDM subcarriers. As we will discover in the following, the DAB+ transmission frame carries two types of data: the main data itself and information-critical data giving all the necessary information to decode the main data (see the FIC or fast information channel block in Figure 7.18). Finally, to be able to recover the OFDM-generated signal at the reception side, synchronization elements also need to be added (also shown in Figure 7.18).

We leave the very high-level view of Figure 7.18 to look at what the standard calls "transport mechanisms." More precisely, the DAB stream combines the three following channels in a 96 ms duration frame for transmission Mode I (see Figure 7.19):

- **Main Service Channel (MSC)**. It is used to carry audio and data service components. The MSC is a time-interleaved data channel divided into a number of sub-channels, which are individually convolutionally coded with equal error protection (EEP) in DAB+. Each sub-channel may carry

one or more service components. The organization of the sub-channels and service components is called the multiplex configuration.
- **Fast Information Channel (FIC)**. It is used for rapid access of information by a receiver. In particular, it is used to send the multiplex configuration information (MCI) and service information. The FIC is a non-time-interleaved data channel with fixed EEP of rate $R \approx 1/3$.
- **Synchronization Channel**. This is used internally within the transmission system for basic demodulator functions, such as transmission frame synchronization, automatic frequency control, channel state estimation, and transmitter identification (TII).

More precisely, looking at Figure 7.20, the MSC is made up of a sequence of four CIF. A CIF is a data field of 55296 bits, transmitted every 24 ms. The smallest addressable unit of the CIF is the **capacity unit (CU)**, having a size of 64 bits. Integral numbers of CUs are grouped together to constitute the basic transport unit of the MSC, called a sub-channel. The MSC constitutes therefore a multiplex of sub-channels. As shown in Figure 7.20, a CIF is made up of 18 OFDM symbols of 1536 subcarriers, each carrying 2 bits. Consequently, the MSC data rate is equal to $MSC_r = (4 \times 1536 \times 2 \times 18)/96 \cdot 10^{-3} = 2304$ kbits/s. In transmission mode I, there are 12 Fast Information Blocks (FIBs) per transmission frame of 96 ms. The 12 FIBs of the three FIC symbols are divided into four groups that are each assigned to one CIF (see Figure 7.20). FIBs are once again divided into fast information groups (FIG) having different types. The most important is FIG type 0 as it carries the MCI field and sub-channels and services information (SI). FIG carry all the necessary information to decode MSC data, such as the error correction profile used by the transmitter. Thus, DAB+ Mode I defines four EEP profiles denoted EEP-1A to EEP-4A with four protection levels corresponding to the code rates 1/4, 1/8, 1/2, and 3/4, respectively. We will not go into more details at this time, particularly concerning the energy dispersal and the time and frequency interleavers, which we will address in the receiver section of this chapter.

Let us now turn to the elements that compose the synchronization channel. First, the DAB transmission frame starts with a null symbol of duration $T_{NULL} \approx 1297$ μs or 2656 samples taken at the reference sample time $T = 1/2\,048\,000$ seconds. Concerning the time reference, the standard states that "time information carried in the FIC shall be taken to be the time of transmission of the start of the null symbol in the transmission frame carrying the time information." Second, to be able to decode the DQPSK symbols contained in the DAB frame, one must have a phase reference. This is the role of the PRS, which is placed right after the null symbol. Moreover, the PRS has a sharp

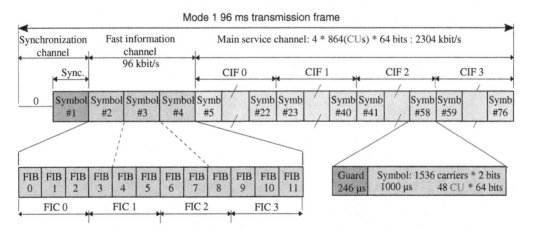

**Figure 7.20** DAB Mode 1 transmission frame detailed.

**Table 7.2** DAB time transmission parameters for Mode I with $T = 1/2\,048\,000$ s.

| System parameter | Mode I |
| --- | --- |
| Frame duration $T_F$ | 196 608 T |
|  | 96 ms |
| OFDM symbols per frame $L$ | 76 |
| Number of OFDM subcarriers $K$ | 1536 |
| Null symbol duration $T_{NULL}$ | 2656 T |
|  | $\sim 1297$ μs |
| Guard interval duration $T_G$ | 504 T |
|  | 246 μs |
| Useful symbol duration $T_U$ | 2048 T |
|  | 1 ms |
| Total symbol duration $T_S$ | 2552 T |
|  | $\sim 1246$ μs |

auto-correlation function, which makes it suitable for time synchronization and integer frequency offset estimation [Zielinski, 2021]. Finally, Table 7.2 summarizes the important time parameters of DAB Mode I.

As a practical way to summarize what we have seen before, it can be interesting to acquire a DAB+ signal with GNU Radio at a sampling frequency of $fs = 2.048$ MHz and look at it in the time and frequency domains. After all, this is simply the starting point for the decoding of the signal, which is going to be performed in the receiver section of this chapter. The result can be observed in Figure 7.21 where we can easily recognize the null symbol in the time domain and an OFDM spectrum occupancy of 1.536 MHz.

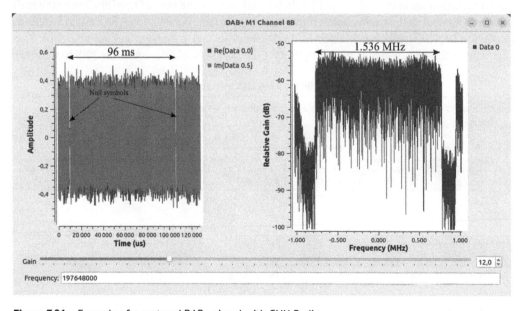

**Figure 7.21** Example of a captured DAB+ signal with GNU Radio.

## 7.3 Building a DAB+ Transmitter

The open-source DAB/DAB+ transmitter we plan to set up is built upon the foundation of ODR-mmbTools, a fork dating back to 2014 from the original CRC-mmbTools. These tools are rooted in the principle of software-defined radio, where the Radio Frequency (RF) signal is not generated by specialized hardware but rather by software running on a standard PC. The development of these tools traces back to 2002 when the Communications Research Centre (CRC) Canada initiated their creation, culminating in the release of the CRC-DabMux DAB multiplexer in September 2009 under the General Public License (GPL) open-source license. Additionally, CRC produced a DAB modulator capable of interfacing with Ettus Universal Software Radio Peripherals (USRPs) through a GNU Radio wave player script written in Python, which was released as CRC-Dabmod in 2010 under the GPL license. The support for SFN was integrated into these tools in 2012 by Matthias P. Brändli during his master's degree at EPFL [Brandli, 2012]. Subsequently, as CRC's support began to wane, the tools were officially forked in February 2014, evolving into ODR-mmbTools, now under the stewardship of the Opendigitalradio association (https://www.opendigitalradio.org/).

Figure 7.22 presents the DAB+ transmitter chain using four different modules included in the ODR-mmbTools. We could assemble and broadcast more than one service, but for the sake of simplicity, we will only consider one DAB+ audio service. Let us describe the key elements of the four tools used.

- **ODR-AudioEnc**. It includes a TooLAME-based MPEG encoder and uses the fdk-aac library as an external dependency to encode DAB+ HE-AACv2 audio. This tool is able to handle almost any audio source using VLC as a plugin and convert it to a DAB+ compatible format by dealing with the framing and adding error correction, thus ensuring an Encapsulation of DAB Interfaces (EDI) protocol compliance. ODR-AudioEnc also allows the insertion of the PAD generated by the ODR-PadEnc tool.
- **ODR-PadEnc**. This tool, which interfaces with the previous one, supports reading and encoding of dynamic label segment (DLS) from a text file and reads images from a folder for multimedia object transfer (MOT) protocol slideshow. These two elements constitute the PAD giving textual and graphical information about the ongoing program.

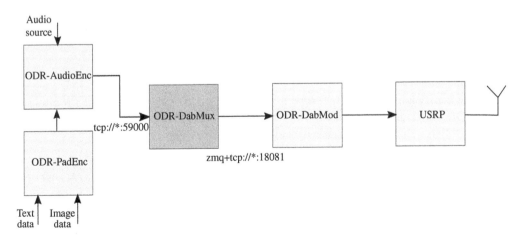

**Figure 7.22** ODR-mmbTools DAB+ transmission chain.

- **ODR-DabMux**. This is the multiplexer which creates the ensemble by combining all input sources. When this program is called, all multiplex parameters (name, programs, bit rates, etc.) must be specified. The user must therefore complete a specific configuration file with the appropriate parameters. The output is an ensemble transport interface (ETI) stream, which already contains the error-corrected data at a constant rate of 2.048 Mbps and that can be saved to a file. An ETI frame contains all the data required to construct all elements of the MSC and FIC and ancillary data (such as Multiplex Network Signalling Channel [MNSC]) of a DAB multiplex signal. It is standardized in the specification ETS 300 799 [ETSITS300799, 1997]. The stream can be transported through a pipe if the multiplexer is on the same computer as the modulator, over IP networks, or even over special interfaces. The same is true for the input: ODR-DabMux can read input audio or data from files, FIFOs (also called "named pipes"), or from a network connection. This particular way used to create an ensemble with multiple sources and sinks local or distant is characteristic of the way broadcasters work today by using the internet. Indeed, a TV or radio program can be created and assembled in Germany in the morning, sent to France using the internet, and be assembled with other programs to be broadcast the same evening.
- **ODR-DabMod**. It is a software-defined DAB modulator that receives or reads ETI streams or from files, and generates modulated I/Q data encoded as complex floats (32 bits per complex sample), which can be used for transmission. The generated I/Q data can be saved to a file or sent to different hardware like USRPs (using The USRP Hardware Driver [UHD]) or to other SDR devices using the SoapySDR library. The output can also be sent to a GNU Radio flowgraph using a ZMQ network connection.

Before our DAB+ radio can go on air, the four previously presented tools have to be installed on the computer. You will find them on Opendigitalradio at [Opendigitalradio, 2024]. The installation instructions are precisely described for the four tools. Be careful about the necessary dependencies that also have to be installed. To be able to follow easily the instructions which are going to be given next, I advise you to `git clone` the source code of the four tools in a dedicated directory called ODR.

The other important things to care about are the audio input and the I/Q output. Concerning the input, we like using internet radio streams whose addresses you can find at [fmstream.org, 2024]. Dealing with the output, we can save the I/Q data to a file, but it is more interesting to transmit a real-life radio signal using an official DAB+ radio channel to be able to listen to the music immediately with a DAB+ commercial radio receiver. Doing so, we were able to transmit with a B210 USRP and an Adalm-Pluto. Anyway, you will be able to try either ways as the provided configuration file is filled to allow these two possibilities.

To go on air, you will have to perform the following steps:

- From the book companion GitHub, copy the configs and PAD directories into the previously created ODR directory. They contain the ODR-DabMod and ODR-DabMux configuration files and the DLS (text) and MOT (images) data to be used by ODR-PadEnc.
- Open a terminal, go to the ODR directory, and run the following command:

```
$ odr-padenc -o GRBookRadio -t PAD/srv1/DLS/DLS1.txt -d PAD/srv1
  /MOT/
```

This command creates the PAD data reading the DLS1.txt text file and the images for the MOT slideshow located in the MOT directory. These images have to be .png or .svg 320x240 pixel images.

- Open a second terminal and run the following command:

  ```
  $ odr-audioenc -v https://icecast.radiofrance.fr/
  francemusiquebaroque-hifi.aac -C 200 -b 88 -o tcp://localhost
  :59000 -D -L --audio-resampler=samplerate -A --sbr -1 -V -P
  GRBookRadio -p 128
  ```

  This command encodes the https://icecast.radiofrance.fr/francemusiquebaroque-hifi.aac audio flux (`web radio`) in stereo using sbr (`-A --sbr`) with a bit rate of 88 kbits/s (`-b 88`). The ETI stream is output on tcp://localhost:59000, and PAD is allocated 128 bytes and added to the stream (`-P GRBookRadio -p 128`).

- Open a third terminal and type the following command:

  ```
  $ odr-dabmux configs/GRBookDab.mux
  ```

- Finally, open a fourth terminal and type the following command:

  ```
  $ odr-dabmod -C configs/dabmod.ini
  ```

  The default modulator configuration uses a B210 USRP and channel 5C (178.352 MHz). If you choose another output, modify the `dabmod.ini` file accordingly.

  Figure 7.23 shows the result of a transmission using the excellent Qt-DAB [Lazy Chair Computing, 2024] receiver software connected to an RTL2832 dongle. Of course this works also perfectly well with my commercial DAB+ radio receiver!

  And now, a few advices in case of problems. DAB OFDM detection algorithms are quite sensitive to carrier frequency offset (CFO). A few kilohertz should not be a problem. Anyway, you can compensate for it in the `dabmod.ini` file, knowing that this value can be measured by the Qt-DAB software. Remember that this CFO value accounts for the shift between the transmitter and the receiver hardware and that it can reach values up to 50 kHz with cheap hardware! Last but not least are the TX and RX antennas. We have had the best results with 1-meter telescopic antennas.

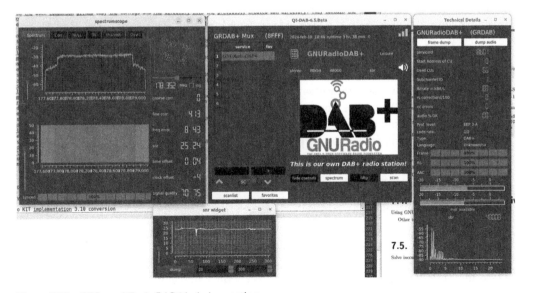

**Figure 7.23** ODR mmbTools DAB Mode I reception.

244 | *7 A Digital Communication Standard: The DAB+ Radio Broadcasting System*

After all that we have explored about this remarkable technology, it is time for a brief evaluation. Are you able to explain why our radio service uses a number of capacity units, CU = 66? Indeed, this is quite easy! We know that a CIF is made up of 864 CUs and that a HE-AACv2 sound frame has a 24 ms duration. Moreover, we have chosen an audio stereo data rate of 88 kbits/s and EEP-3A protection level with a code rate $R = 1/2$. Our radio channel will therefore occupy $(88 \times 10^3 \times 2 \times 0.024)/64 = 66$ CUs! This means that there is room in the CIF for 12 more similar audio channels! Indeed, these are the values adopted for the French Metropolitan M1 and M2 DAB+ multiplexes.

At this point of our discussion, one may ask, "OK, we have built a fully functional DAB+ transmitter but where is GNU Radio gone?" There exist a GNU Radio DAB OOT implementation originally designed by Andreas Müller during his master's degree at ETH Zurich in 2008. The original code has been the subject of a Google Summer of Code 2017 project by Moritz Luca Schmid, a former master's student of the Karlsruher Institut für Technologie (KIT) and mentored by Felix Wunsch and Marcus Müller. Although this OOT is supposed to have a transmitter part, we have not been able to make it work. Moreover, the latest version of this OOT is only compatible with GNU Radio 3.8. Nevertheless, we have performed the conversion to version 3.10, which works quite well for reception purposes. This is precisely the subject of the Section 7.4.

## 7.4 Building a DAB+ Receiver with GNU Radio

The DAB+ receiver we are going to build is based on an out-of-tree (OOT) module originally developed by Andreas Müller during his master degree at ETH Zurich in 2008 and improved by Moritz Luca Schmid, a KIT student in 2017 during a Google Summer of Code project. The module we use in the following is derived from Müller [2021]: we do not use the fork from KIT, which we found less stable. As Andreas Müller's module worked only with GNU Radio 3.8, we had to convert it to version 3.10 [Boeglen, 2024].

### 7.4.1 Basic Usage of `gr-dab`

As can be noticed in Figure 7.24, building a DAB+ receiver with gr-dab is quite simple: only two blocks are required, namely the "DAB OFDM Modulator" and the "DAB+ Audio Decoder" blocks. The source will be an SDR hardware (e.g. an RTL-SDR dongle) or a file, and the sink will be the sound card represented by the audio sink block. The remaining blocks deal with optional signal visualization. In fact, it is not so easy to make the receiver work as the "DAB+ Audio Decoder" block requires the appropriate setting of several parameters that cannot be obtained from the GNU Radio Companion (GRC) interface. Fortunately, the module provides a Python script called grdab, which will help us determine these parameters. The parameters to be able to listen to a radio station are determined by applying the following procedure.

- Open a terminal and run the following command:

```
$ grdab adjust
```

This launches the PyQt interface shown in Figure 7.25. It allows to select the DAB frequency band and set the parameters of an RTL-SDR dongle. These parameters are then saved in a file called adjustment.yaml which is located in the .dab subdirectory of the user's home directory. Scanning the DAB channels allows to locate a DAB transmission.
- Once the DAB channel is selected and the reception parameters fixed, run the following command:

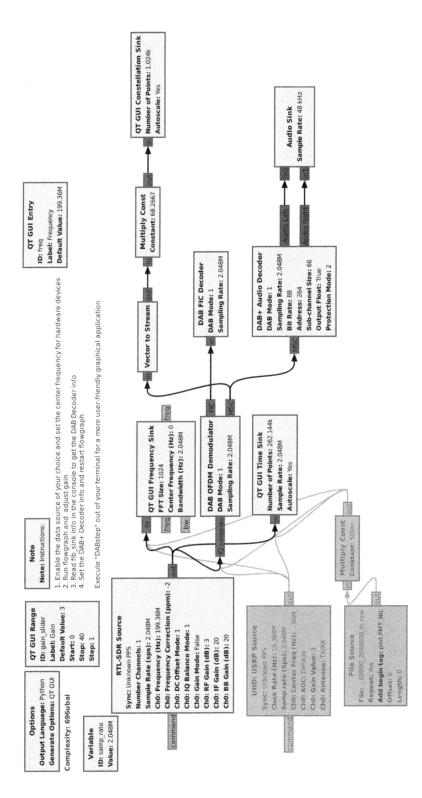

**Figure 7.24** gr-dab DAB+ receiver.

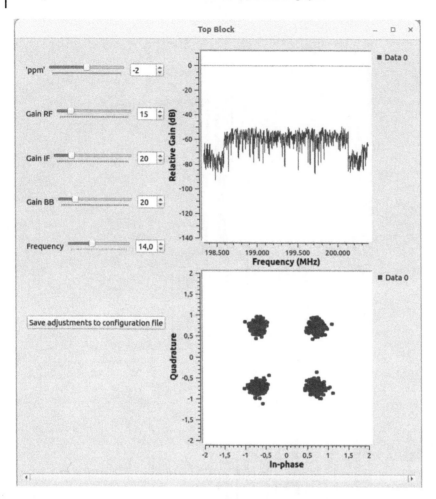

**Figure 7.25** `grdab adjust` GUI.

```
$ grdab info -f 199.36
```

Here we have chosen channel 8B. The information about the radio channels in the multiplex are displayed (see Figure 7.26). They can now be used to parameterize the "DAB+ audio decoder" block in GNU Radio Companion.

- The reception of the chosen radio station can be tested directly by running the following command:

```
$ grdab receive -f 199.36 --bit_rate 88 --address 264 --
  subch_size 66 --protect_level 2 --audiorate 48000
```

### 7.4.2 `gr-dab` in More Details

As shown in Section 7.4.1, the setup of a fully functional DAB+ receiver with gr-dab is easy. However, the `DAB OFDM Modulator` and the `DAB+ Audio Decoder` blocks hide the important steps that have to be performed in order to decode the OFDM DAB+ signal. This is the reason why

```
herve@herve-Precision-3581: ~/Documents                    Q  ≡  –  □  ×

herve@herve-Precision-3581:~/Documents$ grdab info -f 199.36
info handler
Setting frequency: 199.360 MHz
gr-osmosdr 0.2.0.0 (0.2.0) gnuradio 3.10.7.0
built-in source types: file rtl rtl_tcp uhd rfspace soapy redpitaya
[INFO] [UHD] linux; GNU C++ version 11.4.0; Boost_107400; UHD_4.5.0.0-0-g471af98f
[INFO] [UHD] linux; GNU C++ version 11.2.0; Boost_107400; UHD_4.1.0.5-3
Found Rafael Micro R820T tuner
Using device #0 Nooelec NESDR SMArt v5 SN: 00000001
Found Rafael Micro R820T tuner
[R82XX] PLL not locked!
[R82XX] PLL not locked!
buffer_double_mapped :warning: allocate_buffer: tried to allocate   5 items of size 12384. Due to
alignment requirements   128 were allocated.  If this isn't OK, consider padding   your structure
to a power-of-two bytes.   On this platform, our allocation granularity is 4096 bytes.
Allocating 15 zero-copy buffers
Channels:
RFM             : (address: 660, subch_size:  66, protect_level: 2, bit_rate:  88, classic: 0)
EUROPE 2        : (address: 726, subch_size:  66, protect_level: 2, bit_rate:  88, classic: 0)
KTO Radio       : (address: 792, subch_size:  66, protect_level: 2, bit_rate:  88, classic: 0)
FRANCE INFO     : (address:   0, subch_size:  66, protect_level: 2, bit_rate:  88, classic: 0)
FRANCE INTER    : (address:  66, subch_size:  66, protect_level: 2, bit_rate:  88, classic: 0)
FIP             : (address: 132, subch_size:  66, protect_level: 2, bit_rate:  88, classic: 0)
FRANCE CULTURE  : (address: 198, subch_size:  66, protect_level: 2, bit_rate:  88, classic: 0)
FRANCE MUSIQUE  : (address: 264, subch_size:  66, protect_level: 2, bit_rate:  88, classic: 0)
MOUV.           : (address: 330, subch_size:  66, protect_level: 2, bit_rate:  88, classic: 0)
RMC             : (address: 396, subch_size:  66, protect_level: 2, bit_rate:  88, classic: 0)
BFM BUSINESS    : (address: 462, subch_size:  66, protect_level: 2, bit_rate:  88, classic: 0)
BFM RADIO       : (address: 528, subch_size:  66, protect_level: 2, bit_rate:  88, classic: 0)
EUROPE 1        : (address: 594, subch_size:  66, protect_level: 2, bit_rate:  88, classic: 0)
herve@herve-Precision-3581:~/Documents$ []
```

**Figure 7.26** `grdab info` results.

we have created the flowgraph of Figure 7.27. It does exactly the same as the first flowgraph, but, this time, it is easy to see the different steps involved. Let us briefly review the most important ones.

- Time synchronization and fine frequency offset compensation. This is done by the DAB OFDM Synchronization block, which acts directly on the time signal output from the RTL-SDR Source block. Time synchronization consists in finding the beginning of the DAB transmission frame by computing a moving sum having the length of the NULL symbol. The time where the energy is minimal is going to be the end of the null symbol, thus the start of the first (Sync) symbol. Concerning frequency offset, it consists of two parts. The first one is called "coarse" as it represents multiples of the subcarrier spacing (i.e. $1/T_U$). The second, called "fine," represents the remaining part of the frequency offset, which is smaller than the subcarrier spacing. For this block, only the fine frequency offset is dealt with. The algorithm implemented is the cyclic prefix correlator described in Van de Beek et al. [1997].
- The signal is converted to the frequency domain by an FFT operation. This can be confusing, but remember that at the transmission side, we had to perform an IFFT operation to go to the time domain. This is performed by the FFT GNU Radio block.
- Coarse frequency offset correction. This could be realized by using the PRS, but it is not what Andreas Müller has done. The coarse frequency offset $\hat{f}_c$ can also be found by the following energy calculation:

$$\hat{f}_c = \underset{n}{\text{arg}max} \sum_{0 \leqslant i \leqslant K, \, i \neq K/2} |X[i+n]|^2 \;\; n \in [0, L_F - K) \tag{7.12}$$

with $X[i]$ being the $i$th entry in the FFT vector, $K$ the number of used subcarriers, and $L_F = T_U/T$ the FFT length.
- Phase differentiation. This is done by multiplying each symbol with the complex conjugate of the previous symbol:

$$y[n] = z[n] \cdot z^*[n-1] \tag{7.13}$$

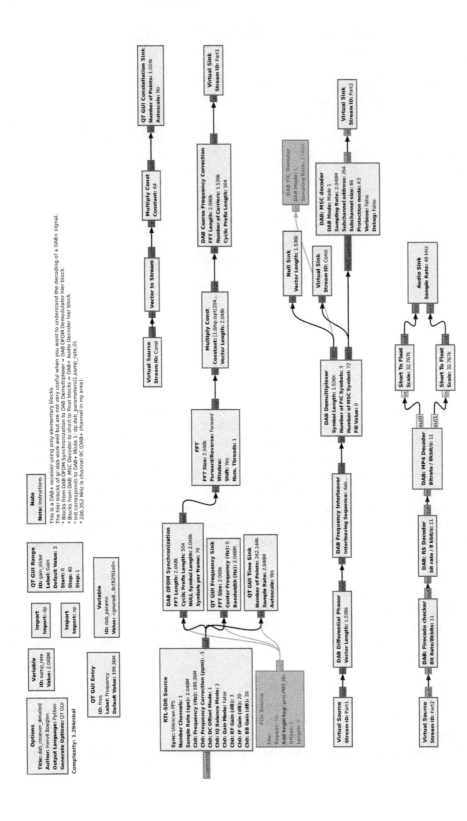

**Figure 7.27** ODR mmbTools DAB Mode I reception.

This is performed by the `DAB Differential Phasor` block.

- Frequency deinterleaving. This is realized by the `DAB Frequency Interleaver` block using the `frequency_deinterleaving_sequence:array` parameter defined in `parameters.py` file.
- The signal is demultiplexed between the 3 FIC symbols and the 72 MSC symbols per frame by the `DAB Demultiplexer` block.
- MSC decoding. The `DAB: MSC decoder` is indeed a "hier" block and performs therefore several operations. First, the CUs concerning the program to be decoded are extracted from the stream followed by symbol demodulation, time deinterleaving, unpuncturing, Viterbi decoding, and energy descrambling.
- Audio decoding. This part starts with error correction performed in two steps. The first step concerns the decoding of a Fire code capable of detecting and correcting most single error burst of up to 6 bits, while the second deals with the decoding of an $RS(120, 110, t = 5)$ code allowing the correction of up to five random erroneous bytes in a received word of 120 bytes. The data finally enters the MP4 decoder block, which produces the left and right channels sent to the audio card.

## 7.5 Conclusion

This chapter has been a broad overview of the DAB/DAB+ system, a modern digital communication standard that summarizes years of research in audio coding and digital communication. HE-AACv2 has reached a high level in audio quality, giving hi-fi rendering at quite low bit rates, while COFDM can cope with time and frequency selective channels with a high level of robustness. Although many of the presented concepts are quite complex to grasp, their comprehension is made easier, thanks to the fact that they have been designed in the digital domain. Indeed, one can easily find open-source software implementations that help to understand a difficult concept using a PC. In particular, this is the case for DAB transmission, where several transmitter and receiver implementations are available. Although we have not been able to cover the subject in all its aspects (this would require a complete book), we expect, however, to have given the main elements which will allow the interested reader to explore the subject further.

## References

H. Boeglen. gr-dab, 2024. https://github.com/hboeglen/gr-dab_amuller.

M. Bosi. How perceptual audio coding has shaped our lives. In *Audio Developer Conference (ADC)*, London, UK, 2018. https://www.youtube.com/watch?v=3BHMykq5PTU.

M. Bosi and R. E. Goldberg. *Introduction to Digital Audio Coding and Standards*. Springer, New York, NY, 2003.

M. Brandli. Single-Frequency Network Support for the CRC mmbTools Open-Source Software-Defined DAB+ Transmitter. Master's thesis, EPFL, 2012.

CEC. COST 207 digital land mobile radio communications, final report, 1989. https://op.europa.eu/fr/publication-detail/-/publication/61fc77e7-bca2-4229-8eb4-77741f0d2ab2.

ETSIEN300401. Radio broadcasting systems; Digital Audio Broadcasting (DAB) to mobile, portable and fixed receivers, 2017. https://www.etsi.org/deliver/etsi_en/300400_300499/300401/02.01.01_60/en_300401v020101p.pdf.

ETSITS102563. Digital Audio Broadcasting (DAB); DAB+ audio coding (MPEG HE-AACv2), 2017. https://www.etsi.org/deliver/etsi_ts/102500_102599/102563/02.01.01_60/ts_102563v020101p.pdf.

ETSITS103466. Digital Audio Broadcasting (DAB); DAB audio coding (MPEG layer II), 2019. https://www.etsi.org/deliver/etsi_ts/103400_103499/103466/01.02.01_60/ts_103466v010201p.pdf.

ETSITS300799. Digital Audio Broadcasting (DAB); distribution interfaces; ensemble transport interface (ETI), 1997. https://www.etsi.org/deliver/etsi_i_ets/300700_300799/300799/01_30_9733/ets_300799e01v.pdf.

fmstream.org. The radio stream directory, 2024. https://fmstream.org.

W. Hoeg and T. Lauterbach. *Digital Audio Broadcasting: Principles and Applications of DAB, DAB+ and DMB*. Wiley, 2009.

ISO/IEC13818-7. Generic coding of moving pictures and associated audio information Part 7: Advanced Audio Coding (AAC), 2006. https://www.iso.org/standard/43345.html.

ISO/IEC14496-3. Coding of audio-visual objects Part 3: Audio, 2019. https://www.iso.org/standard/76383.html.

Lazy Chair Computing. Qt-dab-6, 2024. https://github.com/JvanKatwijk/qt-dab.

W. C. Y. Lee. *Mobile Cellular Telecommunication Systems*. McGraw Hill, 1989.

S. Meltzer and G. Moser. MPEG-4 HE-AAC V2 –audio coding for today's digital media world. *EBU Technical Review*, 305, Jan. 2006. https://tech.ebu.ch/docs/techreview/trev_305-moser.pdf.

A. Müller. gr-dab, 2021. https://github.com/andrmuel/gr-dab.

P. Noll. MPEG digital audio coding. *IEEE Signal Processing Magazine*, 14(5):59–81, Sept. 1997. doi: 10.1109/79.618009

Opendigitalradio. Open source tools for Digital Radio, 2024. https://github.com/Opendigitalradio.

M. Paetzold. *Mobile Radio Channels*. Wiley, 2011.

T. Painter and A. Spanias. Perceptual coding of digital audio. *Proceedings of the IEEE*, 88(4):451–515, Apr. 2000. doi: 10.1109/5.842996

F. A. P. Petitcolas. MPEG psychoacoustic model I for MATLAB –version 1.2.8, 2022. https://github.com/fabienpe/Matlab_MPEG.

G. Schuller. Audio coding video tutorials and Python notebooks, 2024. https://github.com/TUIlmenauAMS/AudioCoding_Tutorials.

J. J. Van de Beek, M. Sandell, and P. O. Borjesson. ML estimation of time and frequency offset in OFDM systems. *IEEE Transactions on Signal Processing*, 45(7):1800–1805, Jul. 1997. doi: 10.1109/78.599949

F. Van De Laar, N. Philips, and R. O. Dubbelink. General–purpose and application–specific design of a DAB channel decoder. *EBU Technical Review*, 257, Nov. 1993. https://tech.ebu.ch/docs/techreview/trev_258-philips.pdf.

T. P. Zielinski. *Starting Digital Signal Processing in Telecommunication Engineering –A Laboratory-based Course*. Springer, 2021.

# 8

# QPSK and CCSDS Packets: Meteor-M 2N Satellite Signal Reception

The signals associated with this chapter are available at https://iqengine.org in GNU Radio SigMF Repo → space → MeteorM2N_180501_17h52_after_resampler and MeteorM2_180502_14h12, both collected using an RTL-SDR receiver fitted with a dipole antenna at a sampling rate of 156.8 kS/s. Polar-orbiting satellite signals are best received at high latitude, where they become visible once every 100 minutes (Figure 8.1).

## 8.1 Introduction

Shifting from analog to digital communication is an unquestionable trend (analog television to Digital Video Broadcast-Terrestrial [DVB-T], analog commercial broadcast Frequency Modulation [FM] to Digital Audio Broadcasting [DAB], telephone), and the satellite radio frequency links are no exception, aimed at optimizing the radio frequency spectrum usage and the quality of the transmitted signals. While the analog automatic picture transmission (APT) protocol of the low-Earth polar-orbiting NOAA satellites is doomed to disappear with the end of this satellite constellation, the succession seems to be taken care of with a protocol using the same bandwidth but digital, allowing the transmission of pictures with improved resolution: low rate picture transmission (LRPT). We shall see that this performance gain is achieved at the expense of a significantly increased processing complexity.

Most readers will probably hardly ever care about LRPT, if only because some functional free open source decoding software is available. Why then spend so much time decoding images transmitted from the Russian Meteor-M 2N satellite [National Research Center Planet, 2020a], the only source currently easily accessible (low-Earth orbiting satellite) transmitting an LRPT data stream (the LRPT emitter of the European Metop-A is broken, and Metop-C is being commissioned while these lines are written)? On the one hand, LRPT is only one example of the general class of digital communication protocols currently in use, with increasingly complex modulation schemes and abstraction levels ranging over all the OSI layers. We hence not only have an opportunity to explore these layers and understand practically their meaning, but other protocols close to LRPT are being used for high-resolution picture transmissions from low-Earth orbiting satellites (including Terra and Aqua from the MODIS constellation) or geostationary orbit [Teske, 2016]. Although we shall depart at some point from the processing chain described in the last reference, the beginning of

*Communication Systems Engineering with GNU Radio: A Hands-on Approach,*
First Edition. Jean-Michel Friedt and Hervé Boeglen.
© 2025 John Wiley & Sons, Inc. Published 2025 by John Wiley & Sons, Inc.
Companion website: www.wiley.com/go/friedtcommunication

**Figure 8.1** The Kongsberg Satellite Services (KSAT) headquarters in Tromsø, Norway, is fitted with multiple parabolic antennas located at a Northern latitude where polar low-Earth orbiting satellites pass most often. This installation is however dwarfed by the extensive parabola field located in Longyearbyen, Svalbard, at 78°N, close to the location where the data analyzed in this chapter were collected.

the acquisition and signal processing chain will be the same for LRPT decoding. Furthermore, if we are to believe the documents provided by the various space agencies over the last two decades, LRPT must be the future of low-bandwidth space communication, even though the Russians are the only ones practically exploiting the protocol in the very high frequency (VHF – 30–300 MHz, hence including the band around 137 MHz dedicated to satellite communications) band currently. Beyond these applied engineering aspects, the techniques we will learn to implement are used in many current digital interfaces used daily, from data communication and storage to television: Forney Jr. [2005] estimated that $10^{15}$ bits/s were being processed using convolutional codes for television alone.

Let us first provide some inspiration to the reader by demonstrating the targeted result (Figure 8.2). Meteor-M 2N transmits on a 137.9 MHz carrier so that current NOAA APT receiver installations are perfectly suited. Analyzing the reception quality for LRPT is a bit more complex than in the case of APT: the digital mode is not suitable for an audio-frequency analysis of the link quality as is the case for APT with its sweet melody at 2400 Hz, and only a constellation diagram observation (see Section 8.5.6) allows for assessing whether the receiver and its antenna are functional. We had to record tens of failed passes before acquiring a usable signal. The challenge with polar-orbiting satellites is that they often fly over the poles and only once every day over any location at our temperate latitudes. We are lucky enough to collect signals from Spitsbergen at 79°N, a latitude at which a polar-orbiting satellite is seen every 100 minutes. If we do not take care, we might even spend most of our time monitoring radio frequency signals from satellites rather than getting work done! An additional bonus is to fetch a direct view of the North Pole, with no geographical interest other than showing that it can be done. The picture shown in Figure 8.3 is aimed at demonstrating how simple hardware is used to receive Meteor-M 2N signals: as for NOAA satellites, two rigid wires and a DVB-T receiver based on the R820T(2) front end and RTL2832U analog-to-digital converter and USB transceiver used as software-defined radio are enough. Such hardware, compact and lightweight, is easily improvised even in remote areas if not readily available.

**Figure 8.2** Meteor-M 2N image acquired from Spitsbergen. Northern Scandinavia is visible on the top right of the image, the North Pole is toward the bottom left of the image.

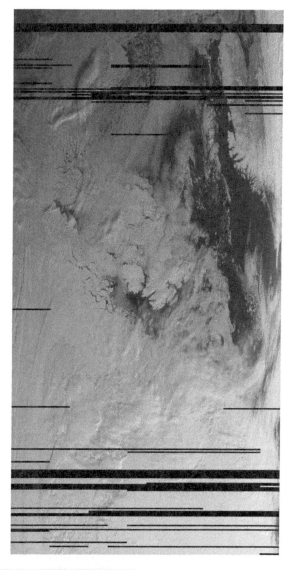

(a)  (b)

**Figure 8.3** (a) Experimental setup. A dipole antenna, a DVB-T receiver, and a computer running GNU Radio are enough to collect data from Meteor-M 2N on a 137.9 MHz carrier or American NOOA POES satellite on the same frequency band. (b) Result after processing the signal received from a NOAA satellite (analog communication) over Spitsbergen. From such a site, low-Earth polar-orbiting satellites reappear every 100 minutes, for passes lasting about 10 minutes. Notice the signal loss around 1/3rd from top of the image when the satellite lies in an antenna null, which cannot be corrected for by forward error correction (FEC) as will be achieved with digital communication. Border limits were added on the raw images by the wxtoimg decoding software. (b, left) is the visible band; (b, right) is the infrared band image.

**Antenna Design**

The careful reader will have noticed (Figure 8.3) that we have replaced a basic dipole antenna with a dipole exhibiting an angle of 120° as advised by https://lna4all.blogspot.com/2017/02/diy-137-mhz-wx-sat-v-dipole-antenna.html. Why this strange angle? For the same reason as the elements of a discone antenna exhibit some angle with respect to the radiating vertical monopole and are not horizontal: in order to match impedance. A quick NEC2 simulation [Marshall, 2002] demonstrates how impedance at resonance is dependent on the angle between the wires:

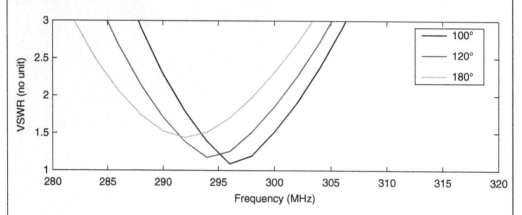

Indeed, an angle of 120° helps us get close to the targeted 50 Ω of the radio frequency connectors and cables. Practically, considering how close ground and other disturbing metallic elements including the snowmobile are, the capacitive parasitic elements are so significant that the antenna is necessarily mismatched. Meeting theoretical expectations can never hurt, nevertheless.

Acquisition and processing are completed as two different steps: GNU Radio is used not only to collect complex data, as in-phase and quadrature ($I$, $Q$), samples from the radio frequency receiver but also to perform pre-processing steps to compensate for frequency offsets between the moving satellite and the ground-based receiver (Section 8.5) and to lock the bit-detection clock on the transitions observed on the digital data stream. By doing so, we reduce the data rate and hence the size of the file storing the data on the hard disk for further processing. The image (Figure 8.2) was decoded using `meteor_decoder` available at https://github.com/artlav/meteor_decoder.git. This software, written in Pascal (`apt-get install fpc`), is trivially compiled with `./build_medet.sh` to generate the program `medet`, which is used with `./medet file.s output -s`, with `file.s` being the file generated from GNU Radio. Following this approach, we have obtained a very nice image, have understood nothing of the processing steps, and remain slaves to an excellent developer who has provided a perfectly functional tool. No interest whatsoever.

Our aim in this presentation is to analyze the detailed operations of `medet`, understand its operating principles, and, without disregarding the comfort of a functional application, benefit from this opportunity of understanding the LRPT protocol to improve our knowledge of "modern" digital communication techniques.

## 8.2 When Will the Satellite Fly Overhead?

The first question to be answered in a project aimed at receiving a low-Earth polar-orbiting satellite signal is to know its flight schedule. Indeed, with an orbit at an altitude of about 800 km from the surface of the Earth, the satellite performs one complete orbit every 100 minutes and is visible from a given location on the surface of the planet for a dozen minutes at most. For historical reasons, our preferred satellite pass prediction tool is `SatTrack`, which, despite its Y2K bug (http://pe1chl .nl.eu.org/SatTrack), remains an excellent, perfectly functional command line software. We fill its `data/cities.dat` configuration file with a new entry, including an identifier and the latitude and longitude (negative toward the east!) coordinates of the site where the receiver is located, as well as the orbital parameters of the satellite in `tle/tlex.dat`: the file fetched from https:// www.celestrak.com/NORAD/elements/weather.txt provides such regularly updated parameters. Identifying the Meteor-M 2N satellite as `METEOR-M 2N`, we obtain a list of passes in the following format:

```
BTS SatTrack V3.1 Orbit Prediction

Satellite #40069  : METEOR-M 2
Data File          : tlex.dat
Element Set Number: 999 (Orbit 21896)
Element Set Epoch : 27Sep18 21:20:50.463 UTC   (2.3 days ago)
Orbit Geometry     : 816.87 km x 823.72 km at 98.593 deg
Propagation Model : SGP4
Ground Station     : NYA    —    JQ58WW
Time Zone          : UTC (+0.00 h)

 Date (UTC)          Time (UTC) of     Duration  Azimuth at  Peak  Vis Orbit
                  AOS      MEL      LOS   of Pass  AOS MEL LOS  Elev
Sun  30Sep18  05:28:54 05:35:40 05:42:25  00:13:31  355  56 117  17.5  DDD 21930
              07:10:16 07:17:34 07:25:00  00:14:44   10  81 153  28.4  DDD 21931
              08:51:18 08:58:56 09:06:38  00:15:20   25 107 187  48.1* DDD 21932
              10:31:55 10:39:41 10:47:31  00:15:37   41 131 220  77.6* DDD 21933
              12:12:20 12:20:02 12:27:48  00:15:28   60 334 250  77.5* DDD 21934
              13:52:24 14:00:07 14:07:53  00:15:28   83   1 276  68.9* DDD 21935
              15:32:29 15:40:15 15:48:01  00:15:33  109  20 299  76.7* DDD 21936
              17:12:46 17:20:32 17:28:22  00:15:37  139 226 318  79.1* DDD 21937
              18:53:39 19:01:17 19:08:59  00:15:20  171 253 334  49.3* VVV 21938
              20:35:17 20:42:35 20:50:01  00:14:44  206 277 349  29.2  VVV 21939
              22:17:44 22:24:29 22:31:19  00:13:35  241 303   5  17.9  VVV 21940

Mon  01Oct18  00:00:55 00:07:00 00:13:09  00:12:14  277 330  23  11.9  VVV 21941
              01:44:06 01:49:51 01:55:39  00:11:33  308 357  46   9.8  VVV 21942
              03:26:49 03:32:46 03:38:46  00:11:58  333  24  76  11.1  VVV 21943
              05:08:47 05:15:20 05:21:58  00:13:11  352  51 110  16.0  DDD 21944
              06:50:13 06:57:23 07:04:41  00:14:28    7  76 146  25.7  DDD 21945
              08:31:14 08:38:53 08:46:31  00:15:16   22 102 180  43.3* DDD 21946
              10:12:00 10:19:42 10:27:32  00:15:33   38 124 213  71.4* DDD 21947
              11:52:25 12:00:07 12:07:57  00:15:33   56 334 244  81.3* DDD 21948
              13:32:33 13:40:15 13:48:02  00:15:28   78 355 271  69.3* DDD 21949
              15:12:38 15:20:20 15:28:06  00:15:28  103  20 295  73.9* DDD 21950
              16:52:51 17:00:37 17:08:27  00:15:37  133 228 314  84.9* DDD 21951
              18:33:32 18:41:14 18:49:00  00:15:28  165 247 331  54.7* VVV 21952
              20:15:02 20:22:28 20:29:54  00:14:52  199 273 346  32.3  VVV 21953
```

**256** | *8 QPSK and CCSDS Packets: Meteor-M 2N Satellite Signal Reception*

**METEOR-M 2N – All Passes**

Search period start: 01 October 2018 00:00

Search period end: 11 October 2018 00:00

Orbit: 820 x 826 km, 98.6° (Epoch: 30 September)

Passes to include: ○ visible only ● all

Click on the date to see the ground track during the pass.

| Date | Brightness (mag) | Start | | | Highest point | | | End | | | Pass type |
|---|---|---|---|---|---|---|---|---|---|---|---|
| | | Time | Alt. | Az. | Time | Alt. | Az. | Time | Alt. | Az. | |
| 01 Oct | - | 00:20:53 | 10° | W | 00:24:30 | 18° | WNW | 00:28:08 | 10° | NNW | visible |
| 01 Oct | - | 02:04:58 | 10° | NW | 02:06:59 | 12° | NNW | 02:09:00 | 10° | N | visible |
| 01 Oct | - | 05:31:12 | 10° | N | 05:32:44 | 11° | NNE | 05:34:16 | 10° | NE | visible |
| 01 Oct | - | 07:12:02 | 10° | NNE | 07:15:20 | 16° | NE | 07:18:36 | 10° | E | daylight |
| 01 Oct | - | 08:52:54 | 10° | NNE | 08:57:23 | 26° | ENE | 09:01:53 | 10° | SE | daylight |
| 01 Oct | - | 10:33:40 | 10° | NNE | 10:38:50 | 43° | E | 10:44:00 | 10° | S | daylight |
| 01 Oct | - | 12:14:17 | 10° | NE | 12:19:43 | 71° | SE | 12:25:08 | 10° | SSW | daylight |
| 01 Oct | - | 13:54:41 | 10° | NE | 14:00:07 | 81° | NNW | 14:05:33 | 10° | WSW | daylight |
| 01 Oct | - | 15:34:52 | 10° | ENE | 15:40:16 | 69° | N | 15:45:39 | 10° | W | daylight |
| 01 Oct | - | 17:14:56 | 10° | ESE | 17:20:20 | 74° | NNE | 17:25:45 | 10° | WNW | daylight |
| 01 Oct | - | 18:55:10 | 10° | SE | 19:00:36 | 85° | SW | 19:06:03 | 10° | NW | daylight |
| 01 Oct | - | 20:35:56 | 10° | S | 20:41:14 | 55° | WSW | 20:46:35 | 10° | NW | visible |

**Figure 8.4** Pass previsions using the website Heavens Above.

We only consider passes with a sufficient elevation for the satellite to be clearly visible (typically about 60° as indicated in the `Peak Elev` column), and the time is given here in Universal time (+1 or +2 hours with respect to French time).

Many users might prefer a graphical interface to get such results. When an internet connection is available, the simplest solution is probably to fetch the information from the Heavens Above (www .heavens-above.com) website, requesting pass predictions for the satellite identified as NORAD 40069 (Figure 8.4). The results are consistent with those of SatTrack, considering that Heavens Above provides results in local time rather than Universal time (hence a two-hour difference on the date of October 1 we are concerned with here), that Heavens Above does not consider passes with a maximum elevation below 10°, and that the schedule for the acquisition of signal (ACQ or AOS) and the loss of signal (LOS) is given for an elevation of 10° and not 0°, inducing an offset of about two to five minutes between the schedules.

We notice, in all cases, the benefits of working at higher latitudes (here 79°N), where passes repeat every 100 minutes, allowing for many more attempts than at the French latitudes.

---

**Satellite track projected with QGIS**

While watching S. Prüfer present *Space Ops 101* at media.ccc.de/v/35c3-9923-space:ops_101#t=1265, we could not resist the urge to reproduce the figures exhibited between 16 and 18 minutes along the presentation illustrating why listening to satellite from polar regions is most favorable. In order to achieve the results depicted later using QGIS (here version 3.4):

1. Install a tool for predicting satellite positions based on the TLE and providing a Shapefile formatted output. We have used https://github.com/anoved/Ground-Track-Generator, which is trivially compiled.
2. Generate the ground trace of the satellite we are interested in. By providing the Meteor-M 2N TLE collected around October 1, 2018, from the Celestrak website as described in the text, we execute `gtg --input meteor.tle --output m2 --start epoch --end epoch+24h --interval 30s` to get the `m2.shp` file, which includes the position, in spherical coordinates (WGS 72), for the satellite as seen from the ground.
3. Download a coast and border database, again in Shapefile format. In our case, we used the archive available at https://www.naturalearthdata.com/downloads/50m-cultural-vectors/50m-admin-0-countries-2.
4. We must always work in a projected Cartesian framework to process geometrical transforms. All spherical coordinates are hence projected to a Cartesian framework in the local tangent planes. For France, WGS84/UTM31N yields good results (EPSG:32631), while https://epsg.io/3576 teaches us that EPSG:3576 will provide an acceptable projection framework around the North Pole.
5. Having clipped (`Vector → Geoprocessing Tools → Clip ...`) the border map to the countries around the Arctic regions (down to about 50° north: this result is achieved by creating a polygon using `Layer → Create Layer → New Shapefile Layer... → Geometry Type: Polygon`), place the receiver location site on the map, for example, by importing an ASCII file including its longitude and latitude.
6. Trace circles of known circumference on the map. This is achieved by saving the receiver site location coordinates in a Cartesian framework (right mouse button on the layer including the symbol of the receiver site, and `Export → Save Feature As ...` by selecting a CRS with the appropriate projection). Then, we will trace an exclusion zone with `Vector → Geoprocessing Tool → Buffer ....` We are left with identifying the circumference of these circles. The case of the angle with respect to the center of the Earth at which the satellite appears over the horizon is trivial since we have a right triangle between the observer, the center of the Earth at radius $R$, and the satellite at a distance $R+r$ from the center of the Earth ($r$ being the flight altitude of the satellite). Hence, the angle between the observer, the center of the Earth, and the satellite is $\vartheta = \arccos(R/(R+r))$, and the length of the arc visible from the satellite is $R\cos\vartheta = R \cdot R/(R+r) = 3050$ km. This yields the radius of the largest circle visible in the figures exhibited later. Practically, we can hardly receive a usable signal from an elevation below $\vartheta' = 15°$. In this case, identifying the radius of the circle defining the visibility zone requires slightly more complex trigonometric relations by replacing the right angle with an angle equal to $(90 + \vartheta')$, but the solution to the problem remains unique for a known given angle (the satellite angle at AOS), a known Earth center-observer distance ($R$), and a known Earth center-satellite distance ($R+r$). The solution for $\vartheta' = 15°$ is a radius of the visibility zone of 1760 km. Finally, we shall not bother facing the Arctic cold if the satellite maximum elevation during a pass remains below $\vartheta' = 60°$. In this case, the radius of the satellite visibility zone for such a minimum elevation is reduced to 400 km around the observer's location. These two circles are concentric and centered on the observation site in the following figures.

*(Continued)*

**(Continued)**

We deduce, by observing these figures, that a single pass at best will yield usable results every day over France with an elevation of at least 60°, while 9 to 10 passes out of the daily 14 will meet this condition over Spitsbergen.

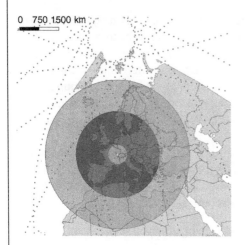 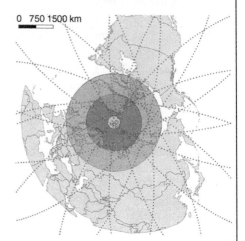

Left: ground trace (dots) of Meteor-M 2N passes as seen from Besançon (France, 47°N) during 24 hours. Right: ground traces of Meteor-M 2N passes as seen from Ny-Ålesund in Spitsbergen (Norway, 79°N). Notice that a single trace lies within the circle defining an elevation of at least 60° during a pass over France, while at least nine orbits meet this condition over Spitsbergen. The left figure is given in WGS84/UTM31N projection, while the one on the right is WGS84/North Pole LAEA Russia. The largest circle centered on each observer site indicates when a satellite becomes visible over the horizon (3030 km radius), the intermediate circle indicates visibility at elevations of at least 15° (1760 km radius), and the smallest circle describes locations at which the elevation of the satellite is at least 60° (400 km radius).

## 8.3 Why Such a Complex Protocol?

APT analog signal emitted from NOAA satellites is trivial [Friedt, 2005]: a dual amplitude (pixel intensity) and frequency (to get rid of Doppler shift as the satellite travels along its orbit) modulation transmits the information decoded on the ground by successive FM followed by Amplitude Modulation (AM) demodulators. Why then leave this simple communication protocol at the expense of a digital communication mode embedding packets in multiple protocol layers (Figure 8.5), which will keep us busy in the pages that follow?

The amount of information transmitted is limited by the allocated bandwidth, but a given bandwidth can be used more or less efficiently to transmit an image with better or worse quality. The radio frequency spectrum is a scarce and busy resource: Meteor-M 2N allows for an improved resolution for a given spectral usage thanks to the spectrum optimization provided by the digital

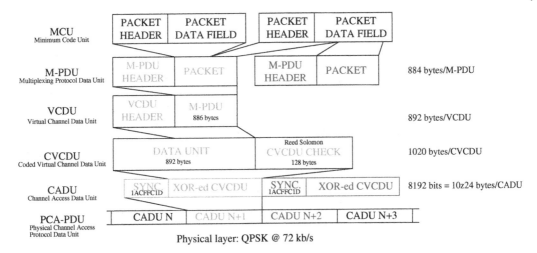

**Figure 8.5** Protocol layers to be addressed to convert the physical signal (bottom) to an image (top): the complexity of the protocol lies in its general purpose and the virtualization of multiple digital channels transmitted by the satellite.

modulation. More importantly, this data encapsulation as packets in successive layers – similar to the OSI layers for ground networks – fits in with a logic of sharing resources needed for space communication, as illustrated in Figure 8.6 (NASA image) following the presentation [Israel et al., 2016].

As an example, a low-Earth orbiting satellite such as the International Space Station (ISS, 400 km altitude) or a weather satellite (800 km altitude) will travel from horizon to horizon in 7–11 minutes under the best circumstances (maximum elevation at zenith). With a period of 90 minutes, this means that for a continuous link with the ISS, 13 stations would be needed along each orbit, or one station every 3000 km, not very practical to implement despite being used in the early days of the space race [Kranz, 2001]. The solution is to communicate through the tracking and data relay satellite (TDRS), a set of geosynchronous satellites acting as relays between the low- and mid-Earth orbiting satellites and the ground. This means that not only is each flying platform fitted with a multitude of instruments, which must share the available bandwidth and hence use a communication protocol for sharing the communication channel more subtly than simply sequentially communicating each instrument's result, as was seen for NOAA [Friedt, 2005], but also that a given satellite might be used to route information from different origins and later from different orbits (e.g. the Moon or Mars [Makovsky et al., 2002]). Such functionalities are, for example, mandatory to fully exploit a satellite such as the Hubble space telescope orbiting at about 500 km from the surface of the Earth, which is nothing more than a spying satellite looking in the wrong direction. All these elements hint at the development of a packaging and routing protocol, which must be robust to the temporary visibility of the satellites from the ground station, with possibly the ability to hand over from one station to another due to Earth's rotation (see, for example, the Deep Space Network and its stations distributed on all continents) without the final user being aware of these successive data sources.

Hence, we find again the same problems of Internet Protocol (IP) packet routing followed by encapsulation of data in TCP or UDP packets but without the robust and user-friendly libraries provided by the various open-source implementations of the networking layers. We must hence

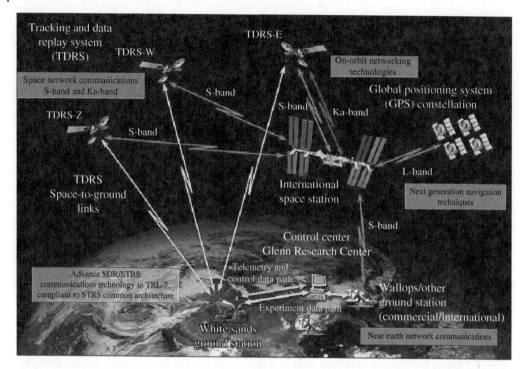

**Figure 8.6** Illustration depicting the complexity of space communications to transmit information acquired around terrestrial orbits, especially from low-Earth orbiting satellites (formerly space shuttle and now ISS at 400 km, Hubble telescope at 550 km altitudes), which could not be efficiently used and monitored by being visible only a few minutes every day from a ground station. Only by using a network of geostationary communication satellites (TDRS in the United States, EDRS in Europe most significantly for the Sentinel satellites) ensures a nearly permanent link between a space agency and its satellites. Source: NASA.

unstack ourselves from the layers of the protocol to understand each subtle operating principle. Luckily, all information is available, assuming we know where to look for them.[1]

## 8.4 How to Tackle the Challenge?

Communicating digital data on a radio frequency communication channel as variable as a space link requires a few data protection strategies to prevent corruption and losses, and even correction capability. These various protocol layers are described in documents published by the CCSDS, the Consultative Committee for Space Data Systems, at https://public.ccsds.org (Figure 8.7).

Reading the document is, to say the least, ... challenging. Let us attempt to ease the challenge by starting from the end (transmitting an image) and reach the signal we have received (the radio frequency wave):

1. An image is split to be transmitted by the satellite as a set of $8 \times 8$ pixel thumbnails.
2. Each one of these thumbnails is compressed (using lossy compression) using JPEG: each image thus exhibits a variable size depending on the amount of details being displayed in each

---

[1] Notice that the author of `libfec` library, which will be used here, Phil Karn KA9Q, is also the author of the TCP/IP stack for MS-DOS that we used during our first steps to discover internet connectivity at the beginning of Linux in 1994/1995, when MS-DOS was still the most common operating system running on personal computers.

## 8.5 From the Radio frequency Signal to Bits

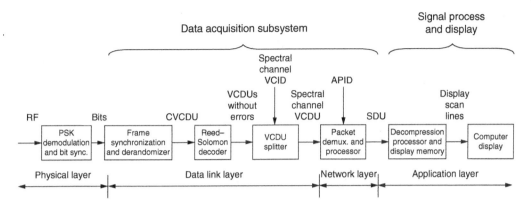

**Figure 8.7** Block diagram of the user ground segment representing OSI layers (bottom) and acronyms (top) used in the literature describing the LRPT image transmission. Source: The Consultative Committee for Space Data Systems (CCSDS)/public domain/https://public.ccsds.org/Pubs/120x0g2s.pdf/Last accessed at 12/04/2024.

thumbnail (few coefficients for a smooth area, many coefficients for areas with many features such as mountains). These steps of image assembly are discussed in Section 8.7.
3. One line of the final image is made up of 196 thumbnails, for a total width of 1586 pixels.
4. Transmitting the images collected at various wavelengths (using various instruments) is alternated by sending the set of 196 thumbnails from one wavelength then the 196 thumbnails from another wavelength. Between image transmissions, a telemetry sentence is transmitted (Section 8.5.7).
5. These variable-size datasets are collected in fixed-size packets. Each packet holds a payload made of 892 bytes followed by 128 bytes for optional transmission error correction, to which a 4-byte header for synchronization is added (total: 1024 bytes). This grouping of bytes into sentences is described in Section 8.6.
6. A convolutional code, which will be described in detail since it is the core theoretical challenge of the whole work, allows for correcting noise distributed uniformly during the transmission. Each bit is doubled to create sentences which are $2 \times 8 \times 1024 = 16\,384$ bits long (Section 8.5.3).
7. The hardware layer is defined as a Quadrature Phase Shift Keying (QPSK or 4-PSK) transmission in which each bit pair is encoded as one of four possible phase states $\{0, 90, 180, 270\}°$. This transmission runs at a rate of 72 kb/s (Section 8.5).

Having described the outline of the encoding, which follows the outline of the OSI layers (Figure 8.7), we must unwrap the problem in the opposite direction to go from the radio frequency signals received by a DVB-T receiver used as an $I/Q$ coefficient source for software-defined radio frequency signal processing.

## 8.5 From the Radio frequency Signal to Bits

Acquiring digital signals following frequency transposition to get to baseband does not involve any significant challenge: the local oscillator of the DVB-T receiver used as an sample source for radio frequency signal processing is tuned to the center frequency of the emitted signal – 137.9 MHz in the case of Meteor-M 2N – and the bandwidth adjusted to be wide enough to collect all spectral components of the signal modulating the carrier. We have discussed at length how phase modulation requires regenerating on the receiver side a local copy of the carrier prior to modulation in order to identify the phase of the signal by mixing and filtering. In the case of binary modulations (binary phase shift keying [BPSK]), we have seen (Section 2.5.2) that the unmodulated carrier was recovered by processing the $(I, Q)$ coefficients with an estimator insensitive to $\pi$ rotations (since the encoding is added to the carrier by rotating the phase to 0 or $\pi$), either by using the arctan $(Q/I)$ function or by squaring the signal in order to double the phase, namely 0 or $2\pi$, hence a carrier having lost its modulation but at double the frequency offset between the emitted signal and the frequency of the local oscillator on the receiver side.

The same principle is exactly transposed to quadrature phase modulation, in which the information is applied as phase rotations of the carrier with values equal to 0, $\pi/2$, $\pi$, or $3\pi/2$ (Figure 8.8, inspired by https://github.com/otti-soft/meteor-m2-lrpt/blob/master/airspy_m2_lrpt_rx.grc). However, instead of simple squaring, getting rid of the phase now requires computing the fourth power of the signal, yielding a beat signal at four times the offset between the emitted frequency and the receiver's local oscillator frequency (Figure 8.9). Hence, for a given acquisition bandwidth, we can only allow for a lower difference between these two frequencies than in the case of BPSK.

On the other hand [EN 50067, 1998], once the carrier has been reproduced (Costas loop), the question of the bit sampling rate remains since the emitter and receiver clocks are not synchronous. Hence, the need to detect transitions from one bit value to another to control the clock sampling the phase. This job is taken care of by clock synchronization blocks such as *Clock Recovery Mueller & Müller* or *Polyphase Clock Sync*, which aim at only providing a single sample for each symbol after controlling the clock sampling the datastream on its transitions (Figure 8.8). This flowchart can be executed on the IQEngine GNU Radio repository files stored in the `space` directory, starting with M2, to generate the soft bit `.s` binary file by processing the raw binary capture from an RTL-SDR dongle to 8-bit integer soft bits by compensating for frequency offsets between the local oscillator and the downlink carrier at nominal frequency 137.9 MHz and symbol synchronization to output one sample for each QPSK symbol.

These two tasks are taken care of by GNU Radio not only because they are perfectly functional in this environment but most significantly to reduce the size of the files stored for further processing. The more the datastream is decimated prior to storage, the smaller the file: in our case, we aim at storing the bitstream as the output of the processing chain, or one byte representing each bit since the recovered values have not yet been discretized to 1 or 0 but remain a probability of maybe 1 or 0. We will see that keeping this uncertainty, considering the clever encoding scheme used during emission, will maximize our chance of recovering the correct value of each bit. This data storage format is named **soft samples**, as opposed to **hard samples**, which have already been discretized to attribute a value of 0 or 1 to each bit [Costello and Lin, 1982, p. 8].

### 8.5.1 Data Format

The first question we might wish to answer is whether our understanding of the data format is correct and if the file is worth processing. Before having the slightest idea about the encoding format, we might simply wonder whether a pattern is repeated. Indeed, when encapsulating messages as packets, it is quite likely that the size of packets is constant and that some pattern such as the header will repeat. Hence, the autocorrelation of the signal must exhibit some peaks spaced by the repetition period of the messages (Figure 8.10).

We observe correlation peaks at every multiple of 16 384 samples (bytes). Getting ahead of the description that will follow, we will learn that each packet transmitted by Meteor-M 2N is 1024 bytes long or 8192 bits and that the coding scheme used (convolutional code [Viterbi, 1967]) doubles the number of bits to 16 384, while in the soft bit format, we have one byte representing each bit of the

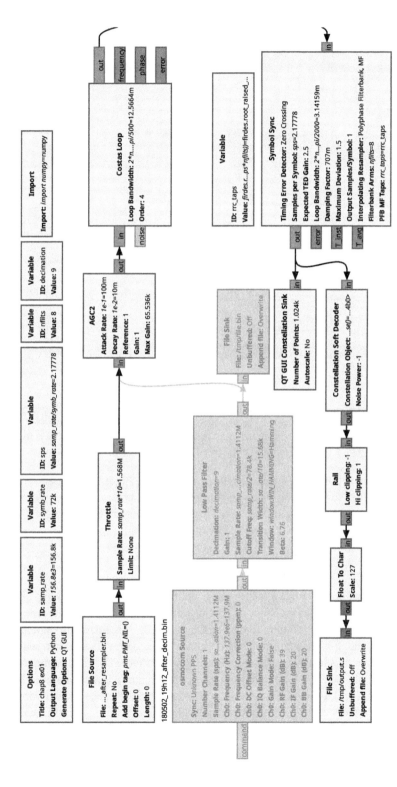

**Figure 8.8** Acquisition sequence aimed at minimizing the size of the file storing data by maximizing the number of processing steps taken care of by GNU Radio, namely creating a copy of the carrier (Costas loop) and the clock synchronizing data sampling, in order to recover the bit sequence and only save one 8-bit sample for each information (bit) transmitted. These samples with be discretized (1 or 0) during post-processing taken care of later. The grayed blocks used during acquisition from the RTL-SDR are commented out and the data are read here from a file available on the IQEngine repository.

# 8 QPSK and CCSDS Packets: Meteor-M 2N Satellite Signal Reception

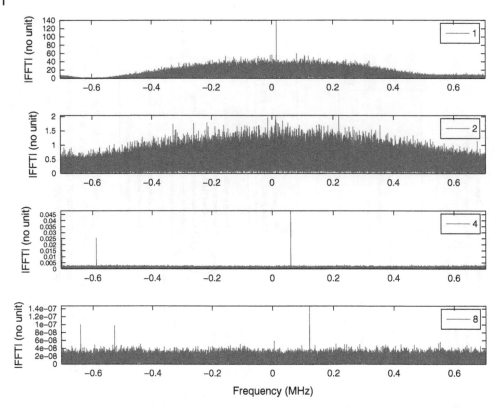

**Figure 8.9** Top to bottom: spectrum of the $I + jQ$ coefficients exhibiting the spectrum spreading; squaring the signal; computing the fourth power and the eighth power. Squaring the signal does not allow for spread spectrum compression to recover the carrier: the modulation is not BPSK. The fourth power allows for removing the modulation and only the carrier is left: the modulation is QPSK.

transmitted message. The position of the correlation peaks observed in Figure 8.10 is hence indeed consistent with the expected shape of the signal: not only have we verified that we understand how to read the file saved by GNU Radio and interpret correctly its content, but we know that the acquired signal contains the information transmitted by the satellite and is worth further analysis.

## 8.5.2 Decoding Data

We have so far obtained a sequence of bytes whose value will most probably be equal to 1 or 0. As with all continuous streams of data, we must have a starting point to know when to start processing the bitstream and assemble them into letters (bytes) and then words, sentences, and paragraphs. The classical approach to identify the beginning of a transmission is to provide a known sequence in the continuous bitstream and search for the occurrence of this pattern. The estimator of resemblance between the successive phases of the signal and the searched pattern is the cross-correlation. Indeed, the technical documentation of LRPT [Ncri, 2000] teaches that all space transmissions are synchronized on the word `0x1ACFFC1D`. The job sounds easy: by cross-correlating this word, we shall find the synchronization as the maximum of the cross-correlation.

Not so fast. First of all, the bits received from a satellite orbiting at more than 800 km from the surface of the Earth are corrupted by noise. We must thus search for some kind of repeated pattern to maximize the chances of properly detecting the transmitted message. A basic approach

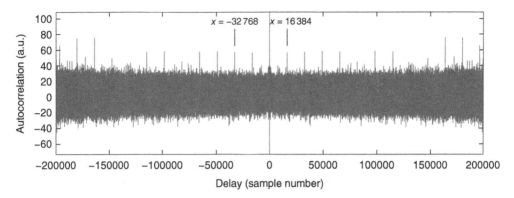

**Figure 8.10** Autocorrelation of 400k samples of soft bits stored as the output of the GNU Radio demodulating chain that was in charge of recovering the carrier frequency and the QPSK clocking rate. Correlation peaks are seen every 16 384 samples.

would consist of simply repeating the message multiple times, but how do we select the good one if two transmissions are not identical? Better, convolutional encoding uses as input a continuous bitstream and creates a new (longer) sequence so that each new bit is a combination of the input bits. This combination is designed to maximize our chances of recovering the initial message: this processing step is convolutional encoding. Bits encoded this way no longer exhibit the synchronization word sequence `0x1ACFFC1D` but its convolutional encoded version, which must be determined and searched for in the received bitstream.

### 8.5.3 Convolutional Encoding of the Synchronization Word

In order to maximize our chances of recovering the value of each bit, a convolutional code spreading the information over time is used to introduce redundancy. While encoding is excessively simple to implement, decoding the most probable sequence following the corruption of some of the bits during transmission is possibly very complex. An optimal approach to the dual problem is to implement decoding using the Viterbi algorithm [Forney, 1973] – named after its author, who is also cofounder of the Qualcomm company – and we must thus master these concepts in order to recover usable bit sequences.

Input data are represented as *soft bits* or sequences of 8-bit values encoding each possible one bit value. The convolution to generate the emitted bits has used each source bit as input and created two output bits as a combination of a given number of input bits: the convolution algorithm is qualified as $r = 1/2$ since it provides twice as many bits on its output as input bits, and $K = 7$ since the shift register used as memory for the input bit sequence is 7 bits long [Neri, 2000, p. 23]. Convolution can be tackled in many ways: one approach, closest to the hardware or programmable logic (FPGA) implementations of the convolutional encoding, consists of a shift register used as a memory, fed by the new bit of the sequence to be encoded, and feeding one or more XOR gates to provide $1/r \geq 1$ output bits for each input bit (Figure 8.11). One way of defining which bits of the shift register feed the XOR gate is to provide the polynomial whose non-null powers match the taps connecting from the shift register to the XOR gate [O'Dea and Pham, 2013] (Figure 8.11, reading from right to left, the binary representation of the byte defining each polynomial coefficient). From this polynomial expression of the convolutional code, we can state the sequence of operations as a matrix operation, as described at http://www.invocom.et.put.poznan.pl/~invocom/C/P1-7/en/P1-7/p1-7_1_6.htm by considering that the register content shift at each time step is achieved by

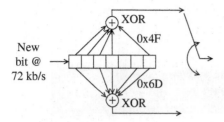

**Figure 8.11** Convolutional coding: bits in the shift register are sampled at the positions whose indices define the polynomials, added as binary values (exclusive OR operation), and the two resulting bits are concatenated as output, inducing an output data rate double of the input data rate.

shifting the polynomial coefficients along the (long) bit sequence to be encoded. This *generator matrix* expression is natural when considering an Octave implementation since it expresses the convolution as a matrix operation and is programmed as follows in this simple case:

```
d=[0 1 0 1 1 1]
G1=[1 1 1]    % 0x7  r=1/2, K=3 (3-bit long shift register)
G2=[1 0 1]    % 0x5
G=[G1(1) G2(1) G1(2) G2(2) G1(3) G2(3)   0     0     0     0     0     0;
      0     0   G1(1) G2(1) G1(2) G2(2) G1(3) G2(3)   0     0     0     0;
      0     0     0     0   G1(1) G2(1) G1(2) G2(2) G1(3) G2(3)   0     0;
      0     0     0     0     0     0   G1(1) G2(1) G1(2) G2(2) G1(3) G2(3);
      0     0     0     0     0     0     0     0   G1(1) G2(1) G1(2) G2(2);
      0     0     0     0     0     0     0     0     0     0   G1(1) G2(1);
]
mod(d*G,2)   % matrix product modulo 2 = XOR
```

with the alternating coefficients of the two polynomials named G1 and G2, which indeed provide a result consistent with the one shown at home.netcom.com/~chip.f/viterbi/algrthms.html: 0 1 0 1 1 1 → 0 0 1 1 1 0 0 0 0 1 1 0.

The creation of the convolution matrix, generated here manually, is generalized to the case we are interested in: a code made of two polynomials applied to a 7-bit-long shift register, with

```
% bytes to be encoded
d = [0 0 0 1 1 0 1 0  1 1 0 0 1 1 1 1  1 1 1 1 1 1 0 0  0 0 0 1 1 1 0 1]; %
    1A CF FC 1D
G1 = [1 1 1 1 0 0 1] % 4F polynomial 1
G2 = [1 0 1 1 0 1 1] % 6D polynomial 2
Gg = [];
for k = 1:length(G1)
  Gg = [Gg G1(k) G2(k)]; % creates the interleaved version of the two
     generating polynomials
end
G = [Gg zeros(1,2*length(d)-length(Gg))] % first line of the convolution
     matrix
for k = 1:length(d)-length(G1)
  G = [G; zeros(1,2*k) Gg zeros(1,2*length(d)-length(Gg)-2*k)];
end
for k = length(Gg)-2:-2:2
  G = [G; zeros(1,2*length(d)-(length(Gg(1:k)))) Gg(1:k)];
end     i          % last lines of the convolution matrix
res = 1-mod(d*G,2);       % mod(d*G,2)
dec2hex(res(1:4:end)*8+res(2:4:end)*4+res(3:4:end)*2+res(4:4:end))'
printf("\n v.s. 0x035d49c24ff2686b or 0xfca2b63db00d9794\n")
```

We observe by executing these few lines that encoding the synchronization word 1ACFFC1D00 with the two polynomials 4F and 6D-defining the taps of the XOR gates to the 7-bit-long shift register, yields the sequence FCA2B63DB00D9794, which is indeed the one given at [Teske, 2016]. Note that it seems that two standards exist within NASA, in which G1 and G2 appear to be swapped. Hence, some of the encoding and decoding software available on the web stating that they use the right code do not yield the expected result. We now understand how to encode the synchronization word using a convolutional code.

### 8.5.4 Convolutional Code Representation as State Machines

We shall need to understand state transitions later, when describing the Viterbi decoding algorithm. Thus, we must not only understand convolutional codes in terms of matrix operations but also view them as a finite state machine with a decision to be taken considering the most probable path from one state to another. How do we express the problem we have just described as a matrix product in terms of a state machine?

The convolutional code we are interested in uses a 6-bit-long shift register to which another new 7th bit is inserted as new input data. The encoder thus has $2^6 = 64$ possible states. We cannot show them all here but will only exhibit the first ones needed to encode the first byte of the synchronization word.

Let us start from a state in which all bits in the shift register are equal to 0 (a state that we shall call "a"). Two outcomes are possible: either the new input value is 0 or 1, so in the previous Octave code, we have the two cases d=[0 0 0 0 0 0 0] or d=[1 0 0 0 0 0 0]. In the former case, the two output bits generated by the convolutional code are mod(d*G1',2)=0 and mod(d*G2',2)=0 or 00, and in the latter case mod(d*G1',2)=1 and mod(d*G2',2)=1 or 11. In the former case, the output state is the same as the input state so that a → a, while in the latter case, the "1" value was input into the register, so we have reached the internal state [1 0 0 0 0 0], which will be called "b." This analysis continues to generate the following table:

| Input bit | Input state | Name | Output bits | Output state | Transition |
|:---:|:---:|:---:|:---:|:---:|:---:|
| 0 | 000000 | a | 00 | 000000 | a → a |
| 1 | 000000 | a | 11 | 100000 | a → b |
| 0 | 100000 | b | 10 | 010000 | b → c |
| 1 | 100000 | b | 01 | 110000 | b → d |
| 0 | 010000 | c | 11 | 001000 | c → … |
| 1 | 010000 | c | 00 | 101000 | c → … |
| 0 | 110000 | d | 01 | 011000 | d → e |
| 1 | 110000 | d | 10 | 111000 | d → … |
| 0 | 011000 | e | 00 | 001100 | e → … |
| 1 | 011000 | e | 11 | 101100 | e → f |
| 0 | 101100 | f | 01 | 010110 | f → … |

| Input bit | Input state | Name | Output bits | Output state | Transition |
|---|---|---|---|---|---|
| 1 | 101100 | f | 10 | 110110 | f → ... |
| ... | ... | ... | ... | ... | ... |
| 0 | 100001 | u | 01 | 010000 | u → c |
| 1 | 100001 | u | 10 | 110000 | u → d |
| ... | ... | ... | ... | ... | ... |
| 0 | 000001 | z | 11 | 000000 | z → a |
| 1 | 000001 | z | 00 | 100000 | z → b |

The ellipsis (...) in the names of the final state represent cases that are not needed when encoding the first byte of the synchronization code (state "c" will not be needed either but is included to make the demonstration clearer). States "u" and "z" are inserted to exhibit loops that will appear when the state machine is followed long enough. We invite the reader to ascertain this result for themselves to be convinced of the relevance of the approach.

This state transition table is exploited to illustrate how the 0x1A byte (the first byte of the synchronization word) is encoded. The first three bits at 0 of the most significant nibble are encoded as 00 and keep the state machine in state "a," while the last bit at 1 is encoded as 11 and leads to state "b." The most significant bit of the A nibble is 1, and we are in state "b" so that we generate 01 and reach state "d." The next bit is 0 (reminder: 0xA=0b1010), and with the current state at "d," we output 01 to reach "e." State "e" takes an input 1 to generate 11 and reach "f," and finally "f," with an input of 0, generates 01. In summary, by encoding 0x1A, we have produced 0b0000001101011101=0x035D, which is indeed the expected value (notice that 0x035D is the opposite of 0xFCA2, which was mentioned earlier as the beginning of the encoding solution – both solutions are the same assuming a 180° phase rotation of the bits).

A state machine representation can hence be given as shown in Figure 8.12, with, on top, the successive states from "a" to "f" as a function of the input bit and, on the bottom, the output bits for each of these transitions.

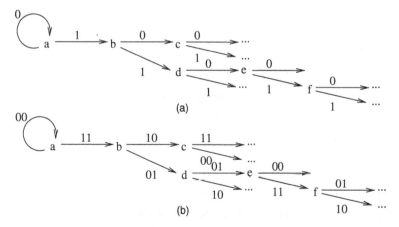

**Figure 8.12** (a) States, named from "a" to "f," and transitions as a function of the input bit value. (b) Output bit values as a function of the transitions. By following the path described in the text, the input sequence 0x1A=0b00011010 is encoded as 0b0000001101011101=0x035D.

*8.5 From the Radio frequency Signal to Bits* | **269**

### 8.5.5 Decoding a Convolutional Code: Viterbi Algorithm

Having described the state machine used to convert an input word to an output word with double length, we now wish to understand the decision sequence that will maximize the probability of reverting the process, considering that some of the received data might have been corrupted during the transmission. Let us consider the result before addressing the explanation.

We will use later a library for efficiently implementing (in C language) convolutional encoding and decoding: libfec (https://github.com/quiet/libfec) is described at [Teske, 2016], and its sample code vtest27.c is used as a starting point to implement the decoder. Alternatively, www .spiral.net/software/viterbi.html provides a code generator to decode, using the Viterbi algorithm, meeting the requirements of LRPT. We start assessing libfec with the simple case of decoding the sentence encoded as FC A2 B6 3D B0 0D 97 94 – which we have already seen as resulting from the convolutional code encoding of the synchronization word – to check whether we are indeed able to recover the initial word:

```c
#include <stdio.h>   // gcc -Wall -o t t.c -I./libfec ./libfec/libfec.a
#include <fec.h>
#define MAXBYTES (4)   // final message is 4-byte long

#define VITPOLYA 0x4F // 0d79: polynomial 1 taps
#define VITPOLYB 0x6D // 0d109: polynomial 2 taps

int viterbiPolynomial[2] = {VITPOLYA, VITPOLYB};
unsigned char symbols[MAXBYTES*8*2] =  // *8 for byte->bit, and *2 Viterbi
    {1, 1, 1, 1, 1, 1, 0, 0,  // fc
     1, 0, 1, 0, 0, 0, 1, 0,  // a2
     1, 0, 1, 1, 0, 1, 1, 0,  // b6
     0, 0, 1, 1, 1, 1, 0, 1,  // 3d
     1, 0, 1, 1, 0, 0, 0, 0,  // b0
     0, 0, 0, 0, 1, 1, 0, 1,  // 0d
     1, 0, 0, 1, 0, 1, 1, 1,  // 97
     1, 0, 0, 1, 0, 1, 0, 0}; // 94

int main(int argc, char *argv[]){
  int i, framebits;
  unsigned char data[MAXBYTES]; // *8 for bytes->bits & *2 Viterbi
  void *vp;
  framebits = MAXBYTES*8;
  for (i = 0; i<framebits*2; i++) symbols[i] = 1-symbols[i];   // flip bits
  for (i = 0; i<framebits*2; i++) symbols[i] = symbols[i]*255; // bit ->
    byte /!\
  set_viterbi27_polynomial(viterbiPolynomial);   // definition of taps
  vp = create_viterbi27(framebits);
  init_viterbi27(vp, 0);
  update_viterbi27_blk(vp, symbols, framebits+6);
  chainback_viterbi27(vp, data, framebits, 0);
  for (i = 0; i<MAXBYTES; i++) printf("%02hhX", data[i]);
  printf("\n");
  exit(0);
}
```

with the only **subtlety of the code**, which takes as input the successive bit values (here hard bits since already saturated as 0 or 1) of the word to be decoded, to encode each bit **on the whole range of a byte (hence 0 to 255 and not 0 or 1)** and possibly flipping the bits (0 ↔ 1) in order not to recover the complement of the word we are looking for (phase rotation by 180° of the initial phase modulation, for example). By executing this code, we indeed recover 1ACFFC1D, which is the word we are looking for, so we understand how to use libfec. Let us now try to understand the underlying algorithm.

The convolutional code implemented during the encoding step was designed to introduce some redundancy in the bit sequence in order to make the transmission robust to random errors that might be introduced by a uniformly distributed noise (as opposed to noise bursts that would corrupt a whole sequence of bits). The problem of decoding hence means going through the state machine proposed in Figure 8.12 the other way around, aiming to find the most probable path considering the received bit sequence. Let us consider that we have received the sequence 0b0000001111011101: what emitted bit sequence has most probably generated the transmitted code? The Viterbi decoder is initialized with all bits set to 0 or state "a" if referring to Figure 8.12. We receive 00, so we stay in state "a" and know that the transmitted bit was 0. Similarly, for the next two 00 sequences that follow, the pair 11 indicates that we have switched to "b" after a 1 was encoded. When receiving 11, we are in state "b," which is not consistent: "b" can only produce 01 (if 1 had been transmitted) or 10 (if 0 had been transmitted). An erroneous bit has thus been received. At the moment, we cannot decide which path to follow, so we continue analyzing the two possible cases, "c" and "d." The next 2 bits are 01, and here we conclude that "c" is hardly possible since "c" can only produce 00 or 11, while "d" is indeed able to produce the observed 01 (emitted bit equal to 0) to yield the state "e." Thus, we give up following the path along "c" in order to continue along "d" and then "e." When in state "e," we receive 11, which is a possible bit state (emitted bit 1), to yield to "f," and finally 01 is a possible value of "f" that allows us to conclude that the last transmitted bit was 0. We have thus been able to correct one erroneous bit and deduce that the most probable transmitted sequence was 0b00011010=0x1A, which should have led to 0b000000110 1011101 in which the erroneous bit is highlighted in bold font (Figure 8.13).

Based on this knowledge, we can now search by correlating with the received sequence the synchronization word once encoded by the convolutional code, with correlation peaks demonstrating our understanding of the message encoding, and the ability to identify the beginning of each transmitted packet. When correlating the encoded word with the $I/Q$ coefficient dataset and converting

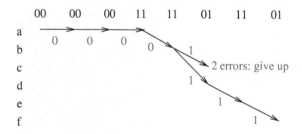

**Figure 8.13** Top: the sequence of received bits, grouped by 2 since the convolutional code generates two output bits for each encoded input bit. On the vertical axis, the states of the machine describing the encoder. In light characters next to each arrow, the numbers of accumulated errors along each path of the decoder. Practically, giving up on one branch would be decided when two paths meet the same state: in this case, the sequence accumulating most errors is left at the benefit of the most probable sequence since exhibiting fewer accumulated errors.

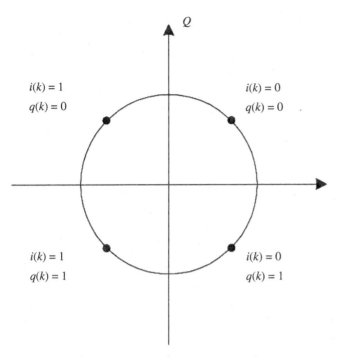

**Figure 8.14** QPSK constellation diagram: each possible phase state encodes 2 bits mapped according to the Gray encoding. Assigning each symbol, from 0 to 3, to the appropriate bit pairs will be part of the decoding challenge.

it to bits through the symbol-to-bit mapping shown in Figure 8.14, the result does not show any correlation peak and is not usable (Figure 8.16, top). Something is still missing in our analysis.

### 8.5.6 Constellation Rotation

The lack of correlation hints at the lack of understanding as to how bits are encoded in the acquired message, assuming that the convolutional code encoded synchronization word is correct, an assumption we will take as true considering the information provided at [Teske, 2016]. When we investigated binary phase modulation (BPSK), we had to consider two possible cases: either the acquired signal was in phase with the local oscillator, and the bit states obtained by comparing the received signal with the local oscillator carrier copy (Costas loop) were those expected, or the signal was in phase opposition and the bits were flipped with respect to their expected value. In both cases, the correlation with a synchronization word accumulates energy along the bit sequence and yields, by considering the absolute value, to a correlation peak, whether we have the right sequence in the collected sentence or its opposite. This simple case is more complex when considering QPSK in which the phase takes one of four possible states, each encoding a pair of bits. Assigning the phase value to a pair of bits is not obvious, but most significantly, any error in the symbol–bit pair mapping yields an erroneous sequence that will not correlate with the word we are looking for. As an example, let us consider the mapping stating that 0° matches the bit pair 10 and 90° to 11 (Gray code in which only a single bit varies between two adjacent phase values). Then, the sequence 0°–90° yields 1011, while the mapping 0° to 11 and 90° to 01 will interpret the same phase sequence as 1101: the two messages generated by the same phase sequence are completely different and have no chance of accumulating the energy needed to generate the correlation peak along the comparison with the reference synchronization word. We have not identified any other scheme for identifying the mapping from phase to bit pairs other than brute force by testing all possible bit combinations, as seen when reading the source code of `m.c.patts[][]` from `medet.dpr` in `meteor_decoder`:

```
1111110010100010101101100011110110110000000110110010111100101000
0101011011111011110100111001010011011010101001001100000111000010
0000011010111010100100111000010010011111111001001101000011010110
1010100100000100001011000110101100100101011011001111100011110100
1111100010100010111100100111110011100000001110011010110110100000
0101011000001000000111001001011100011010101001110011110100111110
0000011101011101000011011000001100011111111000110010100100101110
1010100111110111111000110110100011100101010110001100001011000001
```

provides all possible bit combinations in the sentence encoded by the convolutional code, and the correlations of the received message with all these variations of the code are computed.

However, one issue remains: these bit inversions are easily implemented on the binary values of the reference word by swapping 1 and 0, but how can we perform the same operation with *soft bits* in which the phase of the received message is encoded with continuous values quantified on 8-bit values? Shall we decide from now on which phase value to attribute (*soft → hard bits*), or can we keep on handling raw values? We must interpret bit swapping operations as constellation rotation or symmetry (Figure 8.15). We observe that swapping bits is interpreted as operations between the real and imaginary part, either by rotating or by symmetry along one of the complex axes. Hence, by manipulating the real and imaginary parts of the acquired data, we can achieve the same result as the one obtained by swapping bits, but by keeping the continuous values of *soft bits* and postponing the attribution of each bit (0 or 1) to each phase during the decoding by the Viterbi algorithm.

Identifying the mapping between the four QPSK phase conditions in the constellation diagram (*I* on the *X*-axis, *Q* on the *Y*-axis) and the matching bit pairs hence requires a brute force search, in which all possible combinations are tested. Correlating the phase of the signals with the various combinations of the header word following the convolutional encoder is shown in Figure 8.16. Only one mapping to a given bit pair yields a periodic sequence of correlation peaks (Figure 8.16, bottom): this is the right mapping that will be used throughout the following decoding steps.

This permutation will from now on be applied to all *I*/*Q* sets of the acquired message since we know that doing so will yield the original bit sequence sent by the satellite during the Viterbi decoding applied to the resulting soft bits.

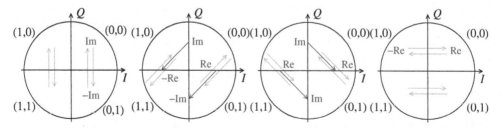

**Figure 8.15** Rotations and symmetry of the constellation, and corresponding result on the real and imaginary axis (*I* and *Q*) of the complex plane on which the raw collected data are represented.

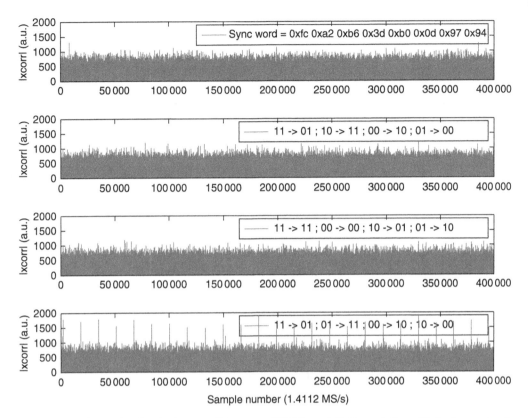

**Figure 8.16** Correlation for the four possible cases of QPSK constellation rotation with the known header word. We observe that only the fourth case – bottom – yields period correlation peaks representative of the beginning of new sentences. This mapping between the four QPSK symbols and bit pairs is the correct one.

### 8.5.7 From Bits to Sentences: Applying the Viterbi Algorithm Decoding

We now have a sequence of phases with values in the set $[0; \pi/2; \pi; 3\pi/2]$ properly organized to become a sequence of bits in which the synchronization word was found, and a unique mapping from the various symbols {00; 01; 11; 10} to each phase value provides a solution exhibiting this correlation. We are now left with decoding to remove the convolutional code encoding and then apply to the resulting bits (which were hence the bits encoded by the satellite prior to the convolutional code) a sequence of XOR (exclusive OR) with a polynomial designed to maximize the randomness of the resulting dataset and hence avoid long repetitions of the same bit state.

We have mentioned the availability of `libfec` for efficiently implementing the decoding of a convolutional encoded signal. We extend the previous basic example to the practical case of decoding full sentences.

Our first idea was to feed the library and decode the whole file of the acquired dataset. By doing so, we hide the initialization and termination issues of the convolutional decoding using the Viterbi algorithm, which assumes that at start-up, all shift register values are equal to 0. This works, as we observe after decoding that, every 1024 bytes, we recover the synchronization word 0x1ACFFC1D.

## 8 QPSK and CCSDS Packets: Meteor-M 2N Satellite Signal Reception

> **Stack Management When Statically Allocating Large Memory Arrays**
>
> Warning: we have met a segmentation fault error when trying to allocate an array large enough to be filled with the whole dataset. Indeed, the file containing the soft bits as one byte for each sample is 11.17 MB large, so we tried to allocate a static array, as any good embedded systems developer would have done, since embedded systems hardly allow dynamic memory allocation through `malloc` due to its excessive resource requirements. However, doing so attempts to allocate the array on the stack, and the default stack size on GNU/Linux is 8192 kB, as shown by `ulimit -s:8192`. Rather than increasing the stack size, we have used the operating system's dynamic memory allocation to save the array on the heap rather than on the stack, hence removing the constraint on the available memory space.

```c
#include <stdio.h>  // from libfec/vtest27.c
#include <stdlib.h> // gcc -o demo_libfec demo_libfec.c -I./libfec./libfec/
    libfec.a
#include <fcntl.h>
#include <unistd.h> // read
#include <fec.h>
#define MAXBYTES (11170164/16)  // file size /8 (bytes-> bits) /2 (Viterbi)

#define VITPOLYA 0x4F
#define VITPOLYB 0x6D
int viterbiPolynomial[2] = {VITPOLYA, VITPOLYB};

int main(int argc, char *argv[]){
  int i, framebits, fd;
  unsigned char data[MAXBYTES], *symbols;
  void *vp;

  symbols = (unsigned char*)malloc(8*2*(MAXBYTES+6)); // *8 for bytes->bits
    & *2 Viterbi
// root@rugged:~# ulimit -a
// stack size            (kbytes,  -s) 8192
// -> static allocation (stack) of max 8 MB, after requires malloc on the
    heap
  fd=open("./extrait.s",O_RDONLY); read(fd,symbols,MAXBYTES*16); close(fd);

  for (i = 1; i<MAXBYTES*16; i+ = 2) symbols[i] = -symbols[i]; // I/Q
    constellation rotation
  framebits = MAXBYTES*8;
  set_viterbi27_polynomial(viterbiPolynomial);
  vp = create_viterbi27(framebits);
  init_viterbi27(vp, 0);
  update_viterbi27_blk(vp, &symbols[4756+8], framebits+6);
  chainback_viterbi27(vp, data, framebits, 0);
  for (i = 0;i<20;i++) printf("%02hhX", data[i]);
  printf("\n");
  fd = open("./output.bin", O_WRONLY|O_CREAT, S_IRWXU|S_IRWXG|S_IRWXO);
  write(fd, data, framebits);
  close(fd);
  exit(0);
}
```

*8.5 From the Radio frequency Signal to Bits* | **275**

L. Teske [2016] tells us, however, that this approach is not optimum since it requires that the whole file is loaded into memory at once. We know that 1024-byte blocks (2048 bytes after encoding) are individually encoded, and hence, rather than decoding the whole file, we might focus on searching for the encoded header synchronization word and decoding the next 2048 bytes starting from this position. For safety and to let the decoder initialize, we will make sure we fetch a few samples before and after the block to be decoded. The resulting `main` function will be as follows:

```
fdi=open("./excerpt.s", O_RDONLY);
fdo=open("./output.bin", O_WRONLY|O_CREAT, S_IRWXU|S_IRWXG|S_IRWXO);
read(fdi, symbols, 4756+8);   // offset
framebits = MAXBYTES*8;

do {
  res = read(fdi, symbols, 2*framebits+50);   // fetches a bit more data
  lseek(fdi, -50, SEEK_CUR);                   // go back
  for (i = 1;i<2*framebits;i += 2) symbols[i] = -symbols[i]; // I/Q
    constellation rotation
  set_viterbi27_polynomial(viterbiPolynomial);
  vp = create_viterbi27(framebits);
  init_viterbi27(vp, 0);
  update_viterbi27_blk(vp, symbols, framebits+6);
  chainback_viterbi27(vp, data, framebits, 0);
  write(fdo, data, MAXBYTES);                  // decoding result, as long
} while (res == (2*framebits+50));             // ... as more data is available
close(fdi); close(fdo);
```

Additionally, this block-based processing will allow to later add Reed–Solomon block error correction (Section 8.A.1). Using this block error correction is optional: the data blocks obtained after convolutional code decoding using the Viterbi algorithm can be used as is, and in the first step, we shall bypass the Reed–Solomon block error correction by only using the first $1024 - 4 - 128 = 892$ bytes in each block (after removing the first four bytes of the synchronization word at the beginning of the block and the last 128 bytes needed for block error correction). This point will be addressed in Appendix 8.A.

Alternatively, the reader who prefers to keep on working with Octave instead of using the C language can use the program provided at https://github.com/Filios92/Viterbi-Decoder/blob/master/viterbi.m, which is also perfectly functional with the following:

```
f = fopen("extrait.s");   % soft bits generated from GNURadio
d = fread(f, inf, 'int8'); % read file
d(2:2:end) = -d(2:2:end); % constellation rotation
phrase = (d < 0)';          % soft -> hard bits
[dv, e] = viterbi([1 1 1 1 0 0 1; 1 0 1 1 0 1 1], phrase, 0);
data = (dv(1:4:end)*8+dv(2:4:end)*4+dv(3:4:end)*2+dv(4:4:end));
```

---

**Forward Error Correcting Code (FEC) Decoding Using GNU Radio**

GNU Radio provides the means for decoding convolutional code since the error-correcting code libraries and `libfec` are implemented in `gr-satellite` at https://github.com/daniestevez/gr-satellites and described in detail in the blog of D. Estévez at www.destevez.net. The `FEC Extended Decoder` requires a decoder object description configured with the CC `Decoder Definition`. The constraint length of $K = 7$ bits and the polynomial description [79, 109] match the $G1 = 0x4F$ and $G2 = 0x6D$ we have just used. The following flowchart

---

*(Continued)*

**276** | *8 QPSK and CCSDS Packets: Meteor-M 2N Satellite Signal Reception*

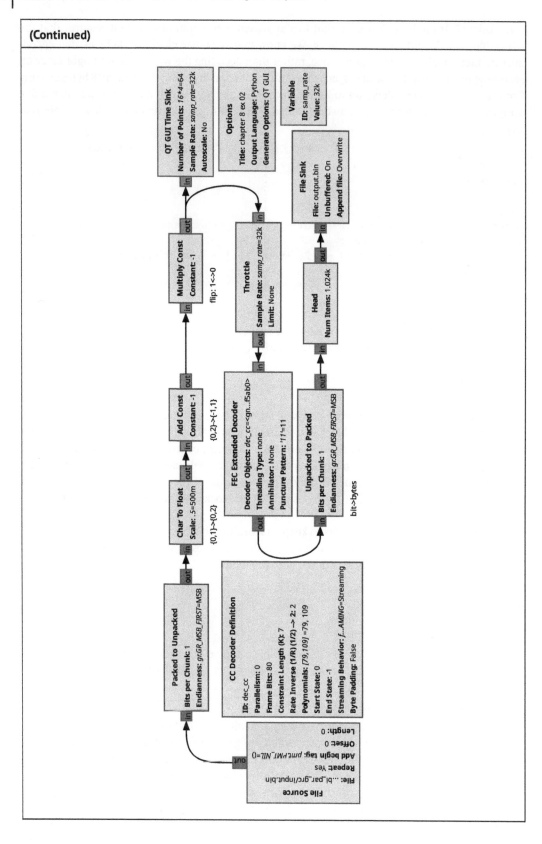

(Continued)

fed with the input file containing

```
$ echo -e -n "\xfc\xa2\xb6\x3d\xb0\x0d\x97\x94" > input.bin
$ xxd input.bin
00000000: fca2 b63d b00d 9794
```

will generate the following time-domain output in which the bits have been inverted and scaled between −1 and +1 (instead of 0 and 1).

**Warning**: while the GNU Radio companion configuration for decoding convolutional codes matches the definition used in `libfec`, the format of the input data differs. Indeed, GNU Radio expects, as hidden in one of the documentations of the `fec_decode_ccsds_27_fb` processing block (GNU Radio companion `Decode CCSDS 27` block) of `gr-fec` block, to be given as input a value between −1 and +1 (and not 0 to +1, as would have been expected for a bitwise representation). We are here using *soft bits* between $\exp(j\pi) = -1$ and $\exp(j0) = +1$ and not *hard bits*. We hence process the bits read from the input file, including the encoded word, multiply by 2 (hence a scale factor inverse of 0.5, a value to be filled in the `Char to Float` block), subtract 1, and possibly switch bit values (multiplication by −1) to achieve the expected result.

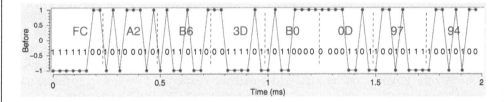

where the input bits have been annotated in small letters (lower row) and the resulting bytes in large letters (upper row), matching the synchronization word definition of FCA2B63DB00D9794. Running this flowchart results in an `output.bin` file whose content is

```
$ xxd output.bin | head -1
00000000: 1acf fc1d 6427 33c0 1acf fc1d 6427 33c0
```

or the repeated pattern – since the input file was allowed to loop – of the decoded synchronization word 1ACFFC1D. The additional bits result from the Viterbi decoder buffer emptying as the input file is allowed to loop. Care must be taken with the initialization state: the Viterbi algorithm assumes that at start-up, all shift register values are equal to 0, which is not necessarily true when the encoded word is repetitively read in a loop. D. Estévez presents at https://destevez.net/2017/01/coding-for-hit-satellites-and-other-ccsds-satellites the various declinations of the decoder configuration to meet the configuration of the various CCSDS norm declinations.

We claim to understand how to decode the messages sent by the satellite, but are we truly sure of the validity of these bits? In order to quickly assess a known bit sequence, we follow up on the strategy described at the beginning of this chapter, namely correlating with a known pattern.

8 QPSK and CCSDS Packets: Meteor-M 2N Satellite Signal Reception

We are told that, without knowing anything about the sentence encoding which will be described in the next section 8.6, the protocol described in [National Research Center Planet, 2020b] in Appendix A includes a frame description of the telemetry claimed to include the date onboard the satellite: this sentence is identified with the PRS-64 code made of the sequence "2 24 167 163 146 221 254 191." Are we able to find this byte sequence in the sentences we have decoded? Having obtained bits that we assume to be stored in an array called dv, we concatenate using Octave bits to nibbles and then nibbles to bytes (data array) and to sentences (see the following fin matrix): array

```
% data is a byte array from Viterbi decoding as discussed previously
for k = 1:24                              % analyze first 24 blocks
 d(:, k) = data(1+(k-1)*2048:k*2048);     % 2048 nibble sentences
 dd(:, k) = d(1:2:end, k)*16+d(2:2:end, k); % nibbles -> bytes
 fin(:, k) = dd(5:end, k);                % remove the synchronization
    header from each sentence
 fin(:, k) = [bitxor(fin(1:255, k)', pn) bitxor(fin(1+255:255+255, k)', pn)
    ...
            bitxor(fin(1+255*2:255+255*2, k)', pn) bitxor(fin
    (1+255*3:255+255*3, k)', pn)];
end
```

We observed that we had to apply the pn code, which was designed to make the bit sequence as random as possible (and hence spread the information), using the bijective function of the XOR mask. This random structure of the sentences avoids long sequences of the same bit value, which would make clock recovery difficult. The 255-byte-long pn sequence is given at https://www.teske .net.br/lucas/2016/11/goes-satellite-hunt-part-4-packet-demuxer, and we just apply this mask to our bytes grouped by 255-element-long packets (the four-byte header, which is not affected by this transform, has already been removed).

Finally, these sentences are analyzed to search for the magic sequence PRS-64 known to prefix the telemetry sentence. Finding this sequence would demonstrate that our procedure is consistent, since not only do we identify the telemetry identification sequence – a set of bytes with little chance of random occurrence – but furthermore, the analysis of the telemetry values provides consistent results with the acquisition time and the output of meteor_decoder (Onboard time: 11:48:33.788) as

```
date_header=final(589:589+7,9)' % found in CADU 9 (out of the 24 processed)
date=final(589+8:589+11,9)'
ans =
     11
     48
     33
     197
```

We have found the time it was onboard the satellite when the image was acquired, demonstrating the validity of the byte recovery from the $I/Q$ stream. Now that we are confident the bit sequence is correct, we are left to analyze how the sentences are assembled to decode the image, an activity closer to computer science than signal processing.

## 8.6 From Sentences to Paragraphs

The bit sequences match the norm described in the technical documentation, so we have finished the decoding investigation. Not completely. The satellite transmits an image, and we have only recovered data. Can we go beyond this basic result?

This is the point where OSI layers appear. An image is a large set of information, too large to be stored in a unique packet transmitted from the satellite. Worse, the satellite sends information acquired in at least three spectral ranges (depending on whether it is on the dayside or nightside, these three ranges change, but what is day and night in Spitsbergen, with its three months of day and three months of complete darkness?) interleaved in various packets. We understand better why the OSI standard separates the various abstraction layers: an image is one large entity split into color layers, which are themselves split into blocks (JPEG encoding), which are further split into packets transmitted to the ground with all the error correction code and redundancy to maximize the chances that the receiver recovers an error-free data stream. We have indeed stated JPEG compression earlier: for this author who was indoctrinated with all the drawbacks of lossy compression and artifacts induced by the JPEG compression, is it possible that images transmitted from satellites are thus encoded? The trade-off lies probably in the available bandwidth around the relatively low-frequency carrier with respect to the large data size of the high-resolution picture to be transmitted. We shall, of course, be careful when processing such images, not to focus on artifacts related to the $8 \times 8$ pixel block encoding, which will be the topic of the forthcoming discussion.

An aspect that was quite unclear to us in the various documentations lies in the definition of the *minimum code unit* (MCU): we are taught that each MCU includes 196 adjacent areas of a picture, made of 14 thumbnails of $8 \times 8$ pixels. The part that was not clear to us in the documentation is that *successive MCUs are independent from each other*. Hence, a picture line is made of 14 MCUs, each including 14 thumbnails of $8 \times 8$ pixels: $14 \times 14 = 196$, and $14 \times 14 \times 8 = 1568$ pixels is indeed the width of an image transmitted by Meteor-M 2N. Thus, our objective is to decode the $8 \times 8$-pixel JPEG thumbnails, concatenate the resulting image, and so on until a full line of the final image is assembled for a given wavelength. The procedure is then repeated for the other two wavelengths before starting again following a maintenance telemetry sentence (identifier 70 in the APID field (*APplication IDentifier*) – the image being defined with APID identifiers in the range from 64 to 69 [National Research Center Planet, 2020b]).

Furthermore, one MCU packet might not completely fit inside the payload of the M-PDU layer protocol: it could be that one M-PDU includes multiple successive MCUs (for example when the JPEG thumbnails are small due to the lack of features in a homogeneous area being imaged) or that one MCU is shared among two M-PDUs. Hence, the VCDU packet header includes a pointer to the index of the starting point of the next M-PDU. Before this pointer, we collect the leftover of the previous M-PDU.

We had stored in the `fin` matrix the successive sentences decoded after the application of the Viterbi algorithm and derandomization: we display the content of the first lines for this matrix to notice the consistency of several patterns – header of the packet – before reaching the random payload. The document best suited to understand how to decode sentences is Fong et al. [2002].

```
octave:28> fin(1:16,:) % see 20020081350.pdf NASA p.9 of PDF
     64    64    64    64    64    64    64    64    64    64    64    64    64    64   Version
      5     5     5     5     5     5     5     5     5     5     5     5     5     5   Type
    140   140   140   140   140   140   140   140   140   140   140   140   140   140   \
    163   163   163   163   163   163   163   163   163   163   163   163   163   163   - counter
     43    44    45    46    47    48    49    50    51    52    53    54    55    56   /
```

8 QPSK and CCSDS Packets: Meteor-M 2N Satellite Signal Reception

```
  0    0    0    0    0    0    0    0    0    0    0    0    0    0 sig. field
  0    0    0    0    0    0    0    0    0    0    0    0    0    0 VCDU insert
  0    0    0    0    0    0    0    0    0    0    0    0    0    0 1 zone
  0    0    2    1    0    0    0    0    0    0    0    0    2    0 5 bits @ 0
 18  142   28   54   18  130   78  226    0   20   70   28   82   32 M_PDU header

 77  166  239  222   73   82   83  199    8   28  232  247  165  183 M_PDU...
133  188  229   42   24   23  220   94   68  117   92  151   87  203 ... 882 bytes
 75  177  221  215    0   48   49  128   13  166    8  218  126  212
 42  138  254   87   12   32  249   87   34  172  247  107    9  142
146  238  236   80  215   96  143  121    0  124   12   89   86  191
179  227   64  144   89   59  240  105  105  251   46   43    0  199
                                        ^
```

which is analyzed as follows, from the first to the last line, beginning with the 6-byte-long *VCDU primary header*:

1. 64 matches the version constant equal to 01 followed by 6 zeros for the most significant bits of the VCDU Id (S/C id). Hence, the first 8 bits are 0100 0000.
2. The identifier of the transmitting satellite follows (field: Type of the VCDU Id): this field is described in [National Research Center Planet, 2020b] as being equal to 5 if the instrument is present and 63 if the instrument is absent. Here, a value of 5 is a positive omen for the next steps of image decoding. Furthermore, Ghivasky [2004, p. 149] and [Le Ber, 2003] indicate that a VCDU Id of 5 (AVHRR LR) is associated with channels APID 64…69, as will be seen later.
3. The three-byte-long VCDU counter is incremented at each new packet, as observed on the last byte (140 163 43…56) of the three bytes matching the sentence counter.
4. All following fields (*signaling field*) are filled with zeros to indicate real-time data transfer, as are the fields VCDU Insert zone and absence of cryptography [Ghivasky, 2004, p. 150], [Le Ber, 2003].
5. Finally, the last two bytes of the header provide the pointer indicating the address of the first packet included in the current sentence. This information is arguably the most important since an M_PDU packet most certainly spans along multiple sentences, and hence knowing where the first M_PDU included in the current packet starts allows for synchronizing the beginning of the decoding process of a new image. The first 5 bits are always equal to 0 [Ghivasky, 2004, p. 147], [Le Ber, 2003], while the last 11 bits provide the address, within the sentence, of the first useful packet. In this processing sequence, the pointer is computed as

   ```
   x=fin(9,:)*256+fin(10,:)+12 = 30 154 552 322 30 142 90 238 12 32 82 40 606 44
   ```

6. The 882 bytes that follow are the M-PDU payload, including the *virtual channel field*. We become convinced that the position of the header computed earlier is correct by

   ```
   for k=1:length(x);fin(x(k),k),end
   ```

   which returns 64 64 64 64 65 65 65 65 68 64 64 64 64 65 which is the list of the virtual channel identifiers we shall analyze later, i.e. the various wavelengths at which the images are collected (APID in the 64 to 69 range [National Research Center Planet, 2020b]).
7. The ninth column is a bit unusual since it includes the first packet of the transmission of the image with APID 68, so it exhibits a header offset equal to 0 with respect to the end of the VCDU header, allowing to start tackling the M_PDU payload format without having to search for the beginning address pointer. We shall thus see that 8 = 0000 1000 is the version (ID = 000/type

indicator = 0/secondary header 1 = present/000 APID). Then APID = 68 is one of the measurement channels [National Research Center Planet, 2020b], and finally the length of the packet (in bytes) is provided by {0 105}.

## 8.7 So Much Text ... Pictures Now

We have identified how to decode the VCDU sentence, so now we have to analyze the M_PDU payload. Multiple M_PDU can be grouped inside one VCDU sentence (for example when the JPEG thumbnail payload is highly compressed), and one M_PDU can be distributed between two successive VCDU – there is no reason for the M_PDU payload size to be a multiple of the VCDU sentence length.

We have previously identified the pointers in the VCDU header toward the beginning of the first M_PDU payload in the VCDU. Displaying the first bytes of each M_PDU, we observe a consistent pattern,

| 8 | 8 | 8 | 8 | 8 | 8 | 8 | 8 | 8 | 8 | 8 |
|---|---|---|---|---|---|---|---|---|---|---|
| 68 | 68 | 68 | 68 | 68 | 68 | 68 | 70 | 64 | 64 | 64 |
| 13 | 13 | 13 | 13 | 13 | 13 | 13 | 205 | 77 | 13 | 13 |
| 34 | 35 | 36 | 37 | 38 | 39 | 40 | 41 | 42 | 43 | 44 |
| 0 | 0 | 0 | 0 | 0 | 0 | 0 | 0 | 0 | 0 | 0 |
| 105 | 47 | 49 | 69 | 81 | 107 | 57 | 57 | 97 | 77 | 79 |
| 0 | 0 | 0 | 0 | 0 | 0 | 0 | 0 | 0 | 0 | 0 |
| 0 | 0 | 0 | 0 | 0 | 0 | 0 | 0 | 0 | 0 | 0 |
| 2 | 2 | 2 | 2 | 2 | 2 | 2 | 2 | 2 | 2 | 2 |
| 136 | 136 | 136 | 136 | 136 | 136 | 136 | 136 | 136 | 136 | 136 |
| 181 | 181 | 181 | 181 | 181 | 181 | 181 | 181 | 186 | 186 | 186 |
| 124 | 124 | 124 | 124 | 124 | 124 | 124 | 124 | 76 | 76 | 76 |
| 0 | 0 | 0 | 0 | 0 | 0 | 0 | 0 | 0 | 0 | 0 |
| 0 | 0 | 0 | 0 | 0 | 0 | 0 | 0 | 0 | 0 | 0 |
| 98 | 112 | 126 | 140 | 154 | 168 | 182 | 2 | 0 | 14 | 28 |
| 0 | 0 | 0 | 0 | 0 | 0 | 0 | 24 | 0 | 0 | 0 |
| 0 | 0 | 0 | 0 | 0 | 0 | 0 | 167 | 0 | 0 | 0 |
| 255 | 255 | 255 | 255 | 255 | 255 | 255 | 163 | 255 | 255 | 255 |
| 240 | 240 | 240 | 240 | 240 | 240 | 240 | 146 | 240 | 240 | 240 |
| 77 | 77 | 77 | 77 | 77 | 77 | 77 | 221 | 77 | 79 | 81 |
| 243 | 186 | 210 | 178 | 136 | 175 | 242 | 154 | 173 | 238 | 235 |
| 197 | 41 | 160 | 177 | 253 | 120 | 216 | 191 | 166 | 148 | 77 |
| 60 | 194 | 210 | 146 | 236 | 9 | 151 | 11 | 88 | 100 | 166 |
| 240 | 156 | 80 | 106 | 84 | 81 | 201 | 48 | 42 | 228 | 208 |
| 105 | 41 | 5 | 65 | 152 | 245 | 135 | 33 | 131 | 208 | 9 |
| 254 | 104 | 41 | 23 | 193 | 172 | 56 | 197 | 38 | 210 | 28 |
| 91 | 52 | 20 | 205 | 1 | 233 | 249 | 0 | 62 | 115 | 118 |

which we analyze [National Research Center Planet, 2020b], by following the first column:

1. "68" is the instrument *packet identifier* onboard the satellite transmitting the information, APID. As the image collected by one instrument spans multiple successive packets, we expect to find

8 *QPSK and CCSDS Packets: Meteor-M 2N Satellite Signal Reception*

the same APID along multiple successive columns. One interesting case is APID number 70 in column 8, which indicates a telemetry sentence. This allowed us to identify earlier the time onboard the satellite when the image was grabbed.

2. "13 34" is made of the first 2 bits indicating whether this is the first packet of a sequence (01) or the follow-up of a transmission (00), followed by the packet number counter encoded on 14 bits, which will be used to check whether we have lost an image thumbnail along a line. We indeed observe that the least significant byte is incremented for each new M_PDU.

3. The next two bytes indicate the length of the M_PDU packet – here 105 bytes.

4. The following 8 bytes encode the date, namely a date on two bytes defined as "0 0," a number of milliseconds within the day encoded on 32 bits "2 136 181 124" valid for all the packets related to a given image, and finally, a date complement in microseconds encoded on 16 bits and fixed, for Meteor-M 2N, to "0 0" [National Research Center Planet, 2020b].

5. The payload description indicates the index of the first MCU, thumbnails whose assembly will create the final image. This MCU index is incremented by 14 between two successive packets, here 98 112 126, since thumbnails are grouped by 14 to improve compression capability [National Research Center Planet, 2020b].

6. Finally, the image header includes 16 bits set to 0 [National Research Center Planet, 2020b] (*Scan Header*) followed by the *segment header*, including an indicator of the presence of the quality factor encoded on 16 bits and set to 0xFF 0xF0 or 255 240 [National Research Center Planet, 2020b], followed by the value of this quality factor which will be used in the quantization levels when decoding the JPEG thumbnail – in our case 77 but can be variable along a line of the final image.

7. The data of the 14 successive MCUs follow, generating 14 thumbnails, each $8 \times 8$ pixels, or 64 bytes in the naming convention of [National Research Center Planet, 2020b] (in the case of the first sentence, this payload starts with 243 197).

This somewhat lengthy description is needed to understand properly the link between the VCDUs and the M_PDU, which finally represent two different abstraction layers of the data stream at different levels of the OSI framework. Once this difference is understood, assembling JPEG thumbnails to create an image is only a matter of rigorous implementation of the standards. With Octave, the bytes representing each MCU forming 14 thumbnails are grouped into individual files by the following piece of software:

```
for col = 1:23                               % column number   =   VCDU frame
    number
first_head = fin(9, col)*256+fin(10, col) % 70 for column 11
fin([1:first_head+1]+9, col)';       % beginning of line 11: 1st header in
    70
fin([1:22]+first_head+11, col)';     % start of MCU of line 11

clear l secondary apid m
l = fin(first_head+16-1, col)*256+fin(first_head+16, col); % vector of
    packet lengths
secondary = fin(first_head+16-5, col);       % initializes header list
apid = fin(first_head+16-4, col);            % initializes APID list
m = fin([first_head+12:first_head+12+P], col);
k = 1;
while ((sum(1)+(k)*7+first_head+12+P)<(1020-128))
    m = [m fin([first_head+12:first_head+12+P]+sum(1)+(k)*7, col)];
```

```matlab
    secondary(k+1) = fin(first_head+16+sum(l)+(k)*7-5, col);
    apid(k+1) = fin(first_head+16+sum(l)+(k)*7-4, col);
    l(k+1) = fin(first_head+16-1+sum(l)+(k)*7, col)*256+fin(first_head+16+
    sum(l)+(k)*7, col);
            % 16 = offset from VDU beginning
    k = k+1;
  end
  for k = 1:length(l)-1 % saves each MCU bytes in a new file
    jpeg = fin([1:l(k)]+first_head+12+19+sum(l(1:k-1))-1+7*(k-1), col);
    f = fopen(['jpeg', num2str(apid(k), '%03d'), '_', num2str(col, '%03d'),
      '_', num2str(k, '%03d'), '.bin'], 'w');
    fwrite(f, jpeg, 'uint8');
    fclose(f);
  end


  k = length(l);                                        % last incomplete
    packet
  jpeg = fin([1+first_head+12+19+sum(l(1:k-1))-1+7*(k-1):end], col);
  first_head = final(9, colonne+1)*256+final(10, col+1);%looks for next VCDU
  jpeg = [jpeg; final([1:first_head]+10, col+1)];
  % we expected 79 bytes in the last packet: 925-892 = 33 are missing
  f = fopen(['jpeg', num2str(apid(k), '%03d'), '_', num2str(col, '%03d'),
    '_', num2str(k, '%03d'), '.bin'], 'w');
  fwrite(f, jpeg, 'uint8');
  fclose(f);
end
```

Having saved the content of the MCUs in individual files named `jpeg*.bin`, we have taken on the task of re-implementing Huffman decoding, Run Length Encoding (RLE), and the discrete cosine transform needed to convert each JPEG-compressed thumbnail into a pixel matrix ready to be displayed. Huffman decoding is especially annoying to implement since it handles data as bit packets whose size is not necessarily a multiple of eight but depends on the size of each information in the encoding binary tree. We have just read and understood the source code of the decoder from `meteor_decoder`, translated to C++ at https://github.com/infostellarinc/starcoder/blob/master/gr-starcoder/lib/meteor, and used small part of the library to validate the content of the bit packets obtained in the preceding processing steps, which we claim to contain JPEG thumbnails. This time, the encoding scheme (and hence decoding) is very well described in CCITT [1993]. The fact that `meteor_decoder` considers the MCUs we feed it as valid and that the resulting thumbnails are consistent since they can be assembled correctly as pictures proves that our decoding of VCDUs and then MCUs is correct.

```cpp
#include "meteor_image.h"
using namespace gr::starcoder::meteor;

int main(int argc, char **argv)
{int fd, len, k, quality = 77;       // fixed quality ...
 unsigned char packet[1100]; // will be provided as argument later
 imager img = imager();
 if (argc < 1) quality = atoi(argv[1]);
```

```
fd = open("jpeg.bin", O_RDONLY);
len = read(fd, packet, 1100);
close(fd);
img.dec_mcus(packet, len, 65, 0, 0, quality);
}
```

is linked with `meteor_image.cc` and `meteor_bit_io.cc` taken from the github archive cited earlier. Exploiting this program by feeding with binary files including the MCU payload, we are provided as output with a 14 × 64 element matrix which we shall name `imag`, each 64-byte-long line being itself an 8 × 8 pixel thumbnail. Reorganizing these 64 elements in Octave using

```
m=[];for k=1:size(imag)(1) a=reshape(imag(k,:),8,8); m=[m a'];end
```

we obtain a 112 × 8 pixel matrix displayed using `imagesc(m)` in order to visualize the image as shown in Figure 8.17 (left). This procedure is repeated for the 14 MCUs each final image line is made of. Figure 8.17 illustrates the concatenation of the first set of thumbnails (left) with the second series, demonstrating how continuous the patterns are. These processing steps are repeated for a whole line of the image acquired at one wavelength by a given instrument (and hence a given APID) before a new APID follows and prompts the processing to restart and so on to create in parallel multiple images acquired at various wavelengths. Notice the excellent compression ratio brought by JPEG on these homogeneous and feature-free areas: only 60 bytes are needed to encode these 14 × 64 = 896 pixel images. Areas exhibiting more features still require bigger MCUs with a few hundred bytes and up to 700 bytes in size.

The result of assembling uncompressed JPEG thumbnails to generate 8 × 8 pixel matrices is shown in Figure 8.18 for the instrument with APID 68. We start seeing some consistent features of

**Figure 8.17** Decoding one MCU (left) made of 14 successive thumbnails each 8 × 8 pixel large, and concatenation with the next MCU (right) to create a picture 28 × 8 = 224 pixel wide and 8 pixel high. The complete final image is thus assembled with small MCU parts. This example is demonstrated on APID 68.

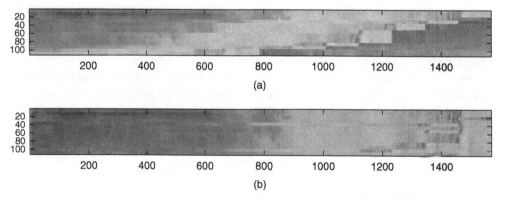

**Figure 8.18** (a) Result of decoding JPEG thumbnails and assembling into a complete picture without considering the counter. We clearly observe that the pattern is shifting from one line to another as packets are missing, resulting in a poor image hardly usable. (b) If the packet counter does not reach the expected threshold of 14 thumbnails/MCU, then some dummy packets are inserted to compensate for the missing thumbnails: this time the image is properly aligned. Here the APID is 68.

an image, but clearly, a few thumbnails are missing at the end of each line since some packets were corrupted and could not be decoded (Figure 8.18).

We compensate for the missing thumbnails, at least temporarily, by duplicating each thumbnail for missing information as indicated by the MCU counter. Rather than recover the missing information, we can at least align along the vertical axis of the picture the adjacent thumbnails and hence achieve a usable picture (Figure 8.19). In this example, we have not used the quality information which modified the quantization coefficients during the JPEG compression depending on the content of the picture, and some discontinuity in the grayscale pattern remains visible.

By integrating the quality factor as an argument provided to the thumbnails decoder of `meteor_decoder` on which our program is linked, the grayscale values become homogeneous to yield a convincing result. Figure 8.20 closely resembles the reference picture fully decoded by `meteor_decoder` (Figure 8.21). Notice that the satellite pass was far from optimal for a

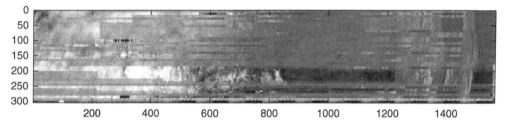

**Figure 8.19** Result of decoding APID 65 while using the counter to identify missing thumbnails, but without exploiting the quality information provided in the JPEG header. The main geographical features are visible, but strong contrast variations exist within the picture, making the result poorly suited.

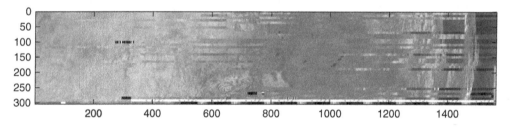

**Figure 8.20** Result of decoding APID 65 while exploiting the counter to identify missing thumbnails and the JPEG quality information. The individual thumbnails become hardly visible and the grayscale evolves continuously along the picture.

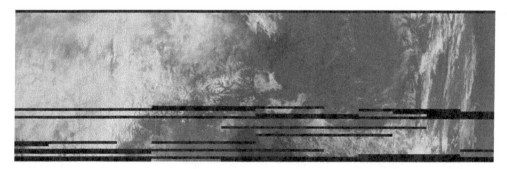

**Figure 8.21** Result of decoding APID 65 with `medet` used as reference picture for comparison with Figure 8.20.

## 8.8 JPEG Image Decoding

We have used `gr-starcoder` without understanding its content. We cannot be satisfied with such an approach: we need to understand the magic behind converting Fourier coefficients into an image of pixel intensity in the spatial domain (since we are in grayscale, we disregard the concept of color: only the intensity of light, or luminance, matters in this processing).

We want to understand how, starting from the MCU provided by decoding the frames, we can retrieve an $8 \times 8$ pixel JPEG image. We know that there is a lossless Huffman coding, with possible elimination of redundancies by copying a repeated value (RLE) and finally converting the Fourier coefficients into the spatial domain through a discrete cosine transform. The JPEG image format, despite being lossy, exhibits excellent **compression** performance and hence allows transferring images with higher resolution than the analog weather satellites (e.g. NOAA POES) in a comparable communication channel.

The main challenge lies in the definition of the Huffman table. Everyone can visit multiple web pages [Dumas et al., 2015] describing the calculation of the binary tree that allows to find a representation of the data that minimizes the number of necessary bits to represent the most frequent information, even if it increases the number of bits required to represent less frequent information. Statistically, this often reduces the size of the file (which is why compressing a text file, with a lot of redundancy, is very effective, while compressing a binary executable, where all bytes are roughly of the same probability, only gives poor results). One aspect that can make the compressed file even larger than the original file is transmitting the correspondence table between bit sequences and decoded data. In the case of LRPT, the choice is made to use a *fixed* table, standardized after analyzing a sufficient number of images to have a statistically representative distribution of the frequency of each symbol. The LRPT standard simply instructs to consult [CCITT, 1993], a somewhat crude shortcut since we have only identified one useful page out of the 186 that the standard contains! The tables that will allow to retrieve the decoding trees (for the DC components – zero frequency – and the AC components – nonzero frequencies – of the Fourier transform of the image) are found on page 158. Therefore, we will make an effort to understand how to read these tables, named DHT for Define Huffman Table, thanks to the explanations from https://www.impulseadventure .com/photo/jpeg-huffman-coding.html and https://imrannazar.com/Let\stquotes-Build-a-JPEG-Decoder (section 3).

We are told that the number of codes made of $N$ bits is

```
X'00 02 01 03 03 02 04 03 05 05 04 04 00 00 01 7D'
```

and that the assignment of the value to each one of these codes follows the sequence

```
X'01 02 03 00 04 11 05 12 21 31 41 06 13 51 61 07' ...
```

Notice how these sequences are those found in `huffman.pas` of `meteor_decoder` (first lines of the `t_ac_0` array). The implicit part is how to create these codes! The first table tells us that there are 0 code with 1 bit, 2 codes with 2 bits, 1 code with 3 bits, 3 codes with 4 bits, 3 codes with 5 bits, two codes with 6 bits, and so on. We must hence assemble this sequence of bits. The operation relies on binary counting, and when we increment the number of bits needed to continue the sequence, we start from the previous sequence filled with 0s (bold characters). Practically, we get

00: first two-bit code
01: second two-bit code
100: unique 3 bit code
1010: first 4 bit code
1011: second 4 bit code
1100: third 4 bit code
11010: first 5 bit code
11011: second 5 bit code
11100: third 5 bit code
111010: first 6 bit code
111011: second 6 bit code
1111000: first 7 bit code

...

and the following table tells us which value is associated with each code:

$00 = 1$
$01 = 2$
$100 = 3$
$1010 = 0$
$1011 = 4$
$1100 = 0x11 = 17$
$11010 = 5$
$11011 = 0x12 = 18$
$11100 = 0x21 = 33$
$111010 = 0x31 = 49$
$111011 = 0x41 = 65$
$1111000 = 6$
$1111001 = 0x13 = 19$

...

We check that we are on the right path by noting that the sum of the elements in the first table is 162, which is indeed the number of elements in the second table; there will therefore be a value for each code. Although the values are between 0 and 255, we observe that 125 of these elements will require 16 bits to be encoded instead of 8. Nevertheless, since the majority of values are grouped around a few elements, these values encoded on 16 bits appear rare enough for the compression to be profitable on average.

An elegant, though not very useful, way to relate to the well-known binary trees of Huffman coding is to represent the code as proposed in Figure 8.22.

The problem is similar but simpler for the DC component, which ranges over fewer values:

```
X'00 01 05 01 01 01 01 01 01'
```

whose content is sequential from 0 to $0xb = 11$. Here again, we observe that there is no 1-bit code, 1 code with 2 bits, 5 codes with 3 bits, then 1 code from 4 to 9 bits. This sequence is hence summarized as follows:

$00 = 0$
$010 = 1$

011 = 2
100 = 3
101 = 4
110 = 5
1110 = 6
11110 = 7
111110 = 8
...

We now have all the information to decode a JPEG image. If we observe the binary sequence at the beginning of an MCU from the previous decoding, we obtain

```
$ xxd -b jpeg.bin | head -2
00000000: 11111011 01010011 11000111 11110110 11110000 11101000 .S...
00000006: 10011111 10110011 00011111 10001101 00111110 00100000 .....>
```

which starts with the DC component of the discrete cosine transform of the image, followed by the 63 AC components (the Fourier transform is bijective, so the 64 pixels of the 8 by 8 thumbnail yield 64 Fourier coefficients, with possibly a certain number of zero values). Each time, we will have the pair "number of bits" followed by "value." Following from the beginning until we reach the first valid DC code, we can see that we obtain 111110 (first zero reached), which corresponds to the value 8 in the second table we just mentioned. This means that the value of the DC component is encoded on the following 8 bits, which is equal to 11 010100 = 212 in decimal. We will see later if this value is correct. The first of the 63 AC coefficients follows: again, we read the bit sequence until we reach the first valid code in the first table mentioned earlier: 11 11000 is the first valid code we encounter, which is associated with the value 6. The first AC coefficient is therefore encoded by the following 6 bits, which is equal to 111 111 = 63 in decimal. Let us continue to discover another subtlety of assigning codes to bit sequences. We resume where we left off reading the bits to find 1011, which corresponds to the value 4 and indicates that we read the next 4 bits, which is equal to 0 111 to determine the value of the next coefficient. Here a simple binary conversion is no longer appropriate, since the representation is not in two's complement but sequential, with the coding of the signed number composed only of 0s representing the minimum value of the range and the coding formed by 1s representing the maximum value. So 0 111 is not +7 as a simple binary to decimal converter would do, but −8, since with 4 bits, we can represent from −15 to +15, excluding the range from −7 to +7, which would have been coded with fewer bits. This is summarized in Table F1 on page 89 of [CCITT, 1993] and interpreted and reproduced in Figure 8.23. The next code is then 100, which is equal to 3, and the following 3 bits are 00 1, which is −6, and we continue like this to decode all the Fourier coefficients or encounter a value of 0 for the number of bits, indicating the end of the MCU with all the other coefficients being zero.

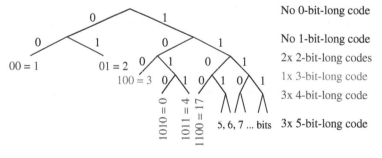

**Figure 8.22** Binary tree representation of the Huffman code. Source: Adapted from CCITT [1993].

| SSSS | DIFF values |
|---|---|
| 0 | 0 |
| 1 | −1,1 |
| 2 | −3,−2,2,3 |
| 3 | −7..−4,4..7 |
| 4 | −15..−8,8..15 |
| 5 | −31..−16,16..31 |
| 6 | −63..−32,32..63 |
| 7 | −127..−64,64..127 |
| 8 | −255..−128,128..255 |
| 9 | −511..−256,256..511 |
| 10 | −1 023..−512,512..1 023 |
| 11 | −2 047..−1 024,1 024..2 047 |

(a)

| Length | Bits | | Value | |
|---|---|---|---|---|
| 0 | | | 0 | |
| 1 | 0 | 1 | −1 | 1 |
| 2 | 00,01 | 10,11 | −3,−2 | 2,3 |
| 3 | 000,001,010,011 | 100,101,110,111 | −7,−6,−5,−4 | 4,5,6,7 |
| 4 | 0000,....,01111 | 1000,....,1111 | −15,....,−8 | 8,....,15 |
| 5 | 00000,....,011111 | 10000,....,11111 | −31,....,−16 | 16,....,31 |
| 6 | 000000,.... | ....,111111 | −63,....,−32 | 32,....,64 |
| 7 | 0000000,.... | ....,1111111 | −127,....,−32 | 32,....,64 |
| 8 | 00000000,.... | ....,11111111 | −255,....,−128 | 128,....,255 |
| ... | ... | ... | ... | ... |

(b)

**Figure 8.23** (a) Original table from CCITT [1993] displaying, as difference magnitude categories for DC coding, the correspondence between the number of bits allocated to a value and the value range. (b) Interpretation by explicitly analyzing the number of bits allocated to a value, the bit sequence, and the value itself. Note that the range covered by a smaller number of bits is excluded from the range represented by a given number of bits.

Once the coefficients are obtained, they are reorganized according to the zigzag pattern described on p. 16 (Fig. 5) of [CCITT, 1993] or specified with the index of each matrix cell on p. 30 (Fig. A.6), and the discrete cosine transform is performed to move from the spectral domain to the spatial domain.

We can convince ourselves of the validity of the analysis by slightly modifying `meteor_ image.cc` from https://github.com/infostellarinc/starcoder/blob/master/gr-starcoder/lib/ meteor to display the coefficients before reorganization and cosine transform: we display the content (floating point numbers!) of `zdct[]` just before the loop `dct[i] = zdct[ZIGZAG[i]] * dqt[i];` which reorganizes the coefficients according to the zigzag pattern. By doing this, we obtain 212 63 −8 −6 −27 −156 52 65 −2 10 3 −1 21 −1 −4 6 −14 0 6 24 10 24 50 16 1 −12 3 −1 0 −3 0 17 3 −2 1 0 27 −8 −4 3 −1 −1 1 −5 −1 0 −1 15 1 −1 −2 1 −1 −1 −1 1 1 0, which indeed starts with the sequence that we have identified and continues with the 64 coefficients. After reorganization according to the zigzag pattern and applying the scaling factor that cancels out a certain number of coefficients during compression, we obtain a vector which we will call `coefficients`: 1484 315 −780 364 −44 108 368 28 −48 −162 390 −9 −168 0 −336 −200 −36 −12 147 0 90 78 224 −104 60 −8 60 52 −23 80 111 145 24 20 34 0 0 −50 47 35 44 0 −75 29 −37 −48 −52 −42 23 0 −72 40 0 −112 −55 46 561 126 −220 −45 52 −46 47 0. These are the 64 coefficients in the spectral domain on which the discrete cosine transform is applied (function `idct2` from Octave's *signal processing toolbox* – page 27 of [CCITT, 1993]) (Figure 8.24):

$$\text{iDCT2}(x,y) = \frac{1}{4} \sum_{u=0}^{7} \sum_{v=0}^{7} C_u C_v S_{vu} \cos\left(\frac{(2x+1)\pi u}{16}\right) \cos\left(\frac{(2y+1)\pi v}{16}\right)$$

with $C_0 = 1/\sqrt{2}$ and $C_{k \neq 0} = 1$, and $S$ being the matrix of Fourier coefficients.

After the iDCT, the decoder provides the sequence of 64 values, which we will call `image` 410 299 332 320 292 508 307 222 304 189 185 283 95 353 290 160 252 474 370 552 558 497 273 109 149 261 215 264 328 340 226 −4 417 289 641 512 644 510 245 95 282 56 407 255 466 390 160 −33 418 131 613 569 551 572 96 59 371 47 498 416 460 478 −45 78 which we reorganize into an $8 \times 8$ pixel image and verify that, within the constant 128 added by `gr-starcoder` (as required by the standard), we obtain the same result between `idct2(reshape(coefficients,8,8))` and `reshape(image,8,8)` (Figure 8.25).

| 0 | 1 | 5 | 6 | 14 | 15 | 27 | 28 |
|---|---|---|---|---|---|---|---|
| 2 | 4 | 7 | 13 | 16 | 26 | 29 | 42 |
| 3 | 8 | 12 | 17 | 25 | 30 | 41 | 43 |
| 9 | 11 | 18 | 24 | 31 | 40 | 44 | 53 |
| 10 | 19 | 23 | 32 | 39 | 45 | 52 | 54 |
| 20 | 22 | 33 | 38 | 46 | 51 | 55 | 60 |
| 21 | 34 | 37 | 47 | 50 | 56 | 59 | 61 |
| 35 | 36 | 48 | 49 | 57 | 58 | 62 | 63 |

**Figure 8.24** Assignment of the position of each coefficient in the zigzag coding of the quantized DCT coefficients maximizing the presence of important coefficients at the beginning of the sequence and eliminating a large number of high-frequency coefficients (toward the right and bottom) during quantization.

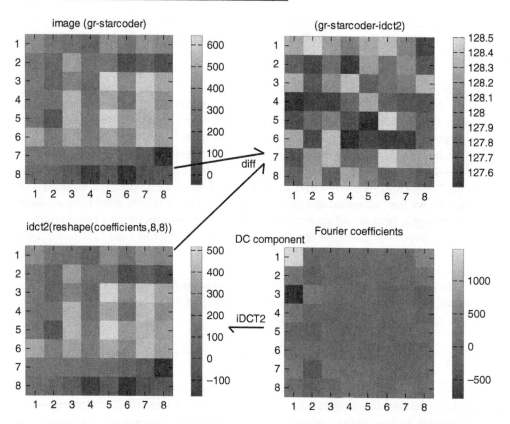

**Figure 8.25** Once the sequence of Fourier coefficients is obtained, the end of the processing is trivial: reorganizing along the zigzag pattern, applying the scaling factor defining the image quality, and finally inverting the discrete cosine transform. Starting from the Fourier coefficients (bottom right, ensuring that the DC component of the signal is in coordinates (1,1) in the top left of the matrix), we compare (top right) the calculation performed by gr-starcoder (a C++ translation of meteor_decoder) and that obtained by inverse 2D discrete cosine transform (idct2 from Octave). We observe that the error is approximately constant and equal to 128 – a constant added by gr-starcoder after calculating the iDCT (top right).

| 16 | 11 | 10 | 16 | 24 | 40 | 51 | 61 |
| 12 | 12 | 14 | 19 | 26 | 58 | 60 | 55 |
| 14 | 13 | 16 | 24 | 40 | 57 | 69 | 56 |
| 14 | 17 | 22 | 29 | 51 | 87 | 80 | 62 |
| 18 | 22 | 37 | 56 | 68 | 109 | 103 | 77 |
| 24 | 35 | 55 | 64 | 81 | 104 | 113 | 92 |
| 49 | 64 | 78 | 87 | 103 | 121 | 120 | 101 |
| 72 | 92 | 95 | 98 | 112 | 100 | 103 | 99 |

**Figure 8.26**  Luminance quantization table provided in CCITT [1993, p. 143].

The scaling factor of each coefficient, which is the source of loss in the JPEG compression algorithm through quantization, is deduced as follows. Starting from the Fourier coefficient quantization table $HTK$ provided in Table K.1 on page 143 of [CCITT, 1993] (Figure 8.26), a quality factor $Q$ is provided with each MCU to indicate the compression factor of the image. From this quality factor $Q$, we deduce $F = 5000/Q$ if $20 < Q < 50$ and $F = 200 - 2 \times Q$ if $50 < Q < 100$. This quantization factor weighs the original quantization table by $PTK = F/100 \times HTK$, and finally, each transmitted coefficient $Sq$ is the result of rounding to the nearest integer the Fourier coefficient $S$ normalized by the adjusted quantization table $Sq = S/PTK$.

## 8.9  Conclusion

This exploration of the LRPT protocol defining digital communications between weather satellites and the ground was the opportunity to address all the OSI layers, from the physical layer (transmission frequency) to the coding (QPSK and convolutional encoding) to packets (words) and images (sentences) broadcast in the messages. This presentation was aimed at demonstrating how a useful tool software-defined radio could be for teaching: each processing step involved a new physical or mathematical principle, each one incredibly boring if tackled independently from a purely theoretical perspective but becoming fascinating in the complex framework of the data transmission concluded with decoding an image.

Hence, we have practically discovered some of the subtleties of QPSK modulation and the various solutions when assigning to each of the four possible phase states a bit pair – a problem that had escaped us when investigating a BPSK modulated signal – and then implementing convolutional code decoding using the Viterbi algorithm. Having demonstrated the consistency of the bit sequence generated by the convolutional code decoding, we have temporarily skipped the Reed–Solomon block code error correction to unstack the various protocol layers encapsulating the thumbnails, which the final image is made. The motivated reader might want to implement manually JPEG decoding, which we simply implemented here without reviewing the theoretical background, emphasizing yet its selection for its high compression capability. Finally, the Reed–Solomon block error decoding algorithm is implemented and its proper operation validated, despite a rather minor benefit in this particular demonstration.

These fundamental principles have been described to provide the basics to decode many other space-borne remote-sensing data streams, as demonstrated by the rich list of applications in the blog written by D. Estévez, who applies his expertise on a multitude of amateur and professional satellite links at www.destevez.net, and most significantly, for example https://destevez.net/2017/01/ks-1q-decoded.

## Appendix 8.A  Reed–Solomon Block Error Correcting Code

Convolutional coding is designed to compensate for noise distributed along the bits due to radio frequency communication noise between the emitted and the receiver, assuming a uniform random noise that might impact each bit independently of its neighbors. Such a coding scheme would be however unable to correct for a corrupted block of data due to a burst of noise: this type of error is taken care of by the block error correcting code, as implemented, for example, by Reed–Solomon. This code is similar to the block-correcting strategy used in the radio data system broadcast over commercial FM bands with the BCH encoding [EN 50067, 1998]. Here, each 255-byte-long packet is made of 223 payload bytes and 32 error-correcting code bytes, allowing to identify transmission errors et possible correct part of them. This type of error-correcting code is thus named RS(255,223) since for 255 transmitted bytes, 223 are data, and the last 32 are error-correcting code bytes. libfec provides the library needed to correct transmission errors using RS(255,223), as described in Teske [2016]. Since the aim of block error-correcting code is to correct a set of erroneous bits, it is wise to distribute information along the sentence in order to minimize the impact of an interference affecting multiple adjacent bits. Hence, instead of splitting a 1020-byte sentence into four neighboring sentences, each 255 bytes long, the data structure interleaves four sequences of data and

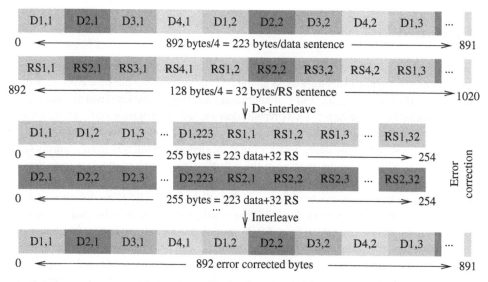

**Figure 8.27**  Organization of data along a 1020-byte long CVCDU sentence (we have already removed the 4-byte-long synchronization word header). The first 892 bytes include the data payload D considered as interleaved for the four sets of 223-byte-long sequences, which can be corrected using Reed–Solomon RS(255,232), and the last 128 bytes include, still interleaved, the four sequences of 32-byte-long error correcting code RS. Applying the correction algorithm requires first de-interleaving the data (top → middle), applying Reed–Solomon to identify and correct errors (middle), and re-interleaving the data (middle → bottom) to put them back in their original position but after correcting some of the bytes possibly corrupted during the radio frequency link.

Appendix 8.A Reed–Solomon Block Error Correcting Code | **293**

their error-correcting code, as illustrated in Figure 8.27, with the error-correcting code located at the end of the sentences.

We can train first to understand how Reed–Solomon is implemented in `libfec`:

```c
#include <fec.h<     // gcc -o jmf_rs jmf_rs.c -I./libfec./libfec/libfec.a

int main()
{int j;
 uint8_t rsBuffer[255];

 uint8_t tmppar[32];
 uint8_t tmpdat[223];

 for (j = 0;j<255; j++) rsBuffer[j] = (rand()&0xff);   // received data
 for (j = 0;j<223;j++)  tmpdat[j] = rsBuffer[j];        // backup data
 encode_rs_ccsds(tmpdat, tmppar,0);                     // create RS code
 for (j = 223;j<255;j++) rsBuffer[j] = tmppar[j-223];// append RS after data
 rsBuffer[42] = 42; tmpdat[42] = 42;                    // introduce errors
 rsBuffer[43] = 42; tmpdat[43] = 42;                    // ... on purpose
 rsBuffer[44] = 42; tmpdat[44] = 42;   // ... to check correction capability
 rsBuffer[240] = 42;tmppar[240-223] = 42;
 printf("RS:%d\n", decode_rs_ccsds(rsBuffer, NULL, 0, 0)); // check that RS
   can correct
 for (j = 0; j < 223; j++)
   if (rsBuffer[j]! = tmpdat[j]) {printf("%d: %hhd -> %hhd; ", j, tmpdat[j],
     rsBuffer[j]);}
 for (j = 223; j < 255; j++)
   if (rsBuffer[j]! = tmppar[j-223]) {printf("%d: %hhd -> %hhd; ", j, tmppar
   [j-223], rsBuffer[j]);}
}
```

In this example, we create a (random) dataset which is encoded, then modify 4 bytes of the payload and 1 byte of the error-correcting code, and test how the decoder can correct these errors. The result

```
RS:4
42: 42 -> 5; 43: 42 -> 23; 44: 42 -> 88; 240: 42 -> 95
```

is consistent with our expectations: four errors were identified and corrected.

Applying this sample program to the 128 bytes at the end of a sentence designed to correct the first 892 bytes at the beginning of the sentence does not work at all! Yet another trick indicated by Lucas Teske, which we had not found in the documentation, is that the implemented algorithm is a *dual basis Reed–Solomon* in which the bytes are once again run through a randomization table as described at https://github.com/opensatelliteproject/libsathelper/blob/master/src/reedsolomon.cpp. Once this transform has been applied, the error-correcting code operates properly, as demonstrated with the following piece of software, which includes the whole decoding sequence: namely Viterbi algorithm deconvolution, application of the polynomial bijective XOR operation to remove the randomization of the data, de-interleaving the data to be grouped with their Reed–Solomon error-correcting code, applying the transposition polynomial, error correction, and removing the transposition polynomial to finally recover the corrected data:

# 8 QPSK and CCSDS Packets: Meteor-M 2N Satellite Signal Reception

```c
#include <fec.h> // gcc -o demo_rs demo_rs.c -I./libfec./libfec/libfec.a

// github.com/opensatelliteproject/libsathelper/blob/master/src/reedsolomon.
    cpp
// dual basis Reed Solomon !
#include "dual_basis.h"

unsigned char pn[255] ={           // randomization polynomial
    0xff, 0x48, 0x0e, 0xc0, 0x9a, 0x0d, 0x70, 0xbc, \
    0x8e, 0x2c, 0x93, 0xad, 0xa7, 0xb7, 0x46, 0xce, \
[...]
    0x08, 0x78, 0xc4, 0x4a, 0x66, 0xf5, 0x58   };

#define MAXBYTES (1024)

#define VITPOLYA 0x4F
#define VITPOLYB 0x6D

#define RSBLOCKS 4

#define PARITY_OFFSET 892

void interleaveRS(uint8_t *idata, uint8_t *outbuff, uint8_t pos, uint8_t I){
  for (int i=0; i < 223; i++) outbuff[i*I+pos]=idata[i];
}

int viterbiPolynomial[2] = {VITPOLYA, VITPOLYB};

int main(int argc,char *argv[]){
 int res, i, j, framebits, fdi, fdo;
 unsigned char data[MAXBYTES], symbols[8*2*(MAXBYTES+6)];// *8 for bytes-
    >bits & *2 Viterbi
 void *vp;
 int derrors[4] = { 0, 0, 0, 0 };
 uint8_t rsBuffer[255], *tmp;
 uint8_t rsCorData[1020];

 fdi=open("./excerpt.s", O_RDONLY);
 fdo=open("./output.bin", O_WRONLY|O_CREAT, S_IRWXU|S_IRWXG|S_IRWXO);
 read(fdi, symbols, 4756+8);   // offset
 framebits = MAXBYTES*8;

 do {
  res = read(fdi, symbols, framebits*2+50);        // 50 additional bytes to
    finish viterbi decoding
  lseek(fdi, -50, SEEK_CUR);                        // go back 50 bytes
  for (i = 1; i<2*framebits; i+= 2) symbols[i] = -symbols[i]; // I/Q
    constellation rotation
  set_viterbi27_polynomial(viterbiPolynomial);
  vp = create_viterbi27(framebits);                // convolution -> Viterbi
```

```
    init_viterbi27(vp, 0);
    update_viterbi27_blk(vp, symbols, framebits+6);
    chainback_viterbi27(vp, data, framebits, 0);
    tmp = &data[4];                                        // rm synchronization
      header
    for (i = 0; i<1020; i++) tmp[i]^ = pn[i%255];   // XOR decode (dual basis)

    for (i = 0; i<RSBLOCKS; i++)
      { for (j = 0;j<255; j++) rsBuffer[j] = tmp[j*4+i];   // deinterleave
        for (j = 0;j<255; j++) rsBuffer[j] = ToDualBasis[rsBuffer[j]];
        derrors[i] = decode_rs_ccsds(rsBuffer, NULL, 0, 0); // decode RS
        for (j = 0; j < 255; j++) rsBuffer[j] = FromDualBasis[rsBuffer[j]];
        interleaveRS(rsBuffer, rsCorData, i, RSBLOCKS);    // interleave
        printf(":%d",derrors[i]);
      }
    write(fdo, data, 4);                                   // header
    write(fdo, rsCorData, MAXBYTES-4);   // corrected frame
  } while (res == (2*framebits+50));
  close(fdi);
  close(fdo);
  exit(0);
}
```

This code is the culmination of the whole phase conversion process from the $I/Q$ coefficients to bits ready to be assembled to recover sentences and the JPEG images, including the two convolutional and block error-correcting codes. The result of this additional correction is illustrated in Figure 8.28 and demonstrates how adding the block error-correcting code allows for extending the range of the JPEG image analysis as the satellite reaches close to the horizon.

Notice that some of the codes have been truncated of the standard input/output library header files to make the sample listings more compact: the reader will not be challenged in recovering the missing header files (stdio.h, stdlib.h, unistd.h ...) which allow one to open, read, write, and close files as well as communicate with the console.

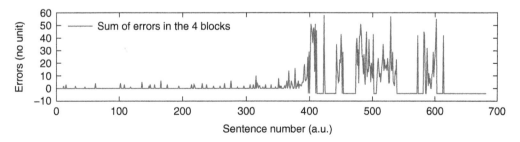

**Figure 8.28** Result of the error correction along the satellite pass. We clearly observe that block correction becomes more efficient as the satellite gets lower on the horizon, degrading the link budget. −1 indicates that too many errors corrupted the payload during the reception to allow for correcting the erroneous bits.

## References

CCITT. Information technology — digital compression and coding of continuous-tone still images — requirements and guidelines, 1993. www.w3.org/Graphics/JPEG/itu-t81.pdf.

D. J. Costello and S. Lin. *Error Control Coding: Fundamentals and Applications.* Prentice Hall, 1982.

J.-G. Dumas, J.-L. Roch, E. Tannier, and S. Varrette. *Foundations of Coding: Compression, Encryption, Error Correction.* John Wiley & Sons, 2015.

EN 50067. Specification of the radio data system (RDS) for VHF/FM sound broadcasting in the frequency range from 87,5 to 108,0 MHz. Technical report, European Committee for Electrotechnical Standardization, 1998.

W. Fong, P.-S. Yeh, S. Duran, V. Sank, X. Nyugen, W. Xia, and J. H. Day. Low resolution picture transmission (LRPT) demonstration system. Technical report, NASA, 2002. https://ntrs.nasa.gov/archive/nasa/casi.ntrs.nasa.gov/20020081350.pdf.

G. D. Forney. The Viterbi algorithm. *Proceedings of the IEEE*, 61(3):268–278, 1973.

G. D. Forney Jr. The Viterbi algorithm: A personal history, 2005. https://arxiv.org/abs/cs/0504020v2.

J.-M. Friedt. Satellite image eavesdropping: A multidisciplinary science education project. *European Journal of Physics*, 26(6):969, 2005.

P. Ghivasky. MetOp space to ground interface specification, doc. MO-IF-MMT-SY0001 rev. 07C. Technical report, EADS/ASTRIUM/METOP, 2004. http://web.archive.org/web/20160616220044; http://www.meteor.robonuka.ru/wp-content/uploads/2014/08/pdf_ten_eps-metop-sp2gr.pdf.

D. J. Israel, G. W. Heckler, and R. J. Menrad. Space mobile network: A near earth communications and navigation architecture. In *2016 IEEE Aerospace Conference*, pages 1–7. IEEE, 2016. https://ntrs.nasa.gov/archive/nasa/casi.ntrs.nasa.gov/20170009966.pdf.

G. Kranz. *Failure is Not An Option: Mission Control from Mercury to Apollo 13 and Beyond.* Simon and Schuster, 2001.

A. Le Ber. MetOP HRPT/LRPT user station design specification, EPS-ASPI-DS-0674. Technical report, EUMETSAT Polar System Core Ground Segment, 2003. www.eumetsat.int/website/wcm/idc/idcplg?IdcService=GET_FILEdDocName=PDF_ASPI_0674_EPS_CGS-US-SPRevisionSelectionMethod=LatestReleased.

A. Makovsky, A. Barbieri, and R. Tung. Odyssey Telecommunications. *JPL, DESCANSO Series on Design & Performance Summary*, 2002. descanso.jpl.nasa.gov/DPSummary/odyssey_telecom.pdf: p. 34 "The command format currently used for deep-space missions, including Odyssey, is defined in the CCSDS standard CCSDS 201.0-B-1."

T. G. Marshall. Numerical electromagnics code, 2002. http://nec2.org/.

National Research Center Planet. "Meteor-M" N.2 satellite, 2020a. http://web.archive.org/web/20200220061824/planet.iitp.ru/english/spacecraft/meteor-m-n2_eng.htm.

National Research Center Planet. Structure of "Meteor-M 2" satellite data transmitted through VHF-band in direct broadcast mode, 2020b. http://web.archive.org/web/20200222233309/http://planet.iitp.ru/english/spacecraft/meteor_m_n2_structure_2_eng.htm.

E. Neri. Single space segment — HRPT/LRPT direct broadcast services specification, doc. MO-DS-ESA-SY-0048 rev. 8, 2000. mdkenny.customer.netspace.net.au/METOP_HRPT_LRPT.pdf.

A. O'Dea and T. T. Pham. Telemetry data decoding, Deep Space Network. Technical report, NASA, 2013. https://deepspace.jpl.nasa.gov/dsndocs/810-005/208/208B.pdf, or http://www.ka9q.net/amsat/ao40/2002paper/.

L. Teske. GOES satellite hunt, 2016. www.teske.net.br/lucas/satcom-projects/satellite-projects/.

A. Viterbi. Error bounds for convolutional codes and an asymptotically optimum decoding algorithm. *IEEE Transactions on Information Theory*, 13(2):260–269, 1967.

# 9

## Custom Source and Sink Blocks: Adding Your Own Hardware Interface

While GNU Radio provides most of the processing features needed for becoming familiar with discrete-time digital signal processing and efficiently analyzing signals, it sometimes happens that some specific functions are missing especially when using field programmable gate arrays (FPGA) [Estévez, 2024] and graphics processing units (GPU) [Azarian, 2024]. These can be at three different layers of integration:

1. Tuning the flowgraph characteristics without accessing the in-phase/quadrature (IQ) stream. This approach was already discussed in Chapter 3 when launching a thread using the `Python Snippet` with the server running as a `Python Module`.
2. Accessing the IQ dataflow within the flowgraph for some basic signal processing that is not designed to be reused in another flowgraph: this feature is provided by the `Python Block`.
3. Providing a new processing block distributed as source code to be compiled or used by others: this feature is provided with out-of-tree (OOT) blocks.

We have mentioned earlier in this document the highlights of software defined radio (SDR), including **stability** (algorithms do not drift and are reproducible from one place to another), **flexibility** (dynamic tuning of the processing parameters, either through remote control or automatically when observing noise as the Kalman filter does), and **reconfigurability** (one hardware, multiple applications). In addition to these features, current networking infrastructures and storage allow for logging and data storage for post-processing despite the huge files involved. Indeed, recording on a dual-channel coherent receiver (e.g. for beamforming) a 10-minute long GNSS Galileo E5 signal sampled at 50 MS/s to cover the whole band with 16-bit resolution complex IQ samples requires $50 \times 10^6 \times \underbrace{2}_{complex} \times \underbrace{2}_{short} \times \underbrace{60 \times 10}_{time} \; \underbrace{2}_{channels} = 2 \times 10^{11}$ or a 200 GB file. Nevertheless, this file can be stored for running multiple processing scripts, possibly decoding different orthogonal signals (e.g. Galileo and Global Positioning System [GPS] signals both broadcasting orthogonal codes in the same frequency band) and many different processing algorithms. Hence, the general outline of our approach to decoding new communication protocols is as follows:

1. Record a few binary files as the emitter was visible from the receiver. This is especially important for intermittent communications such as polar orbiting satellites or planes not constantly visible to the receiver.
2. Assess decoding capability and develop the successive processing sequences, from the lowest level of analyzing the IQ data to recovering soft bits and thresholding to hard bits, and finally assembling bits to bytes and sentences, using high abstraction interpreted processing languages such as Octave or Python and its numerical processing library, numpy. Having access to the

---

*Communication Systems Engineering with GNU Radio: A Hands-on Approach,*
First Edition. Jean-Michel Friedt and Hervé Boeglen.
© 2025 John Wiley & Sons, Inc. Published 2025 by John Wiley & Sons, Inc.
Companion website: www.wiley.com/go/friedtcommunication

**298** | *9 Custom Source and Sink Blocks: Adding Your Own Hardware Interface*

recorded data allows for attempting many algorithm implementations on the **same** data, avoiding the issue of varying source, signal-to-noise ratio, Doppler shift, and making sure the processing algorithm becomes reliable under given conditions.

3. Since interpreted languages are inefficient and slow, convert the high-level interpreted algorithm to lower-level, efficient languages (C/C++), still relying on the recorded signals. By using complete datasets, batches of the appropriate size of samples can be loaded and processed.

4. Once the C/C++ implementation is functional, integrate in the GNU Radio framework where the scheduler provides unknown buffer sizes of data. Batches must be accumulated in a circular buffer until enough samples are available for running the algorithm. e.g. accumulating enough samples in a vector to start processing a fast Fourier transform (FFT). Managing the circular buffer must make sure no sample is lost and all are processed.

5. Once the algorithm has been implemented as the work function of a GNU Radio block, add parameters tunable through callback functions (setters and getters), wrap in the GNU Radio Companion formalism of the YAML interface description and distribute to users.

Various abstraction layers with various difficulties in implementing complex algorithms but also various efficiencies can be considered when implementing a new processing algorithm in GNU Radio, first running on recorded data before tackling live signal analysis. They are described in Sections 9.1 and 9.2.

## 9.1 Python Block

While the Python module does not have access to the IQ stream and can only change parameters of the flowgraph by using setters and getters of the variables defined in the GNU Radio Companion flowchart, it might be desirable to process the IQ stream without getting in the detailed development of a new processing block to be distributed to other users. The Python block provides such a capability.

Various processing blocks are defined within GNU Radio: the synchronous block outputs as many items as inputs, the decimating block decreases by a fixed ratio the data rate, and the interpolating block increases by a fixed ratio the data rate.

Let us consider the generic case where the output rate is unrelated to the input rate, and illustrate this concept with the polynomial fit of the correlation peak for improving the timing resolution by a factor equal to the signal-to-noise ratio. Indeed, when computing the correlation peak between a known signal and time-delayed copies, the basic correlation computation will output a result with a sampling period equal to the acquisition sampling period $T_s$. The timing resolution can be improved by peak-fitting the correlation peak shape and identifying the location of the maximum of the fitting function. This expression becomes analytical in the case of the parabolic fit when the correlation peak located at time delay $\tau$ is characterized by a magnitude $s_0$ while the magnitude of the two neighbors at $\tau - T_s$ and $\tau + T_s$ is $s_{-1}$ and $s_{+1}$. Then the parabolic fit between $(\tau - T_s, s_{-1})$, $(\tau, s_0)$ and $(\tau + T_s, s_{+1})$ leads to a peak position maximum located at

$$\tau + \frac{T_s}{2} \frac{s_{-1} - s_{+1}}{s_{-1} + s_{+1} - 2s_0}$$

This function is implemented in a `Python Block` with, at first, the constructor defining the input parameters as arguments to the `__init__` function as well as the input and output types. This function is only executed once at initialization, and then the `general_work` function loops continuously through the input IQ stream(s) provided in the `input_items` arrays. The output data are filled in `output_items` whose length is returned when exiting `general_work` to inform the scheduler of the number of generated items.

```python
"""
Embedded Python Blocks:
"""

import numpy as np
from gnuradio import gr

class blk(gr.basic_block):  # other base classes are basic_block,
    decim_block, interp_block
    def __init__(self, N=1, cancel=True, cancelP=1, printres=True):
    # arguments show up as parameters in GRC
        gr.basic_block.__init__(
            self,
            name='peak fit',    # will show up in GRC
            in_sig=[np.complex64], out_sig=[np.float32]
        )
        self.N = N
        self.cancel=cancel
        self.bigarray=np.ones((5*self.N),dtype=np.complex64)
        self.posarray=0
        self.printres=printres
        self.cancelP=cancelP

    def general_work(self, input_items, output_items):
        self.consume(0, len(input_items[0]))
        self.bigarray[self.posarray:self.posarray+len(input_items
[0])]=input_items[0]
        self.posarray+=len(input_items[0])
        nbsol=0
        while self.posarray>=self.N:
            if self.cancel == True:
                self.bigarray[self.N//2-self.cancelP:self.N//2+self.
cancelP]=0
            coarse=np.argmax(np.absolute(self.bigarray[0:self.N]))
    # position of the maximum
            s1=np.absolute(self.bigarray[coarse-1])
    # values of neighboring samples
            s2=np.absolute(self.bigarray[coarse])
            s3=np.absolute(self.bigarray[coarse+1])
            if ((s1+s3-2*s2)==0):
                print("ERROR")
            correction=(s1-s3)/(s1+s3-2*s2)/2
    # peak fit
            solution=coarse+correction
```

```
        self.bigarray[0:self.posarray-self.N]=self.bigarray[self
.N:self.posarray]
        self.posarray-=self.N
        output_items[0][nbsol]=solution-self.N/2
        if self.printres==True:
            print(f"{nbsol} {output_items[0][nbsol]:.6f}")
        nbsol+=1
    return(nbsol)
```

The core processing is cancellation of the correlation peak near the 0-delay if leakage between emitter and receiver prevents the time-delayed copies from being detected (argument `Cancel` set to `True` and cancellation of $\pm$`cancelP` samples around the 0-delay) and then peak fitting for returning the fine time delay if enough data have been accumulated, and no item returned otherwise. The associated flowchart for demonstration of the overfitted correlation peak fine time delay detection is illustrated in Figure 9.1. We acknowledge Cyrille Morin (INSA Lyon, France) for helping debug the script.

## 9.2 Out-of-Tree Blocks

OOT is custom signal processing blocks not included in the main GNU Radio software framework. The open-source framework of GNU Radio is prone to promote the development of OOT, but we warn the reader not to jump to this conclusion too quickly and to assess carefully the benefits of writing custom blocks to be maintained over time through the multiple evolutions of GNU Radio, rather than using efficiently existing processing blocks. Some reasons for writing custom blocks rather than assembling existing processing blocks are as follows:

- Hardware peripheral not supported by GNU Radio. In this case, there is no other option than writing custom source or sink blocks, linking low-level drivers with the GNU Radio framework. As an example, radio frequency-grade oscilloscopes are well suited for radio detection and ranging (RADAR) applications when non-contiguous, high-bandwidth signals are needed, with an interface provided at https://github.com/jmfriedt/gr-oscilloscope, arguably the closest solution to the ideal SDR implementation with no frequency transposition to baseband but sampling the whole radio frequency spectrum from direct current (DC) to half the sampling rate.
- Implementing functionalities not available in the GNU Radio framework. But here again, we warn between developing a custom GNU Radio block and streaming the result of real-time SDR processing to external decoding tools, as was described in Chapter 3. As examples of real-time digital communication protocol decoding, the plane-to-ground communication protocol ACARS was implemented at https://sourceforge.net/projects/gr-acars. Similarly, the amateur radio digital protocol was implemented as a GNU Radio encoding and decoding processing block, based on the C library implementing the low-level protocols at https://github.com/M17-Project/gr-m17, emphasizing the need to keep the GNU Radio block in coherence with the library evolutions.
- **Loops are forbidden** in GNU Radio flowgraphs, so any processing that requires some sort of feedback loop – as is the case of the Costas loop – will require custom processing block.

Closed loops are a particularly challenging case since GNU Radio operates with buffers of unknown size, whereas a control loop aims at continuously controlling some actuator. The challenge is even more significant when the control acts on hardware, with uncontrolled latency between input and output. While one might consider that digital processing, with processors clocked in the gigahertz range, might lead to large control bandwidths, the latencies introduced

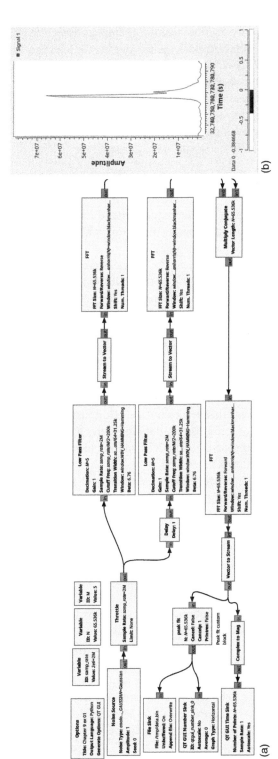

**Figure 9.1** (a) Flowchart with a random sequence recorded on the top reference channel and delayed by a single sample on the bottom channel. Both signals are low-pass-filtered before decimation to spread energy on adjacent samples and widen the correlation peak to simulate sub-sampling period delays. The correlation output is fitted by a parabola in the custom Python block. (b) The correlation peak is stable and detected with sub-sampling period resolution. Notice the fractional value of the delay. Without the low-pass filters, the decimated noise samples would be uncorrelated between the two channels.

302 | 9 Custom Source and Sink Blocks: Adding Your Own Hardware Interface

by digital processing will hardly allow for reaching loop bandwidths above a few tens to hundred kilohertz. Consider, for example, only one low-pass finite impulse response (FIR) filter with $N$ taps: the delay introduced by the FIR is $N/2$ clock cycles, limiting the loop bandwidth to a fraction of the sampling frequency divided by $N/2$. Thus, SDR will provide utmost stability on the long term, thanks to the stability of the digital implementation of algorithms, affected neither by aging nor by drift, but will find it challenging to compete with analog control loops when high bandwidth – and especially radio frequency bandwidth – is targeted. Such bandwidths remain the field of analog processing and even possibly acoustic wave processing when converting the electromagnetic signal into the slower acoustic wave through the inverse piezoelectric effect [Campbell, 1998; Morgan, 2010].

GNU Radio can manage two types of OOT blocks: Python and C++ programs. In all cases, the OOT block must handle the incoming buffer of data and inform the scheduler of the amount of data produced at the output. The processing block can request from the scheduler a minimum number of samples, as needed for, e.g. computing a Fourier transform, which requires integrating over a given number of input samples.

Similar to the `Python Block` described earlier, OOT blocks are defined as objects with a constructor, a work function called in an infinite loop by the GNU Radio scheduler, and a destructor. The work function is provided with input arguments as an array of vectors holding the input IQ stream and the number of items in the input stream. It will possibly generate some output data unless it is a `Sink block` (end of processing chain).

If the work function needs to remember the state from one execution to the next – e.g. the state of a finite state machine – variables are created as private class variables so that their values are kept between iterations.

A block template is created with `gr_modtool`, where the developer only manipulates the content of the `lib` directory holding the actual processing code (initialization and infinite loop function) in C++, possibly the content of `include` for defining setter and getter prototypes, and the `grc` directory content for defining the GNU Radio Companion block features (input type, output type, parameters, and associated callback functions).

A specific warning about the use of the FFT as implemented in FFTW3: this FFT implementation is explicitly stated as not being thread-safe, meaning that the various threads using FFTW3 must be aware of each other to prevent shared data structure from being corrupted during initialization: https://www.fftw.org/fftw3_doc/Thread-safety.html states, "The FFTW planner is intended to be called from a single thread. If you really must call it from multiple threads, you are expected to grab whatever lock makes sense for your application, with the understanding that you may be holding that lock for a long time, which is undesirable." If multiple instances of an OOT processing block are to be included in a GNU Radio Companion flowchart, using FFTW3 is not an option since each thread associated with each OOT instance will be unaware of other initializations. GNU Radio provides a thread-safe implementation of the FFT with `#include <gnuradio/fft/fft.h>` with the planner initialized as `fft::fft_complex_fwd* myplan=new fft::fft_complex_fwd(N);` whose input dataset is defined with `_gr_complex *inbuf;inbuf=plan->get_inbuf();` for filling with the input samples to be processed. The FFT is then executed with `plan->execute()` and returns `plan->get_outbuf()`. Since the GNU Radio scheduler supervises the thread execution, the consistency of the FFT initialization is preserved, and multiple OOT instances can run in parallel. This implementation was used in the ACARS decoder at https://sourceforge.net/p/gr-acars which will be detailed as a demonstration of the sequence for generating an OOT block for real-time decoding of the messages broadcast by planes toward airports (Section 2.3).

ACARS is based on the amplitude modulation (AM), as are all plane communications, of an audio frequency signal with half a 1200 Hz period or a full 2400 Hz period defining the bit state 0 or 1, respectively. Hence, the decoding will aim at identifying the most probable bit state and by comparing the output of two filters centered on each one of these frequencies and convert the soft bit resulting from AM demodulation to hard bit value, feeding the header detector, and then aggregating bits as bytes and analyzing the sentences. We start by defining a few constants for the various parameters set for proper operation of the decoder, namely the sampling rate of 48 kHz selected as a convenient multiple of both tone frequencies and a classical output of a sound card, the FFT size during the convolution to identify each bit probability, the maximum ACARS message size, and a clock synchronization parameter when searching for a bit transition for synchronizing the sampling clock [Laverenne and Friedt, 2020]:

```
#define fs          48000    // sampling frequency
#define CHUNK_SIZE 1024      // minimum number of samples for trigger this
    processing block
#define MESSAGE    (220*2) // twice the max message size !
#define MAXSIZE    (MESSAGE*8*fs) // 48000/2400=20 symbol/bit & 8 bits*260
    char=41600
#define dN          5        // clock tracking at +/-5 samples
```

The constructor defines the minimum and maximum number of inputs (both set to 1 in this case), the minimum and maximum number of outputs (both set to 0 since the ACARS decoder is a sink block), and the type of the input (real floating point numbers) and output (inexistant in this case):

```
acars_impl :: acars_impl(float seuil1, std :: string filename, bool
    saveall)
    : gr :: sync_block("acars",
            gr :: io_signature :: make(1, 1, sizeof(float)),
            gr :: io_signature :: make(0, 0, 0))
            , _threshold(threshold)
```

Finally, an argument is passed to the function, a threshold value used to detect the beginning of a transmission (threshold value). The constructor then initializes all the structures and variables needed by the main infinite loop function and asks the GNU Radio scheduler to request a number of samples multiple of the FFT size (set_output_multiple()) [Economos, 2019]:

```
{
  int t;
  _Ntot = 0;
  _N = 0;
  _threshold = 0.;
  set_threshold(threshold);
  _d = (float*)malloc(MAXSIZE*sizeof(float));
  _tout = (char*)malloc(MESSAGE*8);  // bits
  _toutd = (char*)malloc(MESSAGE*8); // bits
  _message = (char*)malloc(MESSAGE); // bytes
  _somme = (char*)malloc(MESSAGE);   // bytes
  set_output_multiple(CHUNK_SIZE); // only trigger processing if that amount
      of samples was accumulated
}
```

304 | *9 Custom Source and Sink Blocks: Adding Your Own Hardware Interface*

This function benefits from a callback function, which is implemented so that the user can dynamically change the value of the parameter threshold while the GNU Radio Companion flowchart is being executed. In this case, the setter set_threshold() is defined as

```
void acars_impl::set_threshold(float threshold)
{_threshold = threshold;
}
```

The same type of setter and getter functions would be added to the acars class for all parameters dynamically tunable from the graphical interface.

The main function, running as an infinite loop, is the work function which is fed an array of vectors as inputs, an array of vectors as outputs (except in the case of a sink block, as is the case here), and the number of input and output items.

```
int acars_impl::work(int noutput_items,
        gr_vector_const_void_star& input_items,
        gr_vector_void_star& output_items)
```

The template generator launched with the gr_modtool shell command has already filled the cast from the input array of vectors to the input dataset:

```
{
 const float* in = (const float *)input_items[0];
```

and the decoder starts processing the content of the array in, detecting the beginning of the message as a sudden rise in the power (variance) of the signal:

```
_N = noutput_items;
data = (float*)malloc(_N*sizeof(float));
if (_threshold == 0.)
  {_threshold = remove_avgf(in,data,_N);std = _threshold;}
else std = remove_avgf(in, data, _N); // d = d-mean(m); -> returns std()

if ((std > (_threshold*_threshold)) || (_decompte > 0))   // ACARS sentence
  detected: accumulate
  {for (k = 0;k<_N;k++) _d[_Ntot+k] = in[k]; // _Ntot: current position
  _Ntot += _N;
  _decompte++;if (_decompte == 3) _decompte = 0;
  }
else                                       // NO ACARS detected = > decode
  if we had some data
  {_threshold = std;                        // update threshold
   if (_Ntot > 0)                           // we had some data  = >
  process
    {remove_avgf(_d,_d,_Ntot);
    pos_start = 0;
    while ((_d[pos_start] < (_threshold*_threshold)) && (pos_start <
_Ntot))
           pos_start++;                      // get beginning
     pos_end = _Ntot-1;
     while (_d[pos_end] < (_threshold*_threshold))
           pos_end--;                       // get end
```

```
        if ((pos_end > pos_start) && ((pos_end-pos_start) > 200)) // 200
    since we skip first 200 samples
            acars_dec(&_d[pos_start], pos_end-pos_start);
        _Ntot = 0;                              // finished processing: clear
    buffer
        }
    }
if (data != NULL) {free(data);data = NULL;}
consume_each(_N);
return 0; // noutput_items;
}
```

The core signal processing function is `acars_dec()`, which identifies the most probable bit state by comparing the output of two filters centered on 1200 and 2400 Hz, the audio frequencies representing the two possible bit-states, and converts the input samples as output hard bits:

```
void acars_impl::acars_dec(float *d,int N)
{int fin,k,kcut,i,f,t,n,l,pos2400;
char s[256];
float a = 0., max24 = 0., thresholddyn, max2400;
char c;
FILE *fil;
gr_complex mul;
gr_complex *tmp, *_c1200, *_c2400, *_signal, *_fc1200, *_fc2400, *_fsignal,
    *_ffc1200, *_ffc2400;

/* see gr-qtgui/lib/freq_sink_c_impl.cc and freq_sink_c_impl.h for an
    example of GNU Radio 3.9 usage
*/
fft::fft_complex_fwd* plan_1200 = new fft::fft_complex_fwd(N);
fft::fft_complex_fwd* plan_2400 = new fft::fft_complex_fwd(N);
fft::fft_complex_fwd* plan_sign = new fft::fft_complex_fwd(N);
fft::fft_complex_rev* plan_R1200 = new fft::fft_complex_rev(N);
fft::fft_complex_rev* plan_R2400 = new fft::fft_complex_rev(N); // there
    also exists fft_real_fwd and rev if needed
_c2400 = plan_2400->get_inbuf();
for (t = 0; t < 40; t++)                        // convolution with *2*
    periods
    _c2400[t] = gr_complex{(float)cos((float)t*2400./fs*2*M_PI), (float)sin
    ((float)t*2400./fs*2*M_PI)};
for (t = 40; t < N;t ++)                         // zero padding
    _c2400[t] = gr_complex{0., 0.};

_c1200 = plan_1200->get_inbuf();
for (t = 0;t<40;t++)                             // convolution with *2*
    periods
    _c1200[t] = gr_complex{(float)cos((float)t*1200./fs*2*M_PI), (float)sin
    ((float)t*1200./fs*2*M_PI)};
for (t = 40;t<N;t++)      // zero padding
    _c1200[t] = gr_complex{0., 0.};
```

The convolution is performed in the frequency domain by computing the products of the Fourier transform, since using the FFT, the computational load is lower than performing the calculation in the time domain. The two filters are set at the beginning of the function in _1200 and _c2400 respectively, and the product with the Fourier transform of the signal computed below:

```
_signal = plan_sign->get_inbuf();
for (t = 0; t < N; t++)        // zero padding
    _signal[t] = gr_complex{d[t], 0.};
plan_2400->execute();
plan_1200->execute();
plan_sign->execute();
_fc2400 = plan_2400->get_outbuf();
_fc1200 = plan_1200->get_outbuf();
_fsignal = plan_sign->get_outbuf();
_ffc1200 = plan_R1200->get_inbuf();
_ffc2400 = plan_R2400->get_inbuf();
for (k = 0; k < N; k++)
    {mul = _fc2400[k]*_fsignal[k]; // complex number multiplication in C++
    _ffc2400[k] = mul/(float)N;
    mul = _fc1200[k]*_fsignal[k]; // complex number multiplication in C++
    _ffc1200[k] = mul/(float)N;
    }
// Low pass filter after convolution
kcut = (int)((float)N*3500./(float)fs); // cutoff @ 3500 Hz : df = fs/
    length(sf);fcut = floor(3500/df);
for (k = kcut;k<N-kcut;k++)     // low pass filter in Matlab FFT convention
    {_ffc2400[k] = {0., 0.};     // sf2400f(fcut:end-fcut) = 0;
    _ffc1200[k] = {0., 0.};     // sf1200f(fcut:end-fcut) = 0;
    }
plan_R1200 -> execute();       // result in _c1200
plan_R2400 -> execute();       // result in _c2400
_c1200 = plan_R1200->get_outbuf();
_c2400 = plan_R2400->get_outbuf();
{ // skip first 200 samples: we KNOW that N>200
    for (k = 200; k < N; k++) _c1200[k] = gr_complex{abs(_c1200[k]), 0.};
    max2400 = 0.;
    for (k = 200; k < N; k++) {_c2400[k] = gr_complex{abs(_c2400[k]), 0.};
                        if (_c2400[k].real() > max2400) max2400 = _c2400[k].
    real();
                    }
    k = 200;
    do {k++;} while (_c2400[k].real()>0.5*max2400);   // header as long as
    2400 is strong
    k+ = 10; // k at the beginning of the frame: move to the center of the
    first bit
```

The selection of the hard bit state – 0 or 1 – from the soft bit states as the output of the filters is the result of a comparison.

As was discussed earlier in the symbol synchronization block of the quad phase shift keying (QPSK) modem (Section 5.3.1.3),the sampling rate on the receiver might not be the same as the bit generation rate at the emitter as the oscillators might be slightly off frequency. Hence, detecting the bit transitions and aligning the sampling time on the stable state of the signal, far from the transition, allows for symbol synchronization. In the implementation below, a single bit state is identified, and its maximum is considered as the stable signal state most favorable to sampling:

```
_toutd[0] = 0;
 n = 1;
// now clock recovery ...
  while (k < N-40)
    {k += 20;
     if (_c2400[k].real() > _c1200[k].real()) _toutd[n] = 1; else _toutd[n]
   = 0; // soft -> hard bit
     n++;
     if ((_c2400[k].real() > _c1200[k].real()) && ((_c1200[k+20].real()
   >_c2400[k+20].real())) && (_c1200[k-20].real() > _c2400[k-20].real()))
        {max2400 = _c2400[k-dN].real();pos2400 = -dN;
         for (l = -dN+1; l <= dN; l++)
             {if (_c2400[k+l].real()>max2400)   // [m, p] = max(s2400(pos-3:
   pos+3));
                 {max2400 = _c2400[k+l].real();pos2400 = l;}
             }
         k += pos2400;
        }
     if ((_c1200[k].real() > _c2400[k].real()) && ((_c2400[k+20].real() >
   _c1200[k+20].real())) && (_c2400[k-20].real() > _c1200[k-20].real()))
        {max2400 = _c1200[k-dN].real();pos2400 = -dN;
         for (l = -dN+1; l <= dN; l++)
             {if (_c1200[k+l].real() > max2400)   // [m, p] = max(s1200(pos
   -3:pos+3));
                 {max2400 = _c1200[k+l].real();pos2400 = l;}
             }
         k += pos2400;
        }
    }
```

At this point, the most probable bit state has been set in _tout, array and the successive bits are assembled as bytes into the message array:

```
l = 0;fin = n;
_tout[l] = 1; l++;  // first two "1"s are removed since we sync on 1200
_tout[l] = 1; l++;
for (k = 0;k < fin; k++)
  {if (_toutd[k] == 0) _tout[l] = 1-_tout[l-1]; else _tout[l] = _tout[l
  -1];
   l++;
  }
n = 0;
for (k = 0; k < fin;k += 8)
  {_message[n] = _tout[k]+_tout[k+1]*2+_tout[k+2]*4+_tout[k+3]*8+_tout[k
  +4]*16+_tout[k+5]*32+_tout[k+6]*64;
   _somme[n] = 1-(_tout[k]+_tout[k+1]+_tout[k+2]+_tout[k+3]+_tout[k+4]+
  _tout[k+5]+_tout[k+6]+_tout[k+7])&0x01;
```

## 9 Custom Source and Sink Blocks: Adding Your Own Hardware Interface

```
    n++;
    }
 fin = n; // length of message (should be tout/8)
 n = 0;
```

Once the bytes have been assembled, the payload meaning is analyzed with `acars_parse` and the memory allocated to the temporary variables is freed:

```
 acars_parse(&_message[n], fin-n);
} // end of N<200
delete plan_1200;
delete plan_2400;
delete plan_sign;
delete plan_R1200;
delete plan_R2400;
}
```

The message parser checks the validity of the synchronization header and then assembles the successive bytes in sentences according to the ACARS standard as specified in ARINC Specification 618 [Aeronautical Radio, 2016] and implemented as follows:

```
void acars_impl::acars_parse(char *message,int ends)
{int k;
 time_t tm;
 if (ends > 12)
    if ((message[0] == 0x2b) && (message[1] == 0x2a) && // sync
        (message[2] == 0x16) && (message[3] == 0x16) && // sync
        (message[4] == 0x01))                           // Start Of Heading
    SOH
        {time(&tm);
        printf("\nAircraft=");
        for (k = 6; k < 13; k++) printf("%c",message[k]);
        printf("\n");
        if (ends > 17)
            {if (message[17] == 0x02) {printf("STX\n");} // Start of TeXt
            if (ends >= 21)
                {printf("Seq. No=");
                for (k = 18; k < 22; k++)
                    {printf("%02x ",message[k]);}
                for (k = 18;k < 22; k++)
                    if ((message[k] >= 32) || (message[k] == 0x10) || (
    message[k] == 0x13))
                        {printf("%c",message[k]);}
                printf("\n");
                if (ends >= 27)
                    {printf("Flight=");
                    for (k = 22; k < 28; k++) {printf("%c",message[k]);}
                    printf("\n");
                    if (ends >= 28)
                        {k=28;
                        do {if (message[k] == 0x03) printf("ETX"); // End
    of TeXt
```

*9.2 Out-of-Tree Blocks* | **309**

```
                                       else if ((message[k] >= 32) || (message[k]
     == 0x10) || (message[k] == 0x13))
                                            printf("%c",message[k]);
                                k++;
                                } while ((k < ends-1) && (message[k-1] != 0x03)
     );
                                printf("\n");
                          }
                    }
                }
            }
        }
    fflush(stdout);
}
```

We do not detail the `remove_avgf()`, which removes the mean value of the array provided and returns the standard deviation after removal of the mean value.

To complete the relation between the callback functions provided in the C++ main library and the GNU Radio Companion block definition, the YAML block description in the `grc` directory includes

```
parameters:
- id: threshold
  label: Threshold
  dtype: float
  default: '3'
inputs:
- label: in
  domain: stream
  dtype: float
asserts:
  - ${ threshold > 0 }
templates:
  imports: import acars
  make: acars.acars(${threshold})
  callbacks:
  - set_threshold(${threshold})
```

so that the Python-generated code links the `threshold` parameter to the first argument provided to the constructor, making sure this threshold is a positive value. The YAML file also includes the documentation displayed with the block description.

We shall not describe in the same details the `gr-oscilloscope` source block but the prototype in this case will be opposite – no input and as many outputs as oscilloscope channels selected – and the return value of the work function is the number of items generated and provided in each vector of the output array of vectors. Obviously communication rates between a radio-frequency-grade oscilloscope and the computer running GNU Radio is unable to continuously collect and stream all data, but in the context of passive or multistatic RADAR where one channel records a reference signal and other channels record signals backscattered by targets on surveillance channels,

## 9 Custom Source and Sink Blocks: Adding Your Own Hardware Interface

continuous stream is not needed as the oscilloscope memory depth defines the maximum detection range. The only requirement for computing the correlation and identifying the target range is that all channels are sampled simultaneously, a common feature in radio-frequency-grade oscilloscopes.

## 9.3 Cross-compiling for Running on Headless Embedded Systems

GNU Radio, despite its companion graphical interface helping the assembly of processing blocks, is perfectly suited for running on headless embedded systems once the flowgraph is generated with Options → Generate Options → No GUI. After checking that GNU Radio is properly installed on the embedded system, e.g. by checking that in a Python3 console `import gnuradio` does not return an error, one might want to cross-compile a custom block for the embedded target. We will develop the Buildroot solution we are most familiar with, even though GNU Radio is also supported in the embedded distribution build system OpenEmbedded (https://github.com/balister/meta-sdr).

Buildroot is designed to only select and compile the software needed for a given application and will not generate packages as OpenEmbedded does. Under Buildroot, selecting GNU Radio will add all dependencies, although Python support must be manually activated. Since all OOT blocks rely on `cmake` for compilation, this build framework must be informed to link with the target-embedded libraries rather than the host libraries as would be done otherwise. Hence, instead of `cmake ../` from the `build` directory in the OOT tree structure, we will now have

```
cmake -DCMAKE_INSTALL_PREFIX:PATH=/tmp/tmpdir \
    -DCMAKE_TOOLCHAIN_FILE=$BUILDROOT/output/host/usr/share/buildroot/toolchainfile.cmake\
    ../
```

where the `CMAKE_INSTALL_PREFIX` option informs `cmake` to move all the installed files during `make install` to a directory selected here as `/tmp/tmpdir`, which can be easily copied to the `/usr` directory on the embedded system after cross-compilation, and `CMAKE_TOOLCHAIN_FILE` provides the location of the `cmake` configuration files in the Buildroot tree structure. While GNU Radio has been supported in Buildroot since 2015 and version 3.7, the transition to Pybind11 has also not been without consequences for the cross-compilation. Until Pybind11 version 2.9, the dynamic libraries would include the host architecture in their name and would not be found on the target system, requiring manual renaming of the libraries. This issue has been solved since Pybind 2.10, and newer versions are used in Buildroot since mid-2022.

While these authors prefer the simplicity and efficient build sequence tailored to include only the needed packages proposed by Buildroot [Goavec-Merou and Friedt, 2021], rather than a full distribution as generated by OpenEmbedded, it is worth mentioning that a core contributor to GNU Radio, Philip Balister (https://github.com/balister), is supporting GNU Radio cross-compilation in this framework, as described in detail at https://wiki.gnuradio.org/index.php/OpenEmbedded.

As an illustration of GNU Radio running on embedded boards, the Raspberry Pi 4 connected to a Pluto+ receiver and the Raspberry Pi 5 connected to a B210 both record the signal collected by an antenna tuned to receive Iridium signals (Figure 9.2) – a GPS antenna whose front-end band-pass filter was removed – powered by a bias-T and split between the two receivers. While the Raspberry Pi 5 provides the bandwidth and the computational power to process the full 10 MHz of

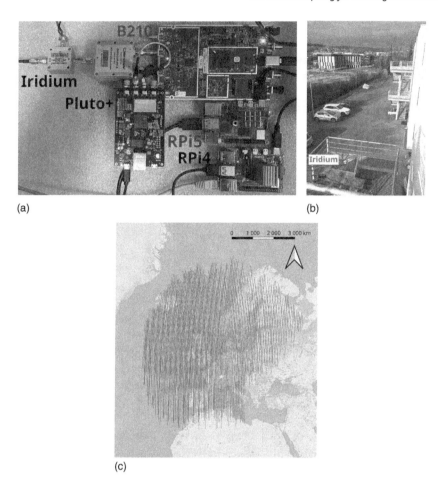

**Figure 9.2** (a) A Pluto+ connected to a Raspberry Pi4 and a B210 connected to a Raspberry Pi5, both receivers fed by a common antenna powered by a bias T and split between the two inputs. The antenna is originally a GPS antenna whose front-end band-pass filter was removed. (b) The antenna facing the sky is hardly blocked except at lower elevation by the building. (c) Location of the satellites whose messages were decoded, as processed by iridium-toolkit, matching the expected 3000 km visibility radius from a satellite flying at an altitude of 765 km. This is the same setup as the one used to introduce gr-iridium in Section 2.2, and the receiver location indicated by a star.

the useful spectrum around 1622 MHz, the Raspberry Pi 4 and the Pluto+ can only record and process 5 MHz bandwidth, already sufficient to receive a large number of frames. Notice that Buildroot sets the processor speed to a conservative "powersave" mode and the higher clock speed (and hence higher power consumption) must be set with echo "performance" > /sys/devices/system/cpu/cpu0/cpufreq/scaling_governor

Furthermore, make sure to run volk_profile to benchmark the default implementation of parallel algorithms and compare with their SIMD instruction implementation to benefit from the optimal solution: despite lower performance of the ARM Neon SIMD instruction sets with respect to those found on Intel-compatible x86 processors, the benefit can remain significant, e.g. on a Raspberry Pi 4.

# 9 Custom Source and Sink Blocks: Adding Your Own Hardware Interface

```
# volk_profile
...
RUN_VOLK_TESTS: volk_16u_byteswappuppet_16u(131071,1987)
generic completed in 501.76 ms
neon completed in 283.372 ms
neon_table completed in 361.166 ms
Best aligned arch: neon
Best unaligned arch: neon
RUN_VOLK_TESTS: volk_32u_byteswappuppet_32u(131071,1987)
generic completed in 746.205 ms
neon completed in 762.513 ms
neonv8 completed in 499.346 ms
Best aligned arch: neonv8
Best unaligned arch: neonv8
...
```

## 9.4 Conclusion

While the reader has been warned that writing a custom OOT module is probably *not* the first objective to be addressed when using GNU Radio, considering the huge number of readily available function blocks, OOT provides the flexibility to add missing functions such as additional hardware (source or sinks) and custom processing functions including decoding unsupported digital communication modes.

Furthermore, this chapter concluded with running GNU Radio on embedded hardware and cross-compiling using dedicated frameworks, including Buildroot and OpenEmbedded.

## References

S. Azarian. Using GPU for real-time SDR signal processing. In *FOSDEM*, Brussels, Belgium, 2024. https://fosdem.org/2024/schedule/event/fosdem-2024-1643-using-gpu-for-real-time-sdr-signal-processing/.

C. Campbell. *Surface Acoustic Wave Devices for Mobile and Wireless Communications, Four-Volume Set*. Academic Press, 1998.

R. Economos. 2019. https://lists.gnu.org/archive/html/discuss-gnuradio/2019-08/msg00188.html.

D. Estévez. Maia SDR: an open-source FPGA-based project for AD936x+Zynq radios. In *FOSDEM*, Brussels, Belgium, 2024. https://fosdem.org/2024/schedule/event/fosdem-2024-1841-maia-sdr-an-open-source-fpga-based-project-for-ad936x-zynq-radios/.

G. Goavec-Merou and J.-M Friedt. Never compile on the target ! GNU Radio on embedded systems using Buildroot. In *FOSDEM*, Brussels, Belgium, 2021. https://archive.fosdem.org/2021/schedule/event/fsr_gnu_radio_on_embedded_using_buildroot/.

Incorporated Aeronautical Radio. *ARINC 618: Air/Ground Character-Oriented Protocol Specification*, 2016.

T. Laverenne and J.-M Friedt. Bitstream clock synchronization in an ACARS receiver: porting GR-ACARS to GNU Radio 3.8. In *Software Defined Radio Academy*, 2020. https://www.youtube.com/watch?v=54URhrJkk28%22 accessed March 2024.

D. Morgan. *Surface Acoustic Wave Filters: With Applications to Electronic Communications and Signal Processing*. Academic Press, 2010.

# 10

## Conclusion

Throughout this book, we have attempted to demonstrate how GNU Radio and its graphical interface, GNU Radio Companion, provide an ideal framework to tackle abstract signal processing concepts and yet reach practical results on real signals recorded by general-purpose software defined radio (SDR) hardware. Starting from such mathematical concepts such as negative frequencies and complex voltages as derived from the analytic signal analysis, intuitively displayed with synthetic signals processed by GNU Radio blocks, we have tackled various digital communication challenges including a quadrature phase shift keying (QPSK) radio modem and decoding a QPSK-based weather satellite signal. However, since SDR provides access to the lowest level of the abstraction layers – hardware – with the raw in-phase/quadrature (IQ) streamed from the analog-to-digital converter, radiofrequency wave analysis such as time of flight (radio detection and ranging [RADAR]) or beamforming/null-steering is possible *before* any digital processing is started. Indeed, once a spoofing signal has reached the initial digital processing steps, a properly tailored signal will no longer be differentiated from the genuine signal. Physics is much harder to spoof, and a constant phase difference between signals broadcast by satellites distributed over the celestial sphere received by two antennas is the sure signature of a unique spoofing source, inconsistent with the varying phase due to the varying directions of arrival. Detecting on the IQ stream, the inconsistency is one result provided by SDR, which might invalidate further processing until the interfering source ceases to emit, but much more powerful is the null-steering capability provided by detecting the contribution of the spoofing signal and subtracting this weighted contribution from the second antenna: once cleaned, the genuine signal is processed and the spoofing source has been cancelled. Interference rejection capabilities comparable with analog controlled radiation pattern antenna (CRPA) results have been demonstrated with this approach [Feng et al., 2021; Friedt, 2024], relying on `gnss-sdr` [Fernández-Prades et al., 2011] for navigation satellite positioning and timing on top of GNU Radio for low-level hardware access and pre-processing, by implementing out-of-tree (OOT) blocks as discussed in the Chapter 9 of this book.

Thus, SDR has been used to address a wide range of applications, from training by addressing basic concepts of discrete time signal processing, to RADAR applications and synthetic aperture beamforming, to digital communication (modem, DAB+), all with our favorite signal processing framework, GNU Radio, and a given hardware. We started this book by stating that the benefit of SDR is stability, reconfigurability, and flexibility. During the investigations leading to the chapters of this document, we have delved into the details of GNU Radio, leading to uncover some of the minute imperfections of the software. As an example, the use of floating point number representation, known to be inaccurate, leads to issues with the trigonometric functions. As described in the issue https://github.com/gnuradio/gnuradio/issues/6800, a simple demonstration

*Communication Systems Engineering with GNU Radio: A Hands-on Approach*,
First Edition. Jean-Michel Friedt and Hervé Boeglen.
© 2025 John Wiley & Sons, Inc. Published 2025 by John Wiley & Sons, Inc.
Companion website: www.wiley.com/go/friedtcommunication

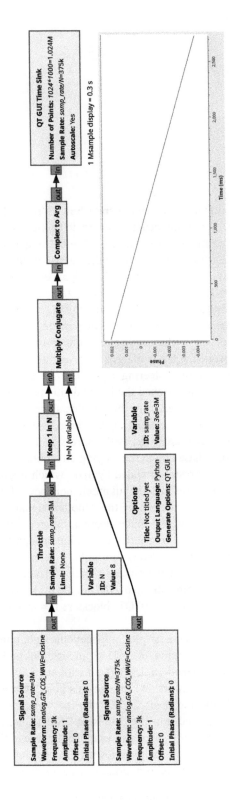

**Figure 10.1** Demonstration of the inaccuracy of the trigonometric function calculation: when selecting a non-unit (1) value of the decimation factor $N$, mixing the decimated sine wave with a sine wave generated at $1/N$th the sample rate leads to a drifting phase when the output of the mixer should remain constant over time.

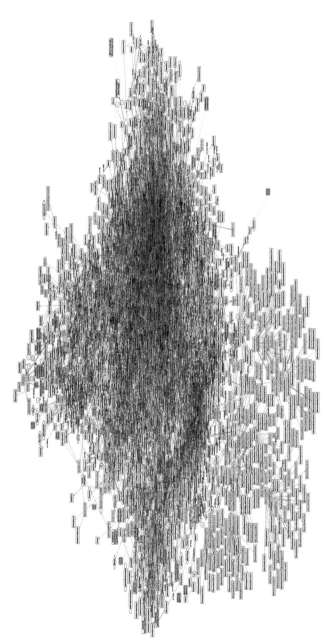

**Figure 10.2** Tree of the 776 dependencies of Debian GNU/Linux SID as of March 2024 needed for installing the binary distribution of GNU Radio.

# 10 Conclusion

with the flowgraph of Figure 10.1 illustrates how inaccurate the trigonometric function calculation is. This demonstration is also valid with the `Xlating FIR Filter`, and its consequences are most dramatic when considering a continuous wave RADAR application, including the case of beamforming when phase stability is of utmost importance.

Furthermore, this conclusion would not be fair without a word of warning about the challenges of keeping up-to-date a complex software framework with a huge number of dependencies. Some of these dependencies are part of the GNU Radio project – VOLK, SigMF – but most are external. Any change to these external libraries, application programming interface (API), will break GNU Radio. While the open-source framework promised long-term stability by avoiding vendor locking and obsolescence of proprietary software that would either no longer run on newer hardware architectures or operating systems, the complexity of modern high-performance computing frameworks makes them prone to excessive dependence to many libraries. Keeping a functional GNU Radio in the long term, with the constant dependencies evolution to improve performance but breaking API, is a time consuming if not a full time activity. The multiple evolutions GNU Radio went through when a developer wanted to install a custom development framework – from manual compilation to PyBOMBS and now conda – are a fine illustration of these challenges. As an illustration of the many dependencies involved in getting GNU Radio to work, under Debian GNU/Linux the command

```
$ debtree gnuradio | cut -d\- -f1 | grep \" | sort | uniq | wc -l
```

returns the number of packages involved: **776**. Any one of these libraries breaks, and the whole GNU Radio framework is broken: keeping this set of tools functioning in harmony is the achievement of the GNU Radio developers (Figure 10.2).

This somewhat pessimistic chapter should not deter the reader from getting acquainted with the awesome field of discrete time digital signal processing applied to radiofrequency signals, using GNU Radio, and reach the level to become a contributor and make the open-source framework faster, more stable, requiring fewer processing resources, and supporting multiple hardware interfaces as the current development team is striving to achieve. The authors are most grateful for their work and achievement.

## References

W. Feng, J.-M. Friedt, G. Goavec-Merou, and F. Meyer. Software-defined radio implemented GPS spoofing and its computationally efficient detection and suppression. *IEEE Aerospace and Electronic Systems Magazine*, 36(3):36–52, 2021.

C. Fernández-Prades, J. Arribas, P. Closas, C. Avilés, and L. Esteve. GNSS-SDR: An open source tool for researchers and developers. In *Proc. 24th Intl. Tech. Meeting Sat. Div. Inst. Navig.*, pages 780–794, Portland, Oregon, Sept. 2011.

J.-M Friedt. 2024. https://github.com/oscimp/gnss-sdr-1pps/ accessed March 2024.

# Index

## a

ACARS   36, 38, 302
active RADAR   113
ambigiuty function   113
autocorrelation   113
automatic gain control   176, 181

## b

Bark scale   221
baseband   5, 6
BER   228
bit error rate   228
Boltzmann constant   67
BPSK   48, 53, 262

## c

CCSDS   260
CGRAN   33
closed loop bandwidth   92
coarse acquisition   167
constellation   57, 156
convolutional encoding   265
corner reflector   139
Costas loop   55
cross-correlation   103
cyclic prefix   235

## d

DAB   219
decibel   55
Digital Audio Broadcasting   219
digital phase locked loop   168
DPLL   168

## f

fading   230
FFTW3   302
File Source   31
fine tracking   167
Forward error correction   236, 292

## g

gnss-sdr   33, 57, 315
GPS   57
gr-iridium   33, 88, 92
gr_modtool   198, 302
GRAVES   16, 63

## h

hard bit   270
hier block   160
hierarchical block   160

## i

Industrial, Scientific, Medical   123, 124
inter-symbol interference   157
IQ   7
Iridium   33, 310
ISM band   123, 124

## j

JPEG   286

## l

Line of sight   228
LO leakage   63
LRPT   251, 264

---

*Communication Systems Engineering with GNU Radio: A Hands-on Approach*,
First Edition. Jean-Michel Friedt and Hervé Boeglen.
© 2025 John Wiley & Sons, Inc. Published 2025 by John Wiley & Sons, Inc.
Companion website: www.wiley.com/go/friedtcommunication

## m

matched filtering    158
MPEG    224
MQTT    93–96
multipath    228

## n

Non-line of sight    228

## o

OFDM    230
OFDM symbol    232
OOT    198, 297
Open Systems Interconnection    75
OSI    75, 251
Out-Of-Tree block    297

## p

passive RADAR    113
phase locked loop    168
PLL    168
POCSAG    50, 70
power delay profile    228
pulse shaping    157
Python Block    297
Python Module    297
Python Snippet    78, 297

## q

QPSK    261
QT Frequency Sink    10
QT Time Sink    9

## r

Radio Data System    32
Raspberry Pi    122, 310

Rayleigh channel    231
RDS    32
Reed Solomon    292
Rice channel    231
roll-off factor    158

## s

S-parameters    55
SigMF    31
Single frequency network    237
soft bit    262
surface acoustic wave    144
Svalbard    252, 253

## t

tags    189
thermal noise    66
Throttle    10
timing error detector    176

## u

Unix pipe    96

## v

Viterbi    265

## x

Xlating FIR Filter    24, 48
XML-RPC    81

## z

ZeroMQ (0MQ)    84
Zynq    33, 122